IET ENERGY ENGINEERING 107

Bifacial Photovoltaics

Other volumes in this series:

Volume 1	**Power Circuit Breaker Theory and Design** C.H. Flurscheim (Editor)
Volume 4	**Industrial Microwave Heating** A.C. Metaxas and R.J. Meredith
Volume 7	**Insulators for High Voltages** J.S.T. Looms
Volume 8	**Variable Frequency AC Motor Drive Systems** D. Finney
Volume 10	**SF_6 Switchgear** H.M. Ryan and G.R. Jones
Volume 11	**Conduction and Induction Heating** E.J. Davies
Volume 13	**Statistical Techniques for High Voltage Engineering** W. Hauschild and W. Mosch
Volume 14	**Uninterruptible Power Supplies** J. Platts and J.D. St Aubyn (Editors)
Volume 15	**Digital Protection for Power Systems** A.T. Johns and S.K. Salman
Volume 16	**Electricity Economics and Planning** T.W. Berrie
Volume 18	**Vacuum Switchgear** A. Greenwood
Volume 19	**Electrical Safety: A guide to causes and prevention of hazards** J. Maxwell Adams
Volume 21	**Electricity Distribution Network Design, 2nd Edition** E. Lakervi and E.J. Holmes
Volume 22	**Artificial Intelligence Techniques in Power Systems** K. Warwick, A.O. Ekwue and R. Aggarwal (Editors)
Volume 24	**Power System Commissioning and Maintenance Practice** K. Harker
Volume 25	**Engineers' Handbook of Industrial Microwave Heating** R.J. Meredith
Volume 26	**Small Electric Motors** H. Moczala *et al.*
Volume 27	**AC-DC Power System Analysis** J. Arrillaga and B.C. Smith
Volume 29	**High Voltage Direct Current Transmission, 2nd Edition** J. Arrillaga
Volume 30	**Flexible AC Transmission Systems (FACTS)** Y-H. Song (Editor)
Volume 31	**Embedded Generation** N. Jenkins *et al.*
Volume 32	**High Voltage Engineering and Testing, 2nd Edition** H.M. Ryan (Editor)
Volume 33	**Overvoltage Protection of Low-Voltage Systems, Revised Edition** P. Hasse
Volume 36	**Voltage Quality in Electrical Power Systems** J. Schlabbach *et al.*
Volume 37	**Electrical Steels for Rotating Machines** P. Beckley
Volume 38	**The Electric Car: Development and future of battery, hybrid and fuel-cell cars** M. Westbrook
Volume 39	**Power Systems Electromagnetic Transients Simulation** J. Arrillaga and N. Watson
Volume 40	**Advances in High Voltage Engineering** M. Haddad and D. Warne
Volume 41	**Electrical Operation of Electrostatic Precipitators** K. Parker
Volume 43	**Thermal Power Plant Simulation and Control** D. Flynn
Volume 44	**Economic Evaluation of Projects in the Electricity Supply Industry** H. Khatib
Volume 45	**Propulsion Systems for Hybrid Vehicles** J. Miller
Volume 46	**Distribution Switchgear** S. Stewart
Volume 47	**Protection of Electricity Distribution Networks, 2nd Edition** J. Gers and E. Holmes
Volume 48	**Wood Pole Overhead Lines** B. Wareing
Volume 49	**Electric Fuses, 3rd Edition** A. Wright and G. Newbery
Volume 50	**Wind Power Integration: Connection and system operational aspects** B. Fox *et al.*
Volume 51	**Short Circuit Currents** J. Schlabbach
Volume 52	**Nuclear Power** J. Wood
Volume 53	**Condition Assessment of High Voltage Insulation in Power System Equipment** R.E. James and Q. Su
Volume 55	**Local Energy: Distributed generation of heat and power** J. Wood
Volume 56	**Condition Monitoring of Rotating Electrical Machines** P. Tavner, L. Ran, J. Penman and H. Sedding
Volume 57	**The Control Techniques Drives and Controls Handbook, 2nd Edition** B. Drury
Volume 58	**Lightning Protection** V. Cooray (Editor)
Volume 59	**Ultracapacitor Applications** J.M. Miller
Volume 62	**Lightning Electromagnetics** V. Cooray
Volume 63	**Energy Storage for Power Systems, 2nd Edition** A. Ter-Gazarian
Volume 65	**Protection of Electricity Distribution Networks, 3rd Edition** J. Gers
Volume 66	**High Voltage Engineering Testing, 3rd Edition** H. Ryan (Editor)
Volume 67	**Multicore Simulation of Power System Transients** F.M. Uriate
Volume 68	**Distribution System Analysis and Automation** J. Gers
Volume 69	**The Lightening Flash, 2nd Edition** V. Cooray (Editor)
Volume 70	**Economic Evaluation of Projects in the Electricity Supply Industry, 3rd Edition** H. Khatib

Volume 72 **Control Circuits in Power Electronics: Practical issues in design and implementation** M. Castilla (Editor)
Volume 73 **Wide Area Monitoring, Protection and Control Systems: The enabler for smarter grids** A. Vaccaro and A. Zobaa (Editors)
Volume 74 **Power Electronic Converters and Systems: Frontiers and applications** A.M. Trzynadlowski (Editor)
Volume 75 **Power Distribution Automation** B. Das (Editor)
Volume 76 **Power System Stability: Modelling, analysis and control** B. Om P. Malik
Volume 78 **Numerical Analysis of Power System Transients and Dynamics** A. Ametani (Editor)
Volume 79 **Vehicle-to-Grid: Linking electric vehicles to the smart grid** J. Lu and J. Hossain (Editors)
Volume 81 **Cyber-Physical-Social Systems and Constructs in Electric Power Engineering** S. Suryanarayanan, R. Roche and T.M. Hansen (Editors)
Volume 82 **Periodic Control of Power Electronic Converters** F. Blaabjerg, K. Zhou, D. Wang and Y. Yang
Volume 86 **Advances in Power System Modelling, Control and Stability Analysis** F. Milano (Editor)
Volume 87 **Cogeneration: Technologies, optimisation and implementation** C.A. Frangopoulos (Editor)
Volume 88 **Smarter Energy: From smart metering to the smart grid** H. Sun, N. Hatziargyriou, H.V. Poor, L. Carpanini and M.A. Sánchez Fornié (Editors)
Volume 89 **Hydrogen Production, Separation and Purification for Energy** A. Basile, F. Dalena, J. Tong and T.N. Veziroğlu (Editors)
Volume 90 **Clean Energy Microgrids** S. Obara and J. Morel (Editors)
Volume 91 **Fuzzy Logic Control in Energy Systems with Design Applications in MATLAB®/Simulink®** I.H. Altaş
Volume 92 **Power Quality in Future Electrical Power Systems** A.F. Zobaa and S.H.E.A. Aleem (Editors)
Volume 93 **Cogeneration and District Energy Systems: Modelling, analysis and optimization** M.A. Rosen and S. Koohi-Fayegh
Volume 94 **Introduction to the Smart Grid: Concepts, technologies and evolution** S.K. Salman
Volume 95 **Communication, Control and Security Challenges for the Smart Grid** S.M. Muyeen and S. Rahman (Editors)
Volume 97 **Synchronized Phasor Measurements for Smart Grids** M.J.B. Reddy and D.K. Mohanta (Editors)
Volume 98 **Large Scale Grid Integration of Renewable Energy Sources** A. Moreno-Munoz (Editor)
Volume 100 **Modeling and Dynamic Behaviour of Hydropower Plants** N. Kishor and J. Fraile-Ardanuy (Editors)
Volume 101 **Methane and Hydrogen for Energy Storage** R. Carriveau and D.S.-K. Ting
Volume 104 **Power Transformer Condition Monitoring and Diagnosis** A. Abu-Siada (Editor)
Volume 108 **Fault Diagnosis of Induction Motors** J. Faiz, V. Ghorbanian and G. Joksimović
Volume 110 **High Voltage Power Network Construction** K. Harker
Volume 111 **Energy Storage at Different Voltage Levels Technology, Integration, and Market Aspects** Zobaa, Ribeiro, Aleem and Afifi (Editors)
Volume 112 **Wireless Power Transfer: Theory, technology and application** N. Shinohara
Volume 115 **DC Distribution Systems and Microgrids** Tomislav Dragičević, Frede Blaabjerg and Pat Wheeler
Volume 117 **Structural Control and Fault Detection of Wind Turbine Systems** Hamid Reza Karimi
Volume 119 **Thermal Power Plant Control and Instrumentation: The control of boilers and HRSGs, 2nd Edition** D. Lindsley, J. Grist and D. Parker
Volume 123 **Power Systems Electromagnetic Transients Simulation, 2nd Edition** N. Watson and J. Arrillaga
Volume 124 **Power Market Transformation** B. Murray
Volume 128 **Characterization of Wide Bandgap Power Semiconductor Devices** F. Wang, Z. Zhang and E.A. Jones
Volume 130 **Wind and Solar Based Energy Systems for Communities** R. Carriveau and D.S.-K. Ting (Editors)
Volume 131 **Metaheuristic Optimization in Power Engineering** J. Radosavljević
Volume 905 **Power System Protection, 4 volumes**

Bifacial Photovoltaics

Technology, applications and economics

Edited by
Joris Libal and Radovan Kopecek

The Institution of Engineering and Technology

Published by The Institution of Engineering and Technology, London, United Kingdom

The Institution of Engineering and Technology is registered as a Charity in England & Wales (no. 211014) and Scotland (no. SC038698).

First published 2018

The Institution of Engineering and Technology
Michael Faraday House
Six Hills Way, Stevenage
Herts, SG1 2AY, United Kingdom

www.theiet.org

British Library Cataloguing in Publication Data
A catalogue record for this product is available from the British Library

ISBN 978-1-78561-274-9 (hardback)
ISBN 978-1-78561-275-6 (PDF)

Typeset in India by MPS Limited
Printed in the UK by CPI Group (UK) Ltd, Croydon

Contents

Acknowledgements

The authors Joris Libal and Radovan Kopecek thank all co-authors for their dedicated writing and reading also during many weekend sessions to finalize this book. We were a great bifiPV consortium agreeing from the beginning on the content – and even if we are working on and believing in slightly different technologies it was never a problem to harmonize the text, so that it was acceptable for all of us. May the better bifacial technology make its way.

In addition, we would like to thank all the dedicated companies that were and are visiting and sponsoring the bifiPV-workshop series from the first bifacial hour on such as PVGS, MegaCell, bSolar, Sunpreme, SolAround, enel, Jolywood, Meyer Burger, Passan, h.a.l.m., DSM and others, so that we were able from the very beginning on to bring the bifacial community together, to discuss about bifacial technology, to work on bifacial standards and to inform investors and banks about this great technology to make it more visible and bankable. These workshops were and hopefully will still be a great fun and success, as the bifiPV community is very active and stands closely behind this future technology. The community is getting bigger and bigger and similar to Mahatma Gandhi's saying, mentioned also in the introduction, 'First they ignore you, then they laugh at you, then they fight you and then you win' bifaciality will certainly win in the end.

Last but not least we would like to thank our families that were very often neglected during evenings and weekends and heard many times 'This weekend I have to finalize the bifacial book'. Thank you Moni, Samuel, Noemi and Frida for being uncomplaining during this time. Saving the world with bifacial PV technology needs some time.

Joris dedicates this bifacial PV book to his family, who always supported him throughout his career in PV as well as when authoring this book – thanks Chiara, Yannick and Alex for being so patient.

Radovan dedicates this book to his parents Alena and Oldrich that convinced me to study Physics instead of Art and to be honest: it's not so bad. Thank you for all your support during my studies in Stuttgart and abroad. You were always there when I needed you. Now enjoy your life and if you have time, read this book before you go to sleep. You definitely will sleep fast and deep. ☺

About the authors

The first bifacial book for PV technology and economy was written by many bifacial believers that support this technology by R&D, writing papers, organising workshops and being involved in technology transfer of new cell and module concepts.

Dr. Bas Van Aken is researcher in the PV Modules and Applications group at the Energy research Centre of the Netherlands (ECN), focusing on fabrication, reliability and performance of n-type bifacial and back-contact modules. Before, he worked on thin film Si solar cells by PECVD. After gaining a Ph.D. in solid state chemistry at the University of Groningen, he worked as a post-doc at Cambridge University and the Max-Born-Institute for Nonlinear and Ultrafast Optics in Berlin. His work has been published by, amongst others, *Nature*, *Nature Materials* and *Physical Review Letters*.

Djaber Berrian was born in Guerrara, Algeria, in 1991, earned his bachelor's degree in physics in 2013 at the University of Science and Technology Houari Boumediene in Algeria. He also received his Master's degree in renewable energy in 2015 at the same university and wrote his master thesis on mathematical modelling of a defective photovoltaic generator. Throughout his study background, he participated in different scientific activities in the field of renewable energy. His passion for renewable energy brought him to ISC Konstanz in February 2017. Currently, he is working as a Ph.D. student on energy yield prediction for bifacial PV systems at ISC. He is also involved in the EU-funded *BiFaLo*-project.

Dr. Chris Deline received the B.S., M.S. and Ph.D. degrees from the University of Michigan, Ann Arbor, in 2003, 2005 and 2008, respectively, all in electrical engineering. Since 2008 he has been a research engineer at the National Renewable Energy Laboratory in Golden, CO, in the photovoltaic performance and reliability group. He manages the US Department of Energy Regional Test Center program at NREL for field assessment of novel PV technologies, and is principal investigator for multiple PV field performance projects including degradation rate assessment and bifacial module power rating and production modelling. Dr. Deline is expert in the field of partially shaded or mismatched PV, with publications on the performance response of distributed power electronics and field performance of PV modules, particularly under mismatched or shaded conditions. His research also includes characterization and energy simulation of advanced PV technologies including thin-film module metastability and bifacial system energy production.

Dr. Matthieu Despeisse is section head in CSEM, Neuchâtel, Switzerland, where he is leading the research activities on crystalline silicon photovoltaics with special focus on bifacial silicon heterojunction, passivating contacts, tandem cells, as well as metallization-interconnection and metrology. He received his degree in electrical engineering from INSA-Lyon, France, in 2002. He then worked until 2008 at the European Organization for Particle Physics (CERN) in Geneva, Switzerland, focusing on novel Silicon radiation sensors (3D-Si, SiPM, MCP, a-Si:H), advanced low-power microelectronics and technology transfer. He obtained his Ph.D. degree in 2006 for his work on new generation silicon sensors. He then joined the IMT PV-lab of EPFL in 2009 in the group of Prof. Ballif, to lead the research team working on multi-junctions thin film silicon photovoltaics technology. In 2013, he started the crystalline silicon and metallization research activities in CSEM.

Dr. Thorsten Dullweber studied physics at the University of Hanover, Germany, and received his Ph.D. degree in 2002 from the University of Stuttgart. From 2001 till 2009 he worked as project leader in the microelectronics industry at Siemens AG and later Infineon Technologies AG. Since 2009, Thorsten is leading the R&D group Industrial Solar Cells at ISFH. His research work focuses on high efficiency industrial-type PERC silicon solar cells and bifacial PERC+ solar cells. Thorsten was awarded with the enercity energy efficiency price in 2015 and with the price of the German Foundation for Industrial Research in 2017 for developing record-efficient PERC and bifacial PERC+ cells. Thorsten is member of the Scientific Committees of the EU-PVSEC and SNEC Conferences and of the Editorial Advisory Board of Photovoltaics International.

Prof. Emeritus Naftali Eisenberg is founder and CTO of Solaround, a company developing advanced p-type PERT Bifacial Cells and Modules. Head of Jerusalem College of Technology Center for Micro-optics and Non-Conventional Optics. Head of Jerusalem College of Technology Silicon PV Solar Cell Laboratory. Former Chief Scientist of the solar energy pioneer company Luz, which installed 360 MWp solar thermal systems in the years 1990 and Founder and CTO of B-Solar, the first company that developed p-type Bifacial Cells and Modules.

Dr. Vahid Fakhfouri obtained his Master in Microengineering from the Swiss Federal Institute of Technology, EPFL in 2004. He obtained his Ph.D. in Nanotechnologies at EPFL in 2008. From 2009, he worked for Pasan, a world reference for measurement equipment in the photovoltaic cell and module manufacturing business, as metrology expert and Head of R&D. In 2014, Dr. Fakhfouri joined the International Electrotechnical Commission IEC as active member and initiated and led the first international standard project for the electrical characterization of bifacial PV devices.

Dr. Eric Gerritsen studied Applied Physics at Twente University in The Netherlands before joining Philips Research Laboratories (Eindhoven, NL) in 1985 to work on ion implantation. Based on his work at Philips, he received a Ph.D. at Groningen University in 1990 on a thesis entitled 'Surface modifications of metals by ion implantation'. He then held various industrial positions within Philips Lighting in Germany and Philips Semiconductors in The Netherlands and France before joining CEA-INES in 2008, as project manager on PV-module technology and applications with a focus on module reliability and bifaciality.

Florent Haffner is studying at Ecole Centrale de Lyon, France, after two years of Classes Préparatoires. Interested by photovoltaics, he decided to make its first intern at INES for four months in 2016. He worked on developing a simulation tool for bifacial systems. Since then he made other interns on wireless power transfer, hydrogen and CSP, each time involving numerical simulations, to gain experience on different topics. After having completed his last year of engineering school, with the focus on Energy and Research and Development, he joined again INES for his last-year internship in April 2018 focusing again on modelling and evaluating performances of photovoltaic systems, including bifacial.

Andreas Halm studied Physics at the University of Konstanz. He obtained his diploma degree in the area of Nanomechanics and -optics in 2006. After volunteering in Guatemala for a healthcare project, he joined ISC in June 2008 as project manager working on industrial solar cells made of SoG-Silicon. Starting 2010, he was part of the n-type group engaged in the development of high efficiency back contacted solar cells. Leading a project on module integration of back contact solar cells running from 2013 to 2015 he moved his focus and joined the module department. In September 2016 he became group leader of the module development group, since September 2017 he is head of the module department at ISC.

Heiko Hildebrandt studied Microsystem Engineering at Freiburg University/Germany. In parallel, he was Co-Founder of Ökostrom Saar GmbH, which is experienced in developing renewable energy projects since 2000. Based on this long-term experience in development, management and (demand-actuated) operation of renewable energy projects, he began working on grid-friendly PV concepts. As a result, the new company Next2Sun GmbH was established in 2015, which is pursuing the commercial launch of vertical East-West bifacial systems. Mr. Hildebrandt is manager and co-owner of Next2Sun and focusing on technical and application development.

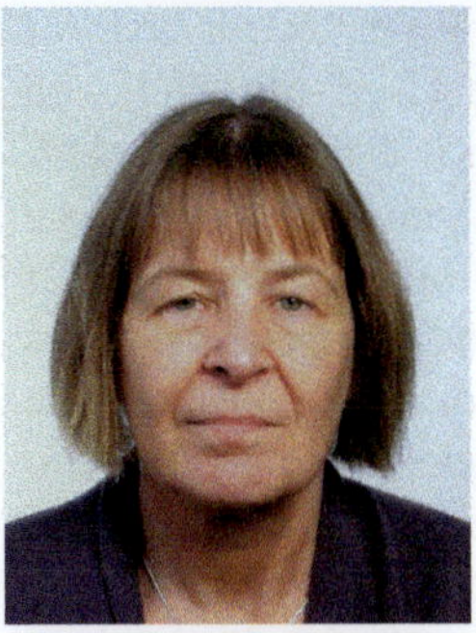

Dr. Gaby Janssen obtained a PhD in Quantum Chemistry from the University of Groningen in the Netherlands. At ECN she has been working as a research scientist on the simulation, characterization and optimization of materials for energy conversion technologies, in particular fuel cells. In 2011 she joined the ECN – Solar Energy department where she focuses on simulation and modelling of photovoltaic cells and modules. In recent years, she has developed at ECN an energy yield prediction model for bifacial systems.

Dr. Wolfgang Jooss received his Ph.D. at the University of Konstanz in 2002 for his work on multicrystalline and back contact buried contact solar cells. The major outcome of the experimental work was a world record of 17.5% for a large area multicrystalline solar cell. In 2002 he joined Sunways AG as R&D engineer and in 2007 he was appointed head of R&D solar cells. When working for the equipment and technology provider centrotherm between 2009 and 2016, his first position was Director Technology Integrated Factories, looking at the complete c-Si value chain. In his second position he was working as Director Technology PV, responsible for the technology development of solar cell equipment. Since April 2016 he is working as Director R&D at RCT Solutions GmbH in Konstanz, Germany.

Dr. Yong Sheng Khoo is the head of PV Module Development group at SERIS. He obtained his Ph.D. degree from Graduate School for Integrative Science and Engineering, National University of Singapore while working as post-graduate researcher at SERIS in the field of solar module. He also holds a B.S. and M. Eng. degrees in Mechanical & Aerospace Engineering from Cornell University, USA. Working on PV since 2010, his research work over the years covered important areas of PV technological development such as investigating outdoor performance of different module technologies in Singapore, modelling of optimal PV module orientation and tilt angle for maximum energy collection by the module, development of novel optical characterization techniques for PV modules, and optimizing the performance of PV modules for tropical conditions. Leading a group of 13 people, his current research focuses on the study of module potential induced degradation (PID), advanced module characterization, and novel module design and fabrication process.

Dr. Markus Klenk studied physics at the University of Constance where he also received his Ph.D. on $CuGaSe_2$ thin film photovoltaics in 2001. He started his work in the PV industry at the Sunways AG where he worked until 2010, first in the R&D, later as Head of Quality Assurance for cells and modules. This was followed by positions as Senior Technologist in the 'integrated factory' – department of the centrotherm PV AG and the rct solutions GmbH. Since 2015 he is working as a research associate at the ZHAW (Zurich University of Applied Science) in Switzerland, where he continues his PV activities.

Dr. Radovan Kopecek obtained the Dipl. Phys. degree at the University of Stuttgart in 1998. In 2002 he completed his Ph.D. dissertation in the field of c-Si thin film silicon solar cells in Prof. Ernst Bucher's group at the University of Konstanz. One of the founders of ISC Konstanz, Dr. Kopecek is since 2007 the leader of the advanced solar cells department dealing with several European, national research projects and technology transfer in the field of silicon feedstock and solar cell development, focusing on n-type devices. Further, Dr. Kopecek has been teaching the basics of PV at the DHBW in Friedrichshafen since 2012. Since 2016 he is in the board of directors at EUREC.

Dr. Lev Kreinin received M.S. in Industrial Energetics from Moscow University of Chemical Engineering Industry; Ph.D. in Energy Conversion from All-Union Scientific Research Institute of Current Sources, Moscow. Career started in solar cells R&D for Russian space program. Developments: fabrication technology of Si cells for first bifacial space solar arrays (in early 1970s) as well as of mono facial space cells based on combination of ion implantation and thermal diffusion. From 1992 – associate professor in Jerusalem College of Technology. 2009 to 2013 – Chief Scientist of bSolar Co., Israel, with main activity in development, indoor and outdoor tests and production of p-PERT bifacial solar cells for terrestrial applications. The same areas of research and development remained in the Chief Scientist position of Solaround Co., Israel. One book, more than 150 published papers and patents.

Dr. Robert Kreiter studied Chemistry and received a Ph.D. on organometallic chemistry and homogeneous catalysis in 2006 at Utrecht University. He then moved to ECN to work on nanostructured organo-silica membranes for energy efficient molecular separation processes. Since 2010 he was involved in research program development and scouting activities for ECN, including solar PV and wind energy. His latest achievement at ECN was the start of a new research program on energy storage and system integration. Since 2017 he moved to Sunfloat where he is responsible for operations. In parallel, he works as program manager System Integration for a national program on Energy and Industry. He is board member of Energy Storage NL, a Dutch platform for energy storage companies.

Dr. Joris Libal joined ISC Konstanz in 2012 where he works as an R&D project manager, focusing on business development and technology transfer in the areas of high-efficiency n-type solar cells and innovative module technology. He received his diploma in physics from the University of Tübingen and a Ph.D. in the field of n-type crystalline silicon solar cells from the University of Konstanz in 2006. He subsequently conducted postdoctoral research at the Università di Milano-Bicocca in Italy, where he investigated the properties of solar grade silicon. From 2008 until 2012, he worked as R&D manager for the Italian company Silfab, being responsible for the company's internal and external research and development projects in the field of solar cells and PV modules. He further coordinated the PV module certification process for the company's production sites in Canada and Croatia.

Jan Lossen studied Physics at the University of Freiburg and the University of Cologne. He graduated in 2003 with a thesis on hot wire chemical vapour deposition of microcrystalline silicon layers. For more than 10 years, he worked in different positions in the production and development of crystalline silicon solar cells at ErSol/Bosch Solar Energy AG. He headed a group that developed a cost effective production process for bifacial nPERT devices based on the use of implantation technology. Since 2014, Jan Lossen has been working at ISC Konstanz as senior scientist and project manager for process transfers. For more than a decade, Jan Lossen researches and promotes the use of bifacial photovoltaics actively.

Dr. Valentin Dan Mihailetchi received the B.Sc. degree in Physics in 2000 from the West University of Timisoara, Romania, and the Ph.D. degree ('cum laude') in 2005 from the University of Groningen, The Netherlands, for his work on device physics of organic solar cells. From November 2005 to June 2008, he worked at the ECN Solar Energy in the Netherlands as a research scientist on crystalline silicon, where he developed n-type–based solar cell processes. He joined ISC Konstanz in July 2008 and is currently a senior scientist and leading the group 'n-Type Solar Cells' in the 'Advanced Cell Concepts' department.

Dr. Hartmut Nussbaumer obtained his Diploma in Physics in 1993 and in 1996 he completed his Ph.D. in the field of silicon photovoltaics. Starting from 1996, he worked as a post doc stipendiary in Italy. From 1998 until 2006, he was head of production and later on head of technology at Sunways AG. From June 2006 to June 2008, he was managing director of centrotherm photovoltaics technology GmbH building up a group of developers for high efficiency silicon solar cell technologies. In July 2008 he joined the RENA group. From 2010 to 2014, he was CTO of RENA GmbH in the photovoltaic business segment. Since 2014 he started as lecturer and head of the group photovoltaic modules at the ZHAW Zurich University of Applied Science at the School of Engineering in Winterthur. He is president of the technical committee Solarpreis Schweiz, member of the scientific committee of the European and IEEE Photovoltaic Conference. He is reviewer in several scientific journals, author, co-author of numerous scientific articles and inventor in several patents in the field of photovoltaics.

André Richter received his diploma degree in electronic engineering (communication engineering, process measuring and control technology and environmental measurement) and operated 11 years own company for electronic education systems. Since 2001 he is working in photovoltaic area: In Conergy AG in development and third level support of solar plants, Conergy Electronics GmbH as CEO and had been Technical Director in Conergy solar plant Frankfurt (Oder). In 2008 he worked at Geneva-based company SES and as a consultant in US to build up module lines. Since 2010 he is working at Meyer Burger Technology AG in technical business development to establish and realize strategic projects.

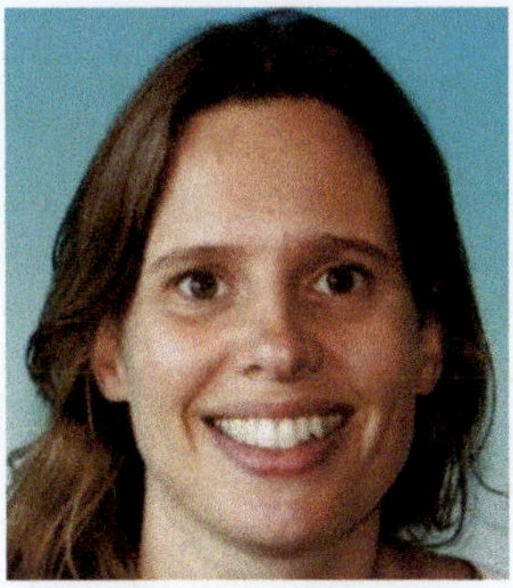

Dr. Ingrid Romijn studied physics at the Leiden University where she received her PhD on metal-insulator transitions in conducting polymers and composite materials. She joined ECN Solar Energy in 2004, where she started working as a researcher and later on as a project leader (2006) in the crystalline silicon group. The research topics involved (o.a.): research on passivating layers, optimization of SiNx deposition systems and advanced p-type solar cell concepts. During 2011, the focus of the work shifted towards the development and industrialization of n-type cell concepts. Since 2012, Ingrid Romijn is program manager for bifacial cells and modules at ECN Solar Energy. From 2014 to present, she also works as team leader and deputy senior manager in the Manufacturing Technology Group.

Prof. Dr. Andreas Schneider obtained his Diploma in Physics from the University of Freiburg in 1999. He obtained his Ph.D. at Professor Ernst Bucher's Department of Applied Solid State Physics at the University of Konstanz in 2004. From 2005, he worked for Day4Energy in Vancouver as head of the company's R&D department and established in 2008 Day4-Energy's quality management department. Subsequently Dr. Schneider joined the American company Jabil where he held the position of advanced engineering manager and was responsible for the scientific supervision of solar panel production. In 2011 Andreas joined ISC Konstanz where he has been responsible for the newly founded Module Development Department. Since 2016 he is full time Professor for Electronic Parts, Circuits and Renewable Energy at the University of Applied Sciences Gelsenkirchen.

Ismail Shoukry obtained his B.Sc. and M.Sc. degrees in the field of renewable energies at the university of Stuttgart as part of the DAAD scholarship programme in 2015. His interest in simulations led him to carry out research activities during his studies on developing physical models and carrying out simulations for various topics in the renewable energy field. His research as part of his Master Thesis at the ISC Konstanz focused on the study of the bifacial gain of stand-alone and in-field PV modules depending on various installation parameters. He is currently pursuing his career at Adaptricity AG as a software engineer, developing and working on a simulation platform for smart electrical distribution grids.

Dr. Jai Prakash Singh is heading Advance PV Module Technology team at Solar Energy Research Institute of Singapore (SERIS). He received his Ph.D. degree in 2015 from National University of Singapore (NUS) for his research on bifacial solar cells and modules. He also holds M.Tech. degree in Energy Science and Engineering from Indian Institute of Technology (IIT) Bombay, Mumbai, and B.Tech. degree in Electrical Engineering from Jamia Millia Islamia University, New Delhi. Dr. Singh has more than 10 years of experience on c-Si solar cell and module. In 2008, he joined Moser Baer Photovoltaic as senior R&D engineer. In 2010, he joined PV module group at SERIS as Research Associate and promoted to Research Fellow position in 2015 after receiving his Ph.D. degree. His research work focuses on bifacial device characterization, simulation, cell-to-module loss analysis, design optimization and reliability study of c-Si–based PV modules.

Dr. Yannick Veschetti obtained his Ph.D. in Physics at Strasburg University, in the field of crystalline silicon PV. He joined CEA-INES in 2005 to develop high efficiency silicon crystalline solar cells. From 2013 to 2015, he was responsible of the homojunction silicon solar cell laboratory, focusing his research work on N-type silicon and bifacial cells. He is currently head of the PV-module division at CEA-INES, addressing topics such as high efficiency and low-cost silicon modules, new modules technologies for specific applications (BIPV, Solar Road, Stratosphere and Space) and OPV/perovskites technologies.

Chapter 1

Introduction

Radovan Kopecek[1] and Joris Libal[1]

Photovoltaics (PV) is becoming, much faster than anybody would have expected, the most cost-effective electricity source on earth. Not only that it is clean, low maintenance, decentralized and scalable – in some cases, the costs per kWh in large systems are already even cheaper than coal energy. In India and other sunny countries, planned coal plants were even cancelled already in 2017 in favour of PV systems [1]. In a couple of years, PV will become an unbeatable electricity source, as there is still huge potential for cost reductions – some of that e.g. offers making full use of bifaciality in many applications as we will see in the coming chapters. In 2017, about 100 GWp additional PV module installations have been added to the existing *ca.* 300 GWp [2] – much faster than anybody would have expected. The most optimistic scenarios forecasted a 100 GWp market in 2022 – which happened now 5 years earlier. In 2020 or 2021, we will have a total of 1 TWp installed PV systems worldwide.

In this chapter, we will sketch a complete picture of PVs status, explain the role of bifaciality and predict what the importance of bifacial PV in future PV systems in terms of reduction of electricity generation costs will be.

1.1 PV 2018 – history, present and future

1.1.1 PV becomes the most cost-effective electricity source

The history of PV is long and extremely dynamic. PV was used for a long time in satellite applications. At the beginning of terrestrial applications only ecologically aware people installed PV systems for their own use at high costs. Everybody else was completely ignoring and later even laughing at this technology.

Figure 1.1 sketches the history, present and a possible future of module costs depending on the demand and production capacity. PV´s history can be matched to Mahatma Gandhi's saying: 'First they ignore you, then they laugh at you, then they fight you and then you win'. At the moment, we are in the middle of 'fighting' against other traditional energy sources and political hurdles, such as e.g. the

[1]International Solar Energy Research Center Konstanz e.V., ISC Konstanz, Germany

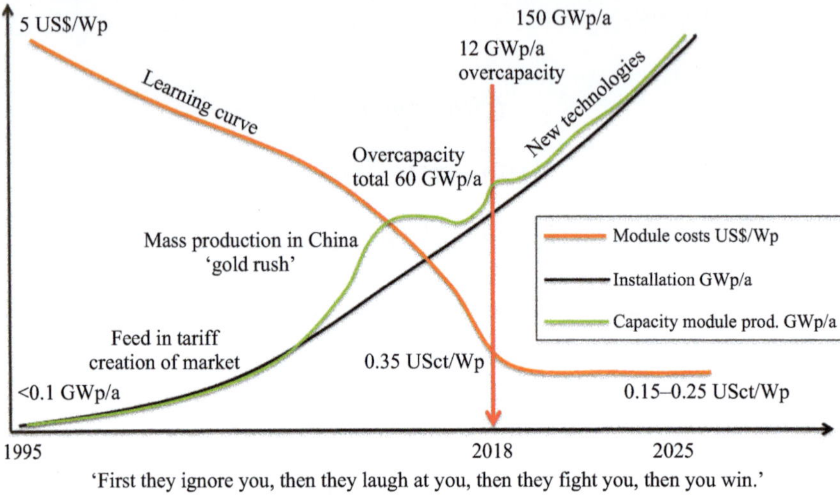

Figure 1.1 History, present and future of PV module prices demand and production capacity. The schematically depicted graphs are showing the trend of each category

introduced taxes for self-consumption in Germany. These taxes were introduced to slow down the energy transition in Germany, so that the large electricity providers have more time to restructure. In the end, PV will win as there is no way anymore to stop this cost-effective technology.

But how did we come so far that PV became that cost effective? The starting point of this success story was the feed in tariff in Germany around 2000 to stimulate the market, followed by the mass production in Germany and later taken over by China as the EU-based manufacturers of production equipment transferred the technology to that flourishing market. At around 2009 (after the silicon feedstock crisis that started around 2005 and lead to a price increase of PV modules), everyone wanted to be a part of the game and to produce cells and modules. This led quite fast to a significant overcapacity (as observed in many other industry branches, such as automotive or electronics industry, before) around 2010. The prices for modules dropped extremely fast at that time, modules were on stock and a big shake-out with many bankruptcies began. After this consolidation phase, the capacity was again close to the demand – however, the prices never went up again remarkably. Actually quite the contrary was the case: the prices even dropped faster.

At the moment we are once more in a small crisis as we have currently once again an overcapacity of *ca.* 12 GWp – mostly because of installations of new, or transformation of old standard Al back surface field (BSF) lines to modern Passivated Emitter and Rear Contact (PERC) lines leading to a higher MW/year for the same line-throughput (cells/h) due to the higher cell efficiency. However, what has changed is that PV is becoming more innovative accepting 'new' technologies on the market such as PERC and nPERT and other cell structures (see Chapter 2). The

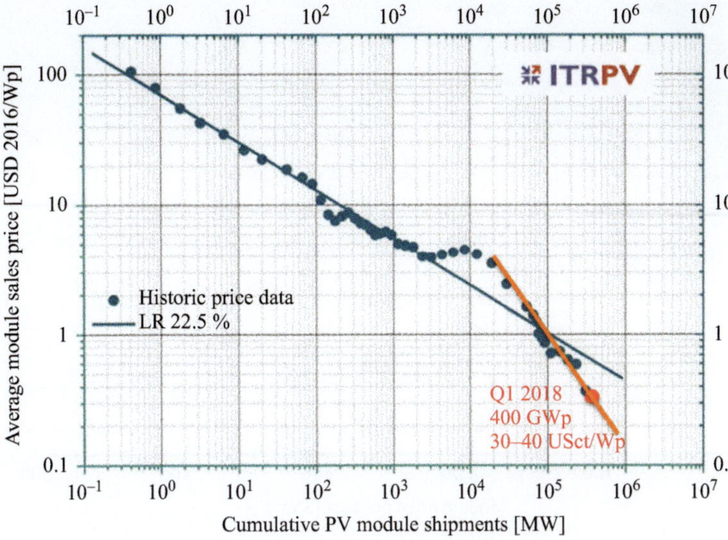

Figure 1.2 Learning curve of PV since 1980 [4]

costs for a standard module at the moment are in average at about 35USct/Wp – and are expected to drop even to the half of this value in the future. The total power of yearly installed PV systems in 2016 is about 70 GWp which is predicted to be almost doubled by 2030 [3]. However, as already mentioned, even Bloomberg did not expect *ca.* 100 GW installations in 2017. Newest studies show that the doubling to 150 GWp could be even reached in 2018 [2]. When we look at the precise learning curve in Figure 1.2 [4], it seems that the already steep curve became even steeper after the feedstock crisis in 2005. At the moment, in Q1 2018, we have a worldwide cumulated power of all PV installations of about 400 GWp at module prices between 30 and 40 USct/Wp.

Even Bloomberg now realized that PV is the future technology for cost-effective electricity [3], and in 12 December 2015 in Paris the UN decided to go in the direction of renewables [5]. Confirming that this was the right decision is not only motivated by the CO_2 reduction plans, but can be seen from the latest Power Purchase Agreement cases in Abu Dhabi, Chile and India, where prices between 2.42 USct/kWh [6] over 2.91 USct/kWh [7] to 5.3 USct/kWh [8] were reached. Currently the lowest bid was given by EDF/Masdar in Saudi Arabia for first time reaching a value below 2 USct/kWh, namely 1.79 USct/kWh [9]. This fact supports our introduction as this low value was reached with bifacial horizontal single axis tracked (HSAT) system – which is the newest trend in bifacial PV. At the end this offer was not selected, which was mostly due to the still low bankability for bifacial systems at that time. This is changing quite quickly as bifacial standards are entering the market and more and more data from large bifacial PV installations are being collected and published supporting the development and improvement of the bifacial system energy yield simulations.

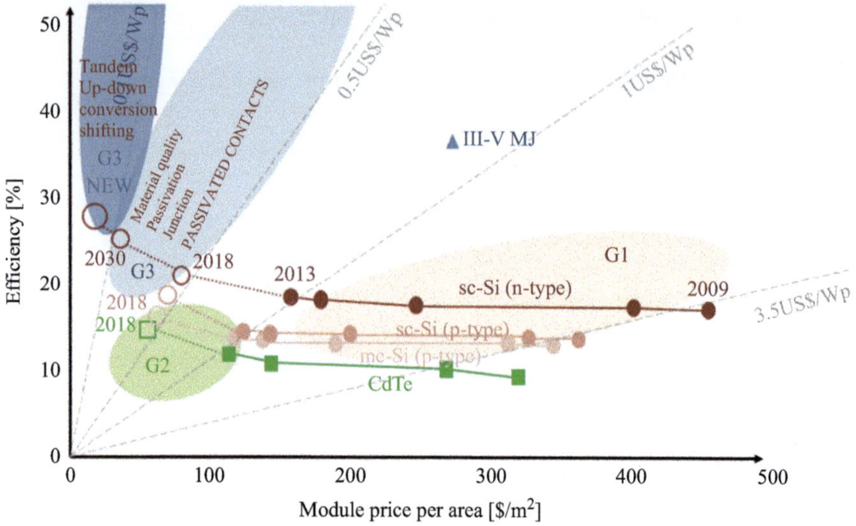

Figure 1.3 Efficiency versus costs in dependence of different technologies. G1: c-Si. G2: thin film. G3: next generation [10]. G3 NEW: combination of c-Si with other technologies

1.1.2 What PV technology will win at the end?

This question very often was raised in the past: what technology will make the race to be the most cost-effective candidate for lowest cost electricity production? Many people were predicting that c-Si (crystalline silicon) cannot make it, as it is an indirect semiconductor and therefore relatively thick wafers are needed (100–150 μm) to absorb the solar spectrum. Therefore, thin film solar cells (direct semiconductors), such as CdTe or CIGS, will be needed, as thicknesses below 10 μs are required. Prof. Martin Green even created a comprehensive graph for this believed fact – which is depicted in Figure 1.3 [10].

He has sketched three regions in an efficiency-price graph and predicted in 2009 that c-Si (generation 1) can be used to a certain extent – after that, for further lowering the costs, thin film (generation 2) and then e.g. multijunction (MJ) III–V structures (generation 3) will be needed. However he, actually nobody, could not predict the unbelievable development of c-Si in respect to maturity and costs. We have updated the graph with newest efficiency and price numbers for c-Si, CdTe and III–V MJ PV until 2018. It is clearly visible that the c-Si technology surpassed the predicted boundary for lowest prices very quickly being well below 100 US$/m^2 and below 40 USct/Wp. CdTe went the same direction, however, at lower efficiencies. III–V MJ cells remained at high costs.

What is important for low costs in large PV systems is not only low module cost but also high power modules, as the balance of system (BOS; installation, material, inverters, etc.) and the land-related cost represent more than 50% of the total costs. The more powerful the modules are, the lower the costs for BOS and land can be. Therefore, the PV industry at the moment is going more and more towards the monocrystalline c-Si technology and also considering more seriously bifacial technologies. This has been observed recently, e.g. at the latest SNEC trade show in Shanghai 2017 [11].

What will happen in the next 15 years in PV is quite easy to predict. The solar cells will be more complex, and monocrystalline c-Si technology at low costs will adapt evolutionary technologies such as selective diffusions, better passivations, improved metallization technologies and maybe carrier selective contacts. The industrial cell will slowly approach the theoretical limit of c-Si and become more and more bifacial. When the limit is reached, it becomes unsure, which technologies could boost c-Si above 30% efficiency. However, we are very sure that it will be a combination of the most powerful and low cost c-Si technology with another technology. We call this 'G3 NEW' (generation 3 new). Maybe we will shift the UV and infrared parts of the solar spectrum towards c-Si band-gap with up and down conversion, or/and additional material in addition to c-Si will be used in tandem configurations such as GaAs nano wires, CIGS or Perovskites.

However, what is sure is that bifaciality will play a more important role in PV in the coming years. Why? As the cell concepts are becoming more complex, the rear side can be only partly metallized which also will lead to cost reduction due to metal savings. Accordingly, bifacial versions of such cell concepts can be easily implemented on an industrial scale. Bifacial PERC+ and PERT cell concepts are the most prominent candidates for this which will be discussed in Chapter 2. In addition, module producers are going to double-glass modules anyhow, due to lower costs and higher lifetimes of the modules. Bifaciality will therefore come naturally on the PV market and serve the large electrical companies (EDF, Enel, Total, Engie, SPIC, Adani, etc.) which already are thinking in terms of costs per kWh rather than in costs per Wp like classical module producers.

1.2 Bifacial PV 2018 – history, present and future

1.2.1 Short bifacial history

The bifacial PV history actually began with the very first solar cell processed at the Bell Labs in 1954. It was an n-type (As-doped) IBC cell with a bifacial character [12]. The movie in the reference shows the very first process and the bifacial nIBC with an efficiency of about 5%. Figure 1.4 shows the bifacial history in few pictures.

In the solar cell Chapter 2, the bifacial cell history is described in more detail. Here we only want to sketch a couple of important incidents and events since the beginning of PV until now. In 1966, bifacial cells were proposed in a Japanese patent, and in the 1970s Russian and Spanish groups were developing bifacial cells – mostly for space applications – which were then used in several Russian satellites. Andreas Cuevas did a detailed study about bifacial terrestrial applications using the albedo from surroundings [13], and Thomas Nordmann was using bifacial double glass modules in noise barriers in Switzerland before 2000. However, at that moment, bifaciality was much more expensive as standard modules technologies.

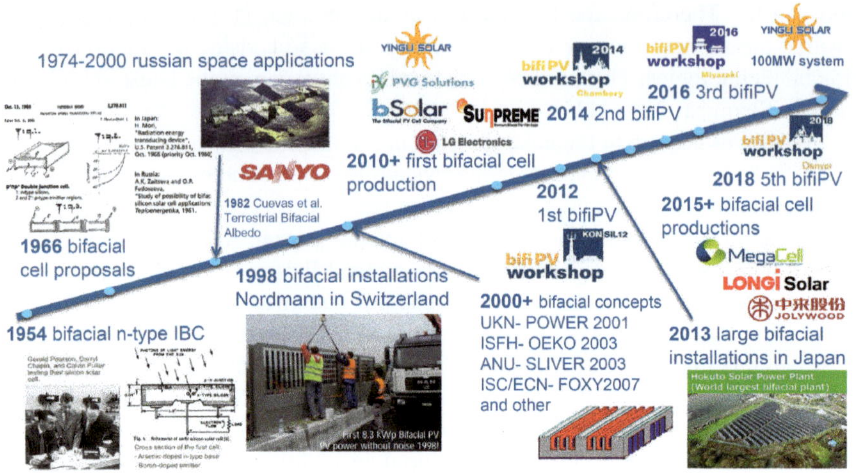

Figure 1.4 Bifacial history in few pictures

Also SANYO was involved in production of costly bifacial modules. Since 2000, many cost-effective bifacial concepts were entering R&D at universities and institutes and later solar cell and module production. Solar Wind, PVGS, Yingli and bSolar were the first producers of standard p- and n-type bifacial solar cells with low-cost processes such as B- and P-diffusions, PECVD passivations and screen printing metallization – followed by more complex processes of LG electronics and Sunpreme. The industrial interest was start of our bifacial workshops, which were initiated in Konstanz in 2012. There the participants identified the most critical issues for bringing bifaciality into the market: standardization, bifacial system gain simulations and bankability. At that time, the largest bifacial systems were only of kW size. After that, in 2013, PVGS created a 1.25 MWp large bifacial PV system in Japan which remained for more than 2 years the largest one and was cited in many publications. Since that time many other producers entered the bifacial PV market, and the system sizes and quantity are monthly increased. The largest bifacial system at the moment of finalizing this book had a size of 100 MWp.

The 'new bifacial history', when low-cost concepts started to enter the PV market, can be shown very briefly in three pictures as depicted in Figure 1.5. As already mentioned, since 2000, couple of scientists were 'preaching' that bifaciality is offering many advantages and that the industry should orientate that way. However, as the PV producers were all in party mood selling all the modules produced, no one was interested to listen and think about the future. Today, in 2018, after the crisis, the direction is set towards innovations, and there are two split groups: the 'mono-facials' that are saying that the bifacial technology is too expensive and the low albedo out there is not worth it to be collected. The 'bi-facials' are convinced that bifacial large systems are the future to generate lowest levelized costs of electricity (LCOEs). The two groups are also considered to be the Wp-group (mono-facials) and the kWh-group (bi-facials). As the 'PV thinking' is

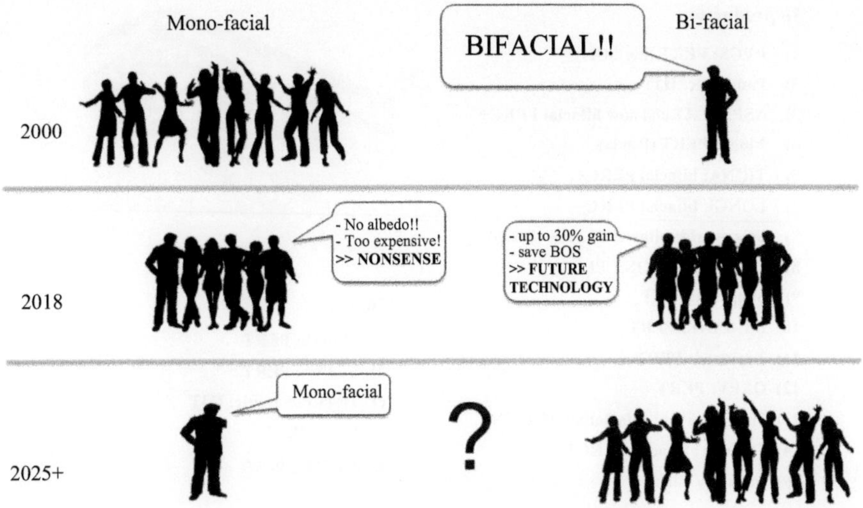

Figure 1.5 Bifacial comic explaining the change of mind of the PV industry [13]

going more and more towards lowest cost per kWh, bifaciality will win at the end. And one of the reasons is also because the new cell types will become bifacial and the modules double-glass anyhow as well.

1.2.2 Bifacial status

As already mentioned before, more and more bifacial cell and module producers are entering the market, and more and more bifacial systems are installed. At the SNEC trade show 2017, many bifacial modules were displayed – even bifacial IBC solar cells [11,14]. Figure 1.6 summarizes the most popular bifacial solar cell producers in a non-exhaustive list.

The most classical and mostly used bifacial cell is nPERT, followed by HJT and bifacial PERC+. These cell concepts, their advantages and drawbacks are discussed in the solar cell Chapter 2 in detail. While MegaCell and Mission Solar had to stop production, Jolywood announced 2.1 GWp nPERT in 2018 and also others like Adani and Linyang are going this way. NSP, LONGi, TRINA and others decided to go the bifacial PERC+ way. As there were *ca.* 35 GWp PERC capacities in 2017, it is expected that bifacial PERC+ solar cells will come on the marked with a high impact soon, even if the bifacial factor is lower than for nPERT.

Figure 1.7 is showing the largest bifacial PV systems on the left and the cumulated power until Q1 2018 on the right. The Japanese 1.25 MWp, first large system, was topped now many times – Yingli having now the largest system with 100 MWp is going online in Q2/2018. There are ambitious plans for 2018 – e.g. EDF wants to install a 90 MWp bifacial system in Mexico [15], which they won in a second auction with 3.28USct/kWh [6]. 8minutenergy plans to install large bifacial systems in the United States, and other big players such as Enel, EDF, Engie and NSP have set their future on bifaciality as well.

In production

1) **PVGS: PERT** (EarthON)
2) **Panasonic: HJT**
3) **NSP: PERT** and now **bifacial PERC+**
4) **Yingli: PERT** (Panda)
5) **TRINA: bifacial PERC+**
6) **LONGi: bifacial PERC+**
7) **Solarworld: bifacial PERC+** (Bisun)
8) **LG: PERT** (NeON/CELLO)
9) **Sunpreme: HJT**
10) **HT-SAAE: PERT**
11) **Jolywood: PERT**
12) **QXPV: PERT**
13) **Shanxi Lu′ an: bifacial mcPERCT+**
14) **JA Solar: bifacial PERC+**
15) and many others

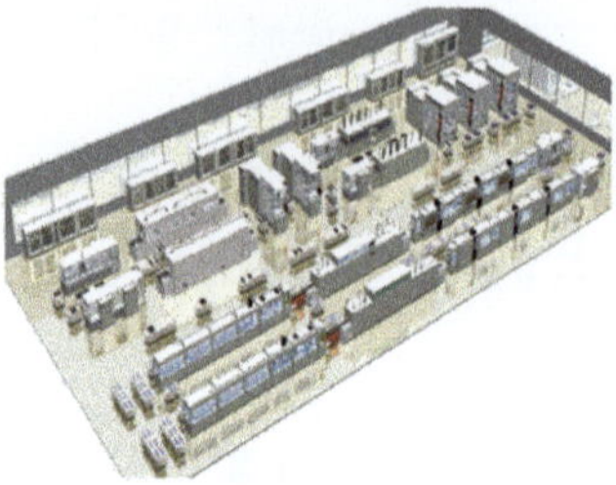

In pilot

a) **Motech: PERT**
b) **TRINA: PERT**
c) **Tesla/Panasonic: HJT**
d) **REC: PERT**
e) and many others

Figure 1.6 List of bifacial cell and module producers [11]

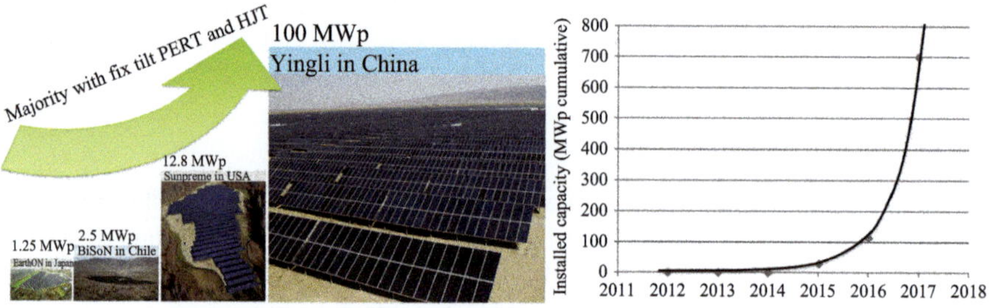

Figure 1.7 Largest bifacial PV systems (left) and cumulated bifacial power (right) [13]

In total, the cumulated power of bifacial installations exceeds 1 GWp in Q2/2018 and is expected to triple by 2019. Chapter 5 will summarize the largest systems until now, their configuration and their benefits in terms of bifacial gain.

1.2.3 Bifacial future

The bifacial future is predicted to be very bright – as e.g. forecasted from ITRPV in Figure 1.8. There it is shown that the world's market share for bifacial modules is getting larger – resulting in about 35% share of 'true' bifacial modules in 2027. True means in this case that the bifacial cells will be used in bifacial modules in bifacial installations. That would be about 75–100 GWp. We will see in the summary chapter that this will be rather more, as on the one hand the

(1) total PV installation in all road-maps is usually underestimated and
(2) the speed of new concepts entering the PV market is considered too low.

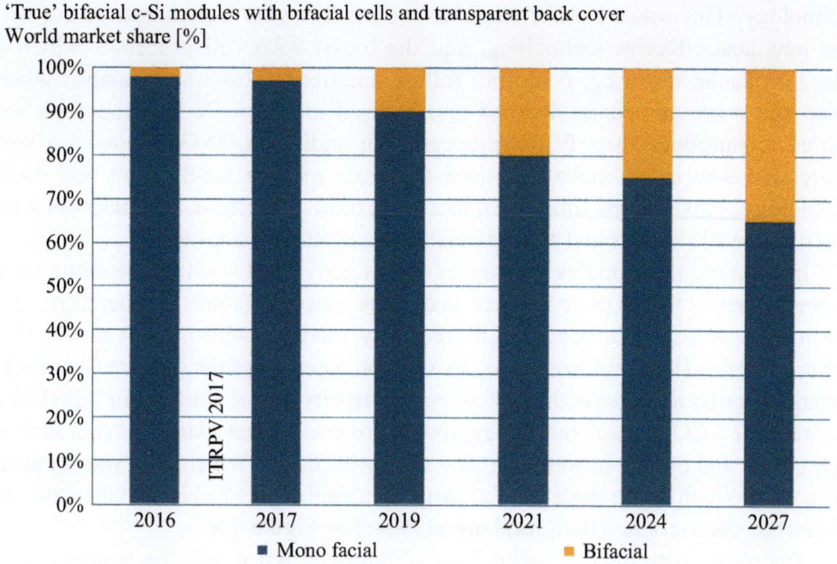

Figure 1.8 Bifacial technology forecast from IRTPV 2017 [4]

In our estimations, it will be rather double the size of the estimated market. So more than 100 GWp – which is more than the total installed module capacity in 2017.

The application areas will be large utility scale PV systems with high albedo – many of them also combined with HSAT or vertical single axis tracking (VSAT) – and flat commercial white roofs like in the case of Sunpreme and their up to now largest bifacial roof top installation of 10 WMp [16].

1.2.4 *Changing to cost per kWh thinking instead of cost per Wp mentality*

The ITRPV did not predict such a fast entering of PERC technology into the PV market in 2018. This was also not foreseen and possible because the PV market is a very complex one. Why this happened so fast is summarized in three most important points and then explained in detail:

- in 2016 Cz-Si wafer prices hit the mc-Si wafer prices (due to diamond wire sawing)
- change from cost/Wp mentality to cost/kWh thinking (as large electricity providers are entering the marked)
- China is pushing high efficiency modules (in their new incentives a 23% solar cell is required)

Beginning of 2016 the PV market started to change slowly with huge impact in 2018. For the first time, since the beginning of the mc-Si era, the Cz-Si wafer costs hit the mc-Si wafer costs because of the introduction of diamond wire sawing technology. This was a game-changer as the mc-Si ingots were harder to cut with this new cost-effective technology, and the mc-Si solar cell process – which is based on acidic texturing, could not follow this trend. This was a catalyst where more and more companies worked and focused more on Cz-Si technology and started to introduce their PERC processes. In addition LONGi followed a very aggressive way to additionally reduce the costs of their Cz-Si ingot and wafer technology. This was the time when more and more powerful Cz-Si modules were entering the PV market and also bifaciality was getting its up-wind.

In addition, the modules became so cost effective that – within the entire costs of the system – the BOS costs were becoming more pronounced. Therefore, it is much more effective to use a bit more costly modules with highest power than lower power modules at lower costs. In addition, because of the same reason, tricks became popular to increase the system power density by e.g. tracking or bifaciality. Because the LCOEs came into the region where also the very large players such as EDF, Total and others showed interest – slowly the cost/Wp mentality was replaced by costs/kWh thinking because the large companies were able to calculate the advantage themselves. These facts are sketched in Figure 1.9.

The most cost-effective technology in terms of Wp is still the simple monofacial mc-Si technology including monofacial mc-Si solar cell with homogeneous Al-BSF and a monofacial white backsheet module technology. What is not included in the COO module calculations is

(1) because of the higher power of the more complex module, the savings of BOS costs, and
(2) the prolonged lifetime when applying a double-glass module.

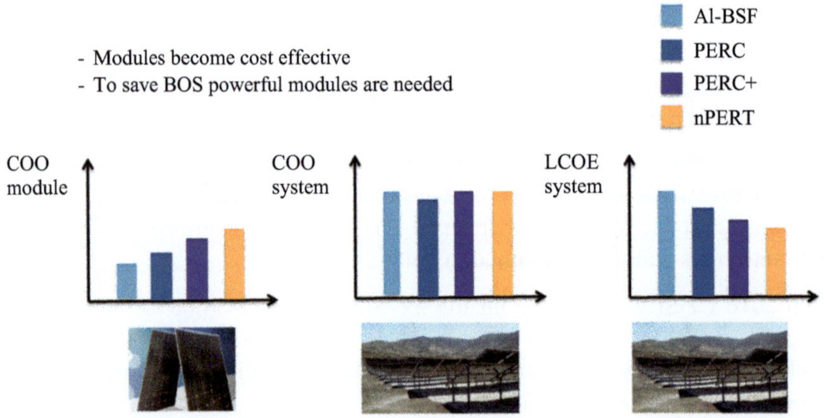

Figure 1.9 Schematic graphs for COO calculations for the most prominent c-Si technologies on module and system level and respective LCOE calculations. The trend in module COO and LCOE of a system is reversed due to savings of BOS and longer lifetime of double-glass modules

The aspect (1) is reflected in COO of the system and bifaciality and (2) in the LCOE. At the end, for most of the cases, the lowest cost electricity prediction is generated by high-power bifacial double-glass modules.

On top of all this the Chinese government is pushing innovations supporting companies with high efficiency technologies. Since 2018, a solar cell producing company in China has to demonstrate a 23% device in order to be supported by the Government.

1.3 Bifacial book 2018

There were many publications in the past, and currently the numbers of publications in this field are even showing an exponential growth. Therefore, we have decided to write a bifacial book to cover and summarize all technical and economical issues.

1.3.1 Latest bifacial publications and presentations

The publications are mostly done at PVSEC and IEEE conferences, in PV international magazine [13,17,18], PV Tech Blogs [19,20] and other magazines also by other authors as well as in various scientific journals. The 'Taiyang News Bifacial Solar Reports 2017 and 2018' also offer great summaries [21,22]. Lately Photon had an issue dedicated to this technology as well, see Figure 1.10.

As already mentioned, bifacial PV workshops take place since 2012 in Konstanz; in 2017 the workshop came back to Konstanz again. The workshop had 130 visitors from 22 countries (see Figure 1.11). In 2018, it will take place in September in Denver co-organized by NREL and Sandia: www.bifiPV-workshop.com. All presentations from the beginning can be found on that webpage and at a webpage

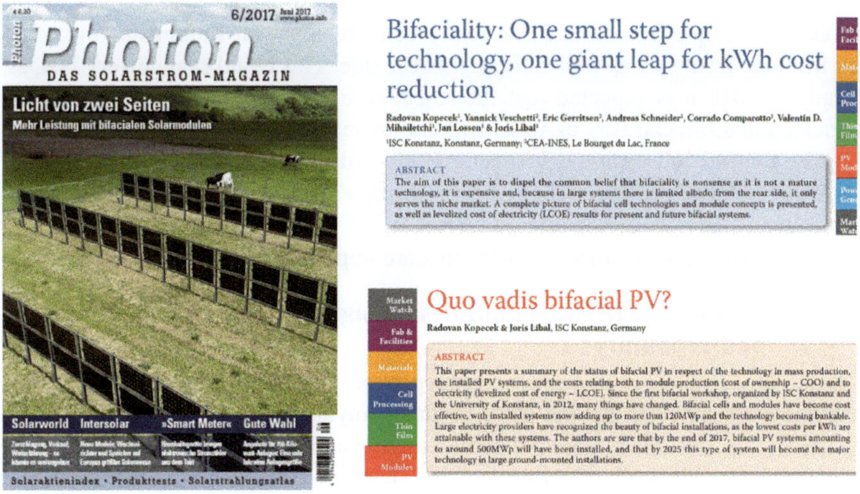

Figure 1.10 Photon6/2017 and PV International 26/2016 and 35/2017

Figure 1.11 4th bifiPV workshop 25/26 October 2017 in Konstanz

from Sandia [23]. At the Intersolar Europe show in June 2017 in Munich many large PV manufacturers announced that bifaciality will be mainstream in the coming years, so e.g. LONGi [24].

1.3.2 Chapters of our bifacial book

The following chapters of the book are structured in the same way as the sessions of bifacial workshops are organized. There are technological and economical parts. The technological parts are always separated into cell, module and system technologies, standardizations and simulations of bifacial energy yield gain. The standardization procedures are included in the cell and module chapters of the book. Then bankability issues are addressed and LCOE calculations are shown and discussed.

1.3.2.1 Chapter 2: Bifacial cells

In 2017, the majority of the PV modules installed are based on p-type silicon solar cells that feature a full-area aluminium rear contact and BSF. Because of their fully covered, opaque rear side, these cells are unable to convert the light that falls on the rear side of the solar cells into electricity. On the other hand, solar cells with a rear side that is only partially covered with metallization (so-called bifacial solar cells) are able to simultaneously and efficiently convert light that illuminates the solar cell from the front side as well as from the rear side. This could change very fast, as already in 2018 it is expected that the so-called PERC technology, which can be fabricated in a bifacial way as well, will dominate the market.

The bifacial cell technology chapter is divided into the history of bifacial cell, properties and measurements. However, the focus is set on the description of different cell technologies – their properties, advantages and disadvantages. This chapter provides an update on all cell technologies that are important for bifacial use.

1.3.2.2 Chapter 3: Bifacial modules: design options, characterization and reliability

Bifacial modules are a logical consequence of PV's evolution. Many module manufacturers move to double glass modules anyhow, due to longer lifetime of the module. As module glasses are getting thinner and more cost effective, the used white back sheet is replaced by rear glass. Therefore, it is also possible to get rid of the costly module aluminium frame which is required for a standard glass/backsheet module to provide a sufficient mechanical stability. For many module manufacturers, it makes also sense to build monofacial double-glass modules. For bifacial applications, many things have to be optimized – so e.g. the shape and placement of a junction box which should not shadow the rear side of the bifacial module. When it comes to bifaciality and additional light penetrating into to module from the rear side, it is obvious that large current densities will be generated in the bifacial devices. In order not the loose electricity by resistive losses, the metallization has to be adapted accordingly. Interconnection technologies and other module properties will be summarized.

In addition, besides the different module technologies which are described, standardization for module measurements is the most important topic. These standards are needed to provide to the end-customers a measure that is comparable with the Wp (at STC) for the monofacial modules, so that the end-customers perfectly know what they get in terms of energy production capacity (kWh/kWp). This is an extreme complex topic, so simplifications are needed.

1.3.2.3 Chapter 4: Energy yield simulations and design rules for bifacial PV systems

The early work that demonstrates power gain exploiting bifacial modules goes back to the late twentieth century. Recently, a significant endeavour has been put to quantify and predict the energy yield of bifacial PV modules for different installation configurations; stand alone, PV plants, east and west orientation. The main reason behind these relevant investigations is to determine the LCOE of bifacial modules and therewith their profitability. Hence, it is necessary to determine the annual energy yield of the bifacial modules.

Compared with standard modules, the calculation of the energy of bifacial modules is more complex, introducing further dependencies, such as the module installation height, the ground reflection coefficient (albedo), the distance between module rows and between neighbouring modules of the same row, and the self-shadowing of the modules on the ground.

To solve this question, mathematical modelling and simulation were found to be a key solution; especially bifacial modules are poorly installed worldwide compared to standard solar panels in present time. At the time of writing this book, commercial software for the energy yield simulation of bifacial modules is just entering the market and still requires extensive validation with experimental results and field data. The availability of such reliable and validated tools will bring on several advantages for bifacial photovoltaic, both academically and commercially. They will allow for better understanding of the performance of bifacial modules and will advance the bankability of projects utilizing bifacial PV modules.

This chapter gives an overview over the challenges related to the energy yield forecast for bifacial PV modules and systems. Different theoretical approaches for modelling of the rear side illumination are shown, and one specific model is used for calculating the expected bifacial gain for some sample scenarios to illustrate the potential of bifacial PV in terms of energy yield and to show the importance of system parameters such as ground albedo, module height over the ground and diffuse fraction of solar irradiance.

1.3.2.4 Chapter 5: Bifacial PV systems and yield data (bifacial gain)

A major motivation for bifacial photovoltaics is an expected additional energy yield, compared with monofacial panels, due to the two-sided light sensitivity. The potential for an improved module power output and energy yield was repeatedly demonstrated by simulations, measurements on stand-alone modules, or installations in various orientations. However, uncertainties concerning the actual output of projected systems still deter possible investors. Even in the PV community, the real quantitative benefit due to bifaciality as well as the best suited technical concepts is still under discussion, as reflected by numerous publications dealing with these topics. For monofacial modules, the clearly defined measurement of the nominal power under standardized conditions (STC) is the accepted classification criteria, and the prediction of the system's energy output with sufficient accuracy is comparatively straightforward with available simulation tools. This is not the case for bifacial devices and systems.

The system chapter describes the application variety of bifacial systems and summarizes the experimental and large bifacial PV systems 'out there'. The most important information about the existing systems is the bifacial gain which is reported for each installation if the data are available.

1.3.2.5 Chapter 6: Impact of bifaciality on the levelized cost of PV generated electricity

The cost of electricity generated by photovoltaic systems is an important criteria that determines the competitiveness of PV in general compared to other – fossil and renewable – methods of electricity generation and that serves also to determine the best choice – from the economic point of view – in terms of PV module technology and system configuration for a given application and location. The reduction of the cost of PV-generated electricity is the driving force behind all research and development activities along the whole value chain of PV manufacturing, starting from the purification of the silicon feedstock and ending with the design and construction of PV systems and their components.

The concept of levelized cost of energy is explained from the theoretical point of view, and in the following, LCOE calculations for monofacial and bifacial PV systems are performed and show the potential of bifacial PV to further reduce the LCOE of PV-generated electricity. In addition, the impact of various factors, such as solar irradiance and additional energy yield by bifacial modules, financing conditions, system price and ground cover ratio, is evaluated by a sensitivity study.

1.3.2.6 Chapter 7: Importance of bankability for market introduction of new PV technologies – bifaciality as example

In this chapter, the concept of bankability for PV power plant projects in general is explained, and examples of important criteria for the related evaluation procedures to be performed by banks and investors are shown and the implications for emerging technologies – such as bifacial PV – are discussed. We will see how difficult it is to bring new technology on the PV market as many criteria are required to be met.

1.3.2.7 Chapter 8: Geographic distribution of bifacial gain

It may be straightforward to get an increased yield from a bifacial system compared with a monofacial system in the same location and having the same orientation. But how much will that extra gain be, and in which geographical regions can we expect bifacial systems to be most advantageous? And where does it make sense to deploy particular configurations such as vertical bifacial systems?

This chapter describes the influence of geographical location on the bifacial gain for interesting installation cases.

1.3.2.8 Chapter 9: Summary

The last chapter will summarize all the findings and describe the future of bifacial PV.

We wish you a lot of fun with all the chapters and we hope that you will become a bifacial believer and if you already are that you will understand in more detail how to apply and/or how to promote bifacial technology.

References

[1] *India cancels plans for huge coal power stations as solar energy prices hit record low* [online]. Available from: https://www.independent.co.uk/environment/india-solar-power-electricity-cancels-coal-fired-power-stations-record-low-a7751916.html

[2] *EGEB: 138GW of solar power in 2018?, California net zero residential after 2020, Trump Tariff attacked, more* [online]. Available from: https://electrek.co/2018/02/22/egeb-138gw-solar-power-2018-california-net-zero-residential-after-2020-trump-tariff-attacked-more/

[3] *Bloomberg's energy outlook* [online]. Available from: https://www.bnef.com/dataview/new-energy-outlook/index.html

[4] *International technology roadmap for PV* [online]. Available from: http://www.itrpv.net/

[5] *The Paris agreement* [online]. Available from: https://www.c2es.org/international/paris-agreement?gclid=CMyMnK2XktQCFQ4R0wodkOkMGg

[6] *Abu Dhabi plant to produce region's cheapest electricity from solar* [online]. Available from: http://www.thenational.ae/business/energy/abu-dhabi-plant-to-produce-regions-cheapest-electricity-from-solar

[7] IRENA, Renewable Energy Auctions, Analysis 2016.

[8] *India's Madhya Pradesh auction nation's lowest-priced solar* [online]. Available from: https://www.pv-magazine.com/2017/02/09/indias-madhya-pradesh-auctions-nations-lowest-priced-solar/

[9] *Saudi Arabia gets cheapest bids for solar power in auction* [online]. Available from: https://www.bloomberg.com/news/articles/2017-10-03/saudi-arabia-gets-cheapest-ever-bids-for-solar-power-in-auction

[10] *Martin Green's graph for 3 PV generations* [online]. Available from: https://depts.washington.edu/cmditr/modules/opv/solar_technologies.html

[11] *SNEC 2017: The next big solar module trend on display* [online]. Available from: https://www.pv-tech.org/editors-blog/snec-2017-the-next-big-solar-module-trend-on-display
[12] *AT&T archives: the bell solar battery* [online]. Available from: https://www.youtube.com/watch?v=3I1JhyOahIw
[13] Kopecek R., and Libal J. Quo vadis bifacial PV? *PV International 35*, March 2017.
[14] *SNEC 2017: World's first full-size IBC bifacial module to be displayed* [online]. Available from: https://www.pv-tech.org/news/snec-2017-worlds-first-full-size-ibc-bifacial-module-to-be-displayed
[15] *EDF wins 90 MW solar project in Mexico's second auction* [online]. Available from: https://www.pv-magazine.com/2016/10/04/edf-wins-90-mw-solar-project-in-mexicos-second-auction_100026345/
[16] *World's largest (10 MW) Bifacial Rooftop Installation from Sunpreme under final completion* [online]. Available from: https://www.youtube.com/watch?v=jnxdHCsVeHk
[17] Kopecek R., and Libal J. Bifaciality: one small step for technology, one giant leap for kWh cost reduction. *PV International 26*, January 2015.
[18] Kopecek R., and Libal J. Cost/kWh thinking and bifaciality: two allies for low-cost PV of the future. *PV International 30*, February 2016.
[19] *Bifaciality: one small step for technology, one giant leap for kWh cost reduction* [online]. Available from: https://www.pv-tech.org/guest-blog/bifaciality_one_small_step_for_technology_one_giant_leap_for_kwh_cost_reduc
[20] *The second summer of new PV technologies – chances for cell and module production outside Asia* [online]. Available from: https://www.pv-tech.org/guest-blog/the_second_summer_of_new_pv_technologies_chances_for_cell_and_module_produc
[21] *TaiyangNews Bifacial Solar Report 2017, Why it's time to produce solar power on both module sides* [online]. Available from: http://taiyangnews.info/reports/taiyangnews-bifacial-solar-report-2017/
[22] *TaiyangNews Bifacial Solar Report 2018, Bifacial Solar Module Technology* [online]. Available from: http://taiyangnews.info/reports/bifacial-solar-technology-report-2018/
[23] *Bifacial PV projects* [online]. Available from: https://pvpmc.sandia.gov/pv-research/bifacial-pv-project/
[24] *Intersolar Europe 2017: bifacial will be mainstream in two years says LONGi* [online]. Available from: https://www.pv-tech.org/news/intersolar-europe-2017-bifacial-will-be-mainstream-in-two-years-says-longi

Chapter 2
Bifacial cells

Ingrid Romijn[1], *Gaby Janssen*[1], *Thorsten Dullweber*[2], *Bas van Aken*[1], *Naftali Eisenberg*[3], *Lev Kreinin*[4], *Matthieu Despeisse*[5], *Valentin Mihailetchi*[6], *Jan Lossen*[6], *Wolfgang Jooss*[7] *and Radovan Kopecek*[6]

2.1 Introduction

In 2017, the majority of the PV modules installed are based on p-type silicon solar cells that feature a full-area aluminum rear contact and back surface field (BSF). Because of their fully covered, opaque rear side, these cells are unable to convert the light that falls on the rear side of the solar cells into electricity.

On the other hand, solar cells with a rear side that is only partially covered with metallization (the so-called **bifacial solar cells**) are able to simultaneously and efficiently convert light that illuminates the solar cell from the front side as well as from the **rear side** (see Figure 2.1).

When bifacial cells are mounted in a module with a reflecting back sheet or with a reflecting material between and behind the cells, transmitted light is reflected back into the cells, resulting in an increased monofacial module efficiency [1,2]. Moreover, if bifacial cells are assembled into a module using a transparent rear cover (i.e. glass or transparent foil) these bifacial modules can lead to a much higher energy output (yield) compared to standard modules under the same conditions. This is due to the additional energy generated by rear irradiance. The additional energy generated by the rear irradiance can range between 5% and 90% of the energy generated by only the front-side illumination. This percentage depends not only on the properties of the module, and on the rear-side conversion efficiency but also strongly on the location, orientation, and immediate surroundings of the

[1]Nederlandse Organisatie voor Toegepast-Natuurwetenschappelijk Onderzoek, Solar Energy Department, The Netherlands
[2]Institute for Solar Energy Research Hamelin GmbH (ISFH), Germany
[3]Lev Academic Center, Israel
[4]SolAround Ltd., Lev Academic Center, Israel
[5]Swiss Center for Electronics and Microtechnology, Switzerland
[6]International Solar Energy Research Center, Konstanz
[7]RCT Solutions GmbH, Germany

module. For equator-facing systems in places with a high direct irradiance, the reflective properties of the ground will determine the rear irradiance, whereas in other climate conditions diffuse light falling on the rear will have a large contribution (see Figure 2.2). The additional energy yield can be particularly high (up to 90%) in those cases where the modules are not perfectly faced towards the equator and therefore also can benefit from direct irradiance on the rear. It has been predicted that at certain locations in the world vertically placed modules with an east-west orientation can have a larger annual energy yield than equator-facing modules [3].

In this chapter, a short review of the history (Section 2.2), physics (Section 2.3), characterization (Section 2.4) as well as a description of the five most common cell architectures of n- and p-type bifacial solar cells (Section 2.5) will be given.

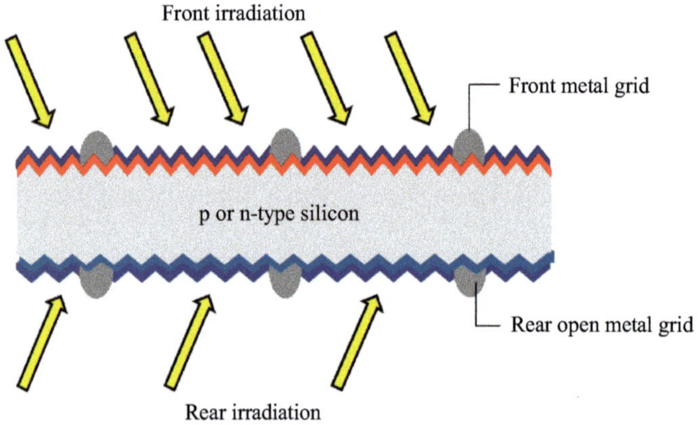

Figure 2.1 Cross section of a bifacial solar cell that can convert light from the front side as well as from the rear side into energy

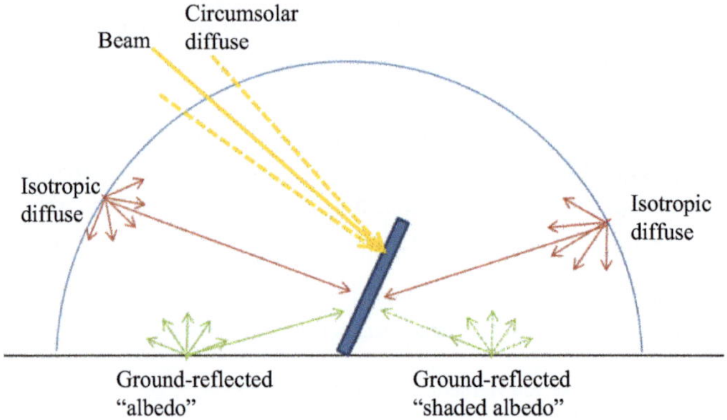

Figure 2.2 Different contributions to the front and the rear irradiation of the bifacial solar cell or module. Reproduced from Duffie & Beckman: Solar Engineering of Thermal Processes [4]

2.2 History of bifacial cells (from 1960 to 2016)

A nice and comprehensive overview of the early history of bifacial solar cells has been published by Cuevas in 2005 [5].

Bifacial solar cells were already made in laboratory environments as early as 1960. The first approach to build a bifacial solar cell was to create a collecting p^+ junction on both surfaces (front and rear) of an n-type silicon wafer, basically forming a p^+np^+ structure. The metal contacts were placed on the sides of this small solar cell, which is described by H. Mori in a patent in 1960 [6] and can be seen in Figure 2.3. In 1961, measurements of solar cell with a similar structure p+np+ structure made of low resistivity Si wafers were reported by Zaitseva and Fedoseeva [7].

Using different materials, but similar design, bifacial GaAs solar cells with front and back p-n junctions were measured and analyzed by Bordina *et al.*, in [8].

The motivation for this double junction design was to improve the collection efficiency for long-wavelength photons, because at that time the efficiency of solar cells was limited by the diffusion length of minority carriers in the silicon bulk. Another reason to work on such a bifacial solar cell was to improve the surface passivation by creating a junction at both surfaces (although diffusion lengths in crystalline silicon wafers at that time were typically too short to feel the surface recombination on the back when illuminated from the front side).

Between 1970 and 1977, many more research groups reported their work on bifacial solar cells [9–12]. The first experimental bifacial devices pioneered by a research group at the UPMadrid in Spain [10] had conversion efficiencies around 7%. In [11,12], it was shown that solar cells based on high resistivity Si, or solar cells with p-i-n structure (characterized by injection level excided doping level), are able to work effectively with both side illumination, i.e. as bifacial cells. Amongst the device structures were transistor-like solar cells (n^+pn^+) [13,14] as well as BSF -type solar cells with the p-n junction on only one (usually front) side of the device and high-low junction (BSF) on second surface. These turned out to be the most usable structures for bifacial cell, and indeed those structures are most common today for both monofacial and bifacial solar cells. This type of bifacial cell design was first proposed by Bordin, and was patented first in Russia in 1970 [15] and later in the United States [16] in 1976. Analysis of these bifacial cells behavior at both

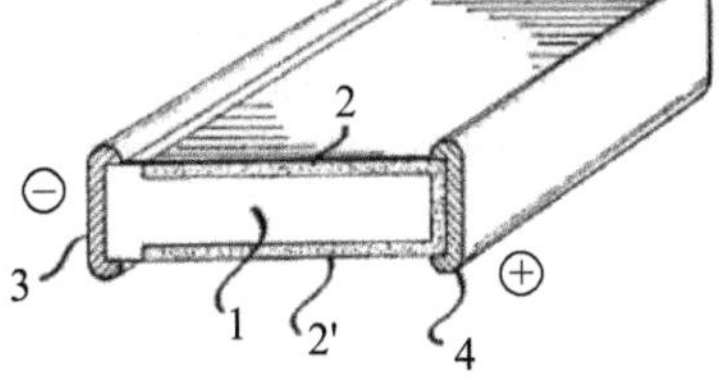

Figure 2.3 After Andres Cuevas [5]: The first published bifacial double junction solar cell, in US patent 3.278.811 [6]. The numbers indicate 1: n-type silicon, 2 and 2': p-type emitter regions. 3, 4 are the metal contacts

sides illumination was published in 1975 [17]. As already mentioned above, the main advantage of the n^+pn^+ (or p^+np^+) structures was in its enhanced collection efficiency for long-wavelength photons and surface passivation.

Alternatively, it was found that a BSF, having the same doping type as the base material, could as well reduce recombination and improve the V_{OC}. The first high-efficiency p+nn+ bifacial solar cells were presented by the UPMadrid group in 1981 [18]. Efficiencies up to 15.7% were obtained on 5 cm^2 bifacial cells under front illumination. Using rear illumination, 13.6% efficiency was reached, indicating a bifaciality of over 85%. Those cells were essentially the same as later fabricated by the company Isofoton in industrial production.

The challenge of collecting carriers generated near the rear surface led to an improved understanding of the physical mechanisms of surface and bulk recombination, and essentially many of the advanced solar cell designs with passivated rear surfaces such as p-type passivated emitter and rear cell (p-PERC), interdigitated back contact cells (IBC) or n-type passivated emitter rear totally diffused cells (n-PERT) that are used and manufactured in mass production today evolved simultaneously with bifacial solar cells. Efficiencies of bifacial cells quickly improved and values above 20% had been reported in several papers by the year 1980 [13] and 2000 [14,19]. Bifacial solar cells were used in space applications at an early stage because bifacial cells have less parasitic infrared absorption, resulting in reduced operating temperatures, as well as increased sunlight collection from the Earth's reflection [20–22]. In fact, the initial development of bifacial solar cells (as well as most solar cell research) for instance in Russia was for space application. Calculations of the anticipated energy gain due to Earth's reflection by bifacial solar cells as an energy source for Earth satellites were proven by space tests in 1974 of bifacial modules [23]. Comparative measurements of optical parameters of bifacial and regular cells (reflection and transmission) and calculation of equilibrium cell temperature in space was described in [24]. Eight and 10 kW bifacial space solar arrays are components of the International Space Station [25]. A 10% to 20% energy generation increase due to substitution of regular cells by bifacial cells was recorded on other space missions.

In 1980, the use of albedo irradiance on earth (albedo is the reflection coefficient from an object or nearby environment) to increase the power output of bifacial solar cells was first realized and described by Cuevas *et al.* [26,27]. Experiments were conducted on bifacial and monofacial cells in the same module setup in a high-albedo site. The ratio of the bifacial to the monofacial short-circuit current J_{SC} was always close to 1.45, in some cases (early morning, cloudy days) even higher 1.55. The module output powers were subsequently calculated, and it was found that a bifacial module would produce about 50% more electrical power than the conventional module [26].

It took until 1989 for the first terrestrial applications of bifacial cells in actual bifacial modules to appear. Some of the earliest bifacial modules were assembled by TNC and mounted in noise barriers along railways in Switzerland [28]. In 2003, researchers from the Institute for Solar Energy Research Hamelin (ISFH) reported on new applications of bifacial cells in shading elements for windows [29].

Between 2004 and 2008, large-scale production of silicon based solar cells and modules took off. The large scale, mainstream production was based on p-type silicon, full aluminum BSF cells (Al-BSF)—i.e. monofacial solar cells and resulting in monofacial modules. The focus of most research institutes and R&D departments of the larger manufactures was on increasing cell efficiency at reduced costs. It soon became clear that two main limitations of p-type Al-BSF solar cells were (1) the rear-side recombination of the BSF and (2) the lifetime in p-type Cz or multi-crystalline material. One direction of research to increase the efficiency therefore was aimed to overcome the recombination on the rear side. This lead to the passivated emitter and rear cell (PERC) solar cells, in which the rear side is passivated by a dielectric layer with only local aluminum contacts [30]. The second option that was pursued was aimed to overcome the limitations in bulk lifetime by the use of n-type base material. Compared to p-type, n-type has higher lifetimes and is less sensitive to common (metal) impurities [31], therefore it is typically used for high-efficiency cell concepts such as n-type heterojunction or IBC solar cells. Both PERC and various n-type cell concepts are naturally suitable for bifacial solar cells, as will also be described in Section 2.5.

In the late 1980s, the Japanese company Sanyo, the first manufacturer worldwide to commercially produce solar cells made of amorphous silicon (a-Si:H), started to integrate such a-Si:H materials into contact layers for n-type crystalline silicon solar cells. Major breakthroughs were achieved in the 1990s with the insertion of a thin intrinsic a-Si:H layer between the doped a-Si:H layer and the silicon wafer, and by applying such heterostructure at the front and at the rear of the solar cells, thus defining the Heterojunction with Intrinsic Thin-layer (HIT®) technology. These HIT® modules entered serial production in 1997, and market introduction in 1998. Still, although the HIT solar cell is bifacial by nature, the first modules were manufactured as monofacial modules. In 2006, Sanyo introduced the bifacial HIT module HIT-Double™, using a glass–glass structure. However, bifacial HIT-Double installations remained niche applications.

The first n-PERT solar cells, which in downstream processing (e.g. soldering and encapsulation) resemble more the prevailing Al-BSF cells, became commercially available around 2010 [32]. In addition, these bifacial cells were firstly assembled in monofacial modules. In 2011, the company PVGS first applied n-PERT cells in a bifacial module [33]. Their publications of the huge increases in energy yield for bifacial modules against reflecting backgrounds (e.g. snowy underground) rekindled the interest of the PV community. Most research institutes started investigations on one or another variation of bifacial solar cell types, see also Section 2.5.2.

Nowadays, bifacial modules based on n-type cells with a passivated emitter and rear totally diffused BSF (n-PERT cells, see also Section 2.5.2) are produced by several companies, including LG, Yingli Solar and Neo Solar Power. In addition, n-type bifacial cells based on silicon heterojunction (SHJ) technology (see also Section 2.5.1) are still being produced by Panasonic/Sanyo (HIT™ technology), but also several new companies (such as Sunpreme) are ramping up production of heterojunction bifacial cells and modules. These cell architectures on n-type silicon base material—either implementation of boron- and phosphorous-doped surfaces passivated with transparent dielectric layers for n-PERT, or ultra-thin n- and p-doped amorphous silicon covered with light-transmitting transparent conductive oxides

(TCOs) for the heterojunction cells—are inherently bifacial. The front and rear metal grids are typically H-patterned screen-printed silver.

Standard Al-BSF p-type cells are not bifacial as they rely on the passivation and contacting of a full-area aluminum layer on the rear side. However, the next generation of p-type solar cells that is based on PERC can be made bifacial and is named PERC+ (see also Section 2.5.4) [34]. The rear side of monofacial PERC and bifacial PERC+ cells is passivated by one or more dielectric layers. In case of PERC+ cells, the rear contacting can be done by screen-printing an aluminum H-pattern finger grid on top of laser contact openings (LCOs) of the rear passivating layer [34], whereas for PERC cells contact is made by screen-print of a full-area aluminum layer.

Converting a production line from monofacial PERC cells to bifacial PERC+ cells requires no additional or other production tools but mainly to change the screen layout of the Al screen-print. As PERC cells already account for around 20% of the global silicon solar cell production capacity, the bifacial PERC+ concept represents an easy to implement bifacial cell concept for many solar cell manufacturers. Accordingly several cell and module manufacturers such as SolarWorld, Neo Solar Power, Trina Solar and LONGi Solar, are now producing bifacial PERC+ solar cells.

At a more prospective development phase, back contacted solar cells also are potentially bifacial, as demonstrated by research institutes, such as ISC Konstanz with bifacial Zebra cells [35]. A more detailed overview of the most common bifacial solar cells types that are currently either in or close to (mass) production will be given in Section 2.5.

2.3 Characteristics of bifacial cells

2.3.1 Bifaciality factor

An important parameter of any bifacial device is its bifaciality factor [36]. This factor describes the ratio between the rear and the front responses of the device under the same circumstances. The bifaciality factor of a module essentially allows to determine the additional energy that can be generated by the rear irradiance.

Here, we will define the bifaciality factor φ for current density J_{SC}, voltage V_{OC}, power and efficiency

$$\varphi_{Jsc} = Jsc_r/Jsc_f \tag{2.1}$$

$$\varphi_{Voc} = Voc_r/Voc_f \tag{2.2}$$

$$\varphi_{Pmax} = \text{Pmax}_r/\text{Pmax}_f \tag{2.3}$$

$$\varphi_{eta} = eta_r/eta_f \tag{2.4}$$

The bifaciality factor is usually expressed as a percentage. For example Jsc_x is the short-circuit current density at standard test conditions (STCs) under single-sided illumination, with index x: f for front and r for rear side. The measurement of the front and rear IV characteristics of bifacial cells can be compromised by stray light falling on the dark side. Section 2.4.1 describes the procedures for correct characterization of the front and rear cell performance.

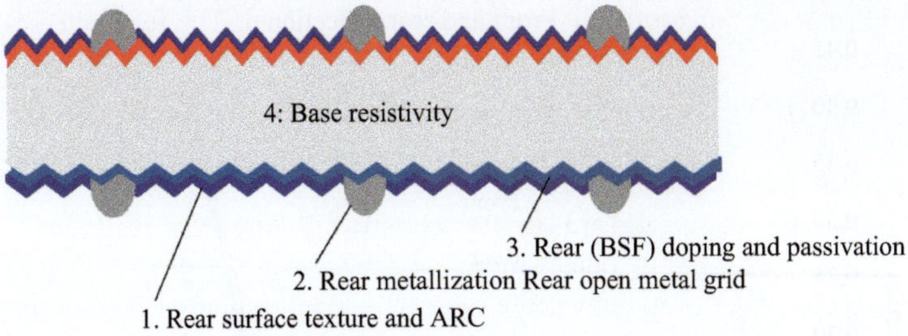

Figure 2.4 Cross section of bifacial cell, with the parameters that influence the cells bifaciality factor φ_{eta} indicated by numbers 1 to 4

2.3.2 Parameters influencing the bifaciality factor φ

Most bifacial solar cells are not symmetrical. They feature for instance an emitter on one side (usually the front) and the BSF on the other side (rear), as well as a different optimization of metal grid on either side. The front has a grid optimized for 1 Sun illumination, whereas at the rear the optimization is typically for reduced illumination, therefore potentially towards a pattern providing more shading on the rear than on the front but with reduced resistive losses. Due to the nonsymmetrical layout, the response of a cell to front or rear illumination is not the same, which is reflected in a bifaciality factor of the cells deviating from unity.

In this section, the most commonly used layout of bifacial cells will be used to describe some of its physics: H-patterned (screen or stencil) printed contacts on both front and rear side, an emitter on the front and optionally a BSF on the rear side (see Figure 2.4).

The main characteristics that determine the bifaciality factor φ_{eta} of a cell with such specifics are:

1. rear surface texture and antireflection coating (ARC)
2. metal coverage on the rear side
3. rear side (BSF) doping and passivation
4. base resistivity and lifetime of the solar cell.

The first two characteristics govern the light coupling into the cell to create charge carriers, while the other characteristics determine if the charge carriers will subsequently recombine or be collected at the electrodes and contribute to electricity generation.

2.3.2.1 Rear surface texture and ARC

The influence of rear surface texture and ARC to the rear J_{SC} is similar to that of the front texture to the front J_{SC}. To optimize the bifaciality factor and increase the rear J_{SC_r}, the texture and ARC can both be tuned for low reflection and optimal light coupling. Ideally, this should be realized without lowering the front-side short-circuit current J_{SC_f} and without compromising on the rear surface passivation. The optimization between the above factors typically results in a (slightly) different texture on front and rear side. In Figure 2.5, reflectance curves of both the front and

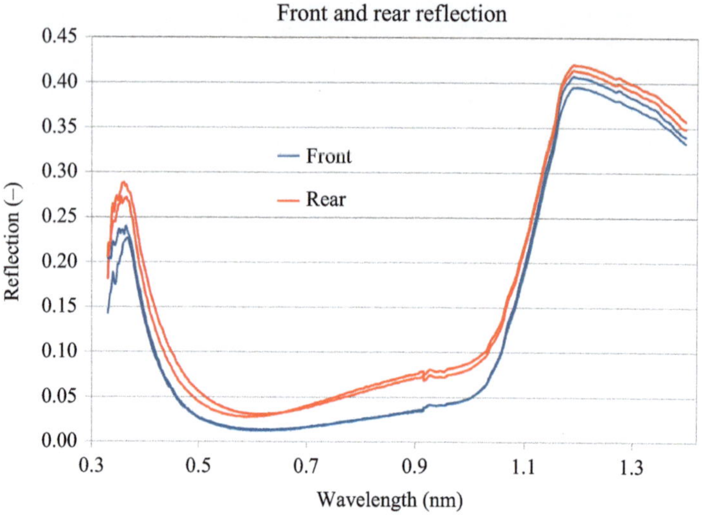

Figure 2.5 Front and rear reflection curves of two bifacial cells with similar front and rear texture and antireflective coatings. The higher absolute rear reflection is caused by the higher metallization fraction on the rear side

Figure 2.6 Photographs of bifacial solar cells on mirror-tables, showing front and rear metallization

the rear side of a bifacial cell are shown. The texture and antireflective coatings are similar. The higher absolute rear reflection is caused by the higher metallization fraction on the rear side.

2.3.2.2 Metallization

Most bifacial cells feature an H-pattern metallization, with several (typically 3–6) busbars and a multitude of fingers (see Figure 2.6). Like for the front side, the number and width of fingers and busbars of the rear-side metal grid is a compromise between recombination losses at the contacts, shading losses for rear-side illumination, resistive losses and silver paste consumption (cost). The rear current J_{SC_r} will be directly influenced by the metallization fraction on the rear, while the fill factor (FF) will be governed by the resistive losses in the grid, the contact resistance and lateral conductivity in the silicon. Both V_{OC} and FF can also be adversely influenced by recombination at the contacts.

Additional lateral conductivity in a highly doped BSF contributes to a good FF despite the open rear-side metallization and increases the tolerance to high substrate resistivity. If bifacial cells do not feature a doped BSF region, as is the case in the p-PERC+ cells (see also Section 2.5.4), the lateral conduction of charge carriers between the rear metal contact has to rely on the base conductivity. To prevent too large resistive losses, the rear metallization pitch—the distance between the metal fingers—has to be restricted, adversely influencing the rear J_{SC_r} unless thinner metal fingers can be printed or deposited.

2.3.2.3 BSF

For application in monofacial solar cells, the BSF is optimized firstly for passivation and secondly for conduction, to complement the lateral conductivity in the base. However, for bifacial application different requirements must be met for passivation and conduction. If the partial metallization at the rear has a larger pitch, the doping level has to be increased to compensate by increased conductivity for the larger path length. On the other hand, the J_{SC_r} is very sensitive to surface recombination at the rear and, particularly in cells with a front-side junction, also to Auger recombination at the rear. This latter effect will be explained below. Moreover, a high doping level will also give rise to a higher free carrier absorption (FCA) as compared to a lowly doped or absent BSF. The FCA reduces both front and rear J_{SC_r}.

The metallization and BSF will therefore have to be optimized together, depending on the type of bifacial solar cell (i.e. n-PERT, p-PERC, or HIT).

2.3.2.4 Base resistivity and lifetime

Comparing the response of a *front-junction* bifacial solar cell to front illumination with that to rear illumination, the main effect is that, when under rear illumination a current is drawn from the cell, the minority charge carriers have to travel by diffusion through the bulk towards the junction. Under front-side illumination majority carriers are driven by the electric field towards the BSF. The results are (1) a larger gradient of the excess carrier concentration Δn over the bulk, and (2) a higher average excess carrier concentration in the bulk under rear illumination, as shown schematically in Figure 2.7. Under rear illumination, this high Δn at the rear results in an enhanced recombination current J_{recomb} at the BSF, as J_{recomb} depends both on the doping concentration N_D and on the excess concentration Δn in the base material close to the BSF:

$$J_{recomb} = J_{0\_BSF} \frac{(N_D + \Delta n)\Delta n}{n_i^2} \tag{2.5}$$

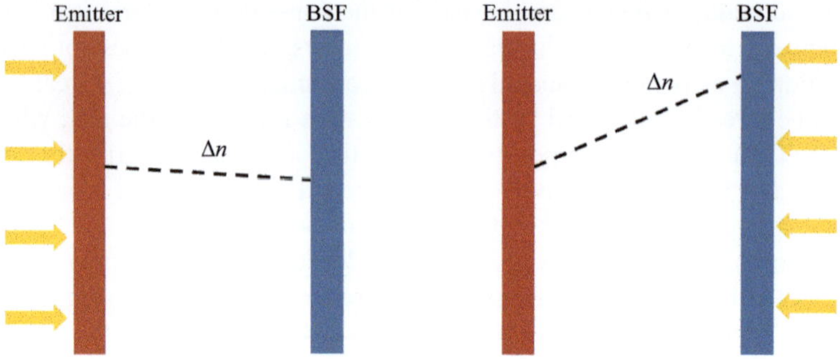

Figure 2.7 Gradient in carrier concentration for a bifacial solar cell subjected to front (left) or rear (right) irradiation

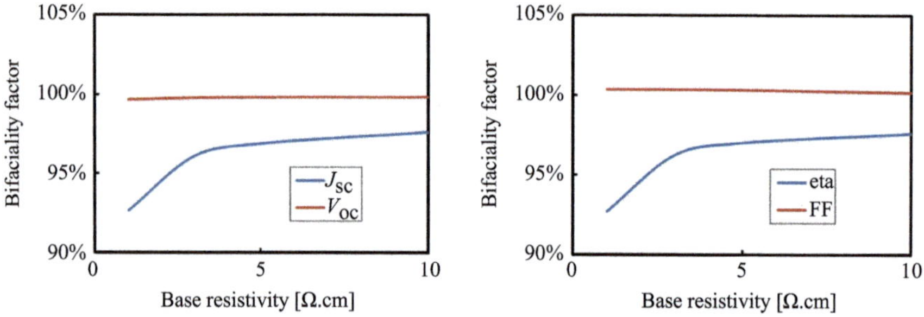

Figure 2.8 Bifaciality factor dependence on base resistivity for the cell parameters J_{SC}, V_{OC} (left) and FF, efficiency (right)

Here J_{0_BSF} is the recombination parameter of the BSF. At the emitter side the recombination will be reduced under rear illumination, but the difference in Δn at the emitter under front or rear illumination is much more smaller than at the BSF. As a lower base resistivity corresponds to a higher doping concentration ($N_D \sim 1/\rho$), and as at short-circuit conditions $\Delta n << N_D$, the recombination increases with decreasing resistivity. This means that almost always $J_{SC_r} < J_{SC_f}$ and that φ_{Jsc} can strongly increase with increasing resistivity, as shown in Figure 2.8 [37]. But note, that a reduction of the recombination parameter $J_{0,BSF}$ will also improve the bifaciality factor, as already mentioned above. Because of the higher average excess concentration for the rear illumination case, enhanced base recombination also reduces the J_{SC_r} compared to J_{SC_f}. A higher Shockley–Read–Hall lifetime will therefore also improve the $\varphi_{J_{SC}}$.

At open-circuit conditions, the excess concentration is much more uniform and much less dependent on the illumination side. Therefore, V_{OC} is hardly dependent on the base resistivity and the illumination side. The FF increases significantly with decreasing base resistivity, especially below 3 Ωcm, both for front and rear illumination. The combined effects on J_{SC}, FF and V_{OC} are that the front-side efficiency slightly increases with decreasing base resistivity whereas the rear-side

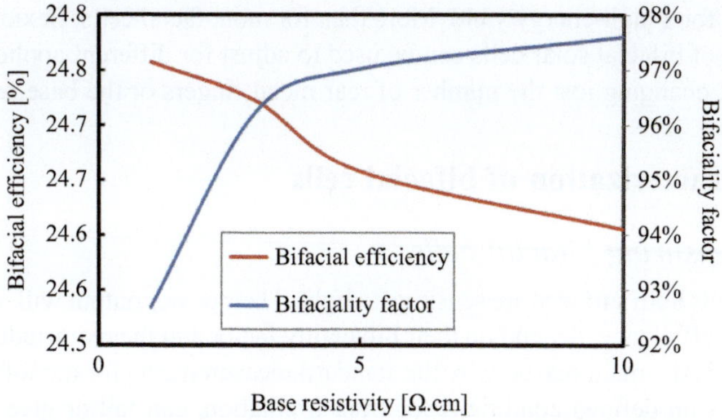

Figure 2.9 Simulation of bifacial efficiency (η_{bifi20}) and bifaciality factor (φ_{eta}) of a n-PERT cell as function of base resistivity. The definition of the bifacial efficiency η_{bifi20} will be further explained in Section 2.4.2

efficiency decreases, dominated by effects on the J_{SC}. This will result in a lower φ_{eta} for cells made on low resistive base material as compared to higher resistive material, as is shown in Figure 2.8 [38].

2.3.3 Design of bifacial cells

Similar to the monofacial cells, the specific design of bifacial solar cells will depend on the base material (p- or n-type) that is used, how the junctions are formed (for instance by diffusion, using heterojunction or local BSFs) and on the contacting methods (printed metallization, transparent conducting oxides (TCOs), plating). This has resulted in bifacial versions of well-known monofacial cell concepts such as heterojunction, n-PERT, PERC+ or IBC cells. Furthermore, for each type of bifacial solar cell the design will depend on the interconnection technology, number of busbars and optimization of costs. Finally, the cell response should be optimized according to the intended application. The cell can either be optimized on highest front-side efficiency only, on the highest efficiency under specific front and rear illumination (bifacial efficiency or η_{bifi}) or on the highest bifaciality factor φ_{eta}—typically these cases lead to different cell designs in terms of doping and metallization. For example in Figure 2.9, both the simulated bifacial efficiency (η_{bifi20}, calculated for 1,000 W/m^2 front and 200 W/m^2 rear irradiation) and the bifaciality factor φ_{eta} of a n-PERT cell are shown as function of n-type base resistivity. Low base resistivity results in a higher bifacial efficiency mainly due to increase in FF due to improved lateral conductivity. High base resistivity on the other hand results in higher a bifaciality factor due to reduced rear recombination effects as was described in Section 2.3.2.

The choice for optimization should depend on the expected front and rear irradiance. Typical equator-facing bifacial modules with moderate albedo would need optimization on the front-side efficiency for highest energy yield. In east-west vertical situations, or at very high albedo a high bifaciality factor will result would be favorable for a high-energy yield. More than for monofacial cells, flexibility in the basic design of bifacial solar cells can be used to adjust for different applications by, for example, changing just the number of rear metal fingers or the base resistivity.

2.4 Characterization of bifacial cells

2.4.1 Measuring bifacial cells

In bifacial cells both surfaces are sensitive to light. Their power output will, in addition to their front efficiency, depend on their bifaciality factor and the rear irradiance. This obvious fact is the main reason why the standard measurements for monofacial cells, that rely *only* on defined conditions for front irradiation, can fail or give erroneous results for bifacial cells [39]. Subsequently, any characterization of bifacial cells has to be done with care, and any stray irradiance on the rear side should be either avoided or well controlled. In this section, the focus will be on standard current and voltage (*IV*) measurements, but naturally similar care will have to be taken to avoid any non-controlled contributions from rear irradiance for, i.e. spectral response measurements, and any other measurements that involve illumination of the solar cells.

The international standards IEC 60904-1 and IEC 60904-9 [40,41] describe measurement procedures and standard test conditions for a correct measurement of solar cells. STCs include a defined temperature of 25 °C, a defined irradiance G of 1,000 W/m^2, and a defined specified spectrum (AM1.5G) [41,42]. This standardized measurement allows for comparison of efficiencies between cell types and between measurements done using different solar simulators. The 60904-1 standard has been defined and written for monofacial solar cells and modules, and needs to be adjusted for bifacial measurements. At the time of writing, an IEC project team is working on the adjustments of this norm for bifacial solar cells as well as modules, which is at the committee draft stage.

To measure the *IV* characteristics of a solar cell, the cell is illuminated with 1,000 W/m^2 from the front side at STC [42]. The current is measured during a voltage sweep (*IV* curve), after which this *IV*-curve is analyzed to find parameters such as the short-circuit current density J_{SC}, the voltage and current density at maximum power point (V_{MPP} and J_{MPP}) and the open-circuit voltage (V_{OC}). From these parameters, the efficiency of the solar cells is calculated via

$$\text{Efficiency} = J_{SC} * V_{OC} * FF/\left(1{,}000\ \text{W/m}^2\right) \tag{2.6}$$

where the FF is defined as

$$\frac{V_{MPP} \times J_{MPP}}{V_{OC} \times J_{SC}} \tag{2.7}$$

Basically two setups are frequent for *IV* measurements of solar cells:

- Measurements on a temperature-controlled chuck (kept at 25 °C during the whole measurement): Temperature stabilization enables the use of multiple flashes, or of a continuous light source. The rear side of the cell is in contact

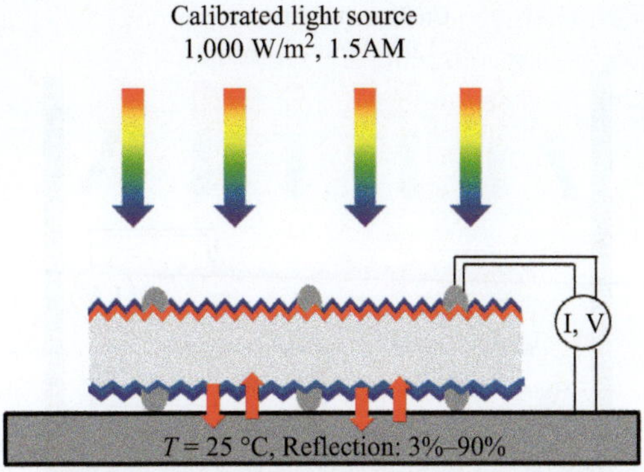

Figure 2.10 Laboratory IV setup where the solar cell is placed on a temperature controlled measurement chuck. The reflection depends on the material of the chuck (typically brass, copper coated with a gold layer or anodized aluminum)

with a chuck that controls the temperature and is used for rear contacts. The *IV* curve can be measured slowly, so that capacitance of the cell does not play a role, or be measured in several parts when multiple flashes are used. Measurements on temperature controlled chucks are typically restricted to use in laboratories, as the required test time is not suitable for mass production.

- Flasher measurements with rear contacts bare. This setup it typically used for inline measurements in a cell production line. The cell is contacted on both sides with multiple pins on the busbars, so in fact the rear side of the cell is open to the surroundings. Because the cell is not in (full) contact with a measurement chuck, the temperature cannot be controlled as well as in the laboratory case and typically only 1 flash of duration some tens to hundreds of milliseconds (also calibrated to 1.5AM) is used so as not to heat up the cell during testing. The cell temperature during measurement is recorded and the measured IV values are corrected to 25 °C, using prior determined temperature coefficients of the cell type.

In Figures 2.10 and 2.11, examples are given for the irradiation situation in typical measurement setups in laboratory and inline testers using bifacial cells. Both measurement configurations pose specific challenges for bifacial solar cells.

For standard test configurations, only the front irradiation of the tested cells is considered and calibrated. Any stray irradiation that falls on the rear side of the cells is not an issue for monofacial cells. However, on bifacial cells any stray rear irradiation will also generate current and will subsequently also influence the J_{SC}, V_{OC}, and FF of the solar cells. This can result in higher, but not well-defined efficiencies for bifacial cells. In laboratory environments, the reflection from the measurement chuck can be significant. Light that is not absorbed by the solar cell (typically infrared light) will exit the cell at the rear surface, and will be reflected

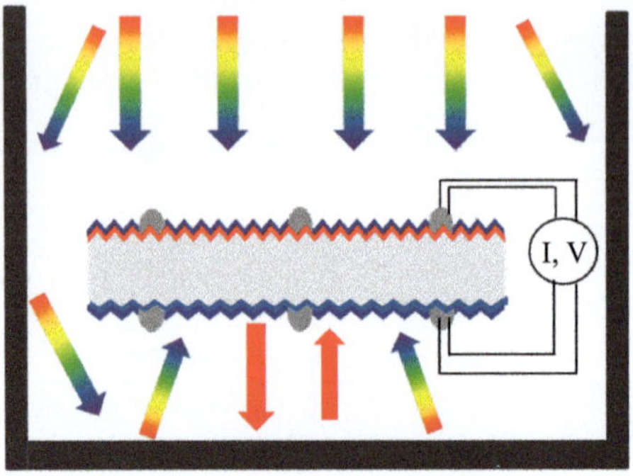

Figure 2.11 Inline IV measurement setup. The solar cell is contacted by probes on front and rear side. In order to avoid illumination from the surrounding, the setup is typically placed in a housing. However, reflections from the part below the cell are often not controlled

back causing an increase in J_{SC} that can be up to 4%. Furthermore, if the chuck is also conducting, the open rear metallization can be shorted resulting in an increase in FF. When the bifacial cells are mounted in a bifacial module and subsequently measured against a black background, the too high measured cell J_{SC} and FF result in apparently increased cell-to-module losses (see also Chapter 3.1.2, "Cell-to-module loss analysis in bifacial PV modules").

In order to generate a correct measurement for a bifacial solar cell, a nonreflecting and nonconducting chuck should be used. Nonreflecting means that the contribution from reflected or any stray irradiation remains below 3 W/m^2.

Inline IV measurements in cell production lines are done by contacting the front and rear busbars with probes. While the front side of the cell is illuminated, the rear of the cell is exposed to stray light reflected from the sides and the bottom of the measurement setup. These reflection can be very high (even higher than in the case of a rear chuck), resulting in unrealistically high currents that are sometimes reported for bifacial solar cells.

By illuminating the rear side, the rear *IV*-characteristics of a bifacial cell can be measured in either set-up, but preferentially using the laboratory setup with a non-conductive and non-reflective chuck (any stray irradiation should remain below 3 W/m^2). Spectral mismatch correction to counterbalance for different spectral responses of the front and rear of the bifacial cell under investigation, and the used reference cell, should be applied to the measurement of both J_{SC_f} and J_{SC_r}, according to IEC 60904-7 [40].

The bifaciality factors (2.1)–(2.4) can then be calculated from the front and rear characteristics.

Short-circuit current bifaciality coefficient φ_{Isc} is the ratio between the short-circuit current generated exclusively by the rear side of the bifacial device and the one generated by the front side. Both currents are measured at STC (1,000 W/m^2, 25 °C, with the IEC 60904-3 reference solar spectral irradiance distribution):

$$\varphi_{Isc} = \frac{Isc_r}{Isc_f} \tag{2.8}$$

where φ_{Isc} is the short-circuit current bifaciality coefficient. It is usually expressed as a percentage, Isc_x is the short-circuit current at STC under one-sided illumination, with index x: f for front and r for rear side.

However, the efficiency rating of single-side measurements with a nonreflecting and nonconducting chuck will of course not fully to reflect the bifacial properties of the solar cells. A new procedure to unambiguously rate the bifaciality factor and the bifacial efficiency of solar cells for simultaneous front and rear side illumination is currently under construction by the IEC workgroup and will be described in Section 2.4.2.

2.4.2 IV measurements under bifacial irradiation

Typical outdoor conditions involve rear irradiance (G_{rear}) levels between 5% and 30% of the front irradiance, in many cases between 10% and 20%. The exact value naturally depends on the installation as well as on diffuse irradiance at the site. Thus, a good and realistic assessment of the gain in generated power of bifacial cells is to measure the *IV* characteristics with simultaneous G_{front} 1,000 W/m^2 and G_{rear} 100 W/m^2 or 200 W/m^2 illumination. However, as this would involve two light sources, or a very well-defined reflector behind the solar cell, another method is being proposed for indoor measurements:

The new standard for measurements of bifacial cells and modules will based on the equivalent irradiance (G_{E}) (also called the compensated current) method [43]. It considers that bifacial cells operate at a higher total irradiance and with a different generation profile due to the higher illumination. Thus, this method takes into account the higher V_{OC} that is achieved due to the additional irradiance as well as the related increased resistive losses. In order to perform these measurements, a standard solar simulator with adjustable irradiance levels for one-side illumination can be used. The irradiance on the rear side should of course remain below 3 W/m^2.

As first step in this procedure, the *IV* characteristics of both the front and the rear sides must be measured at STC ($G = 1{,}000$ W/m^2). A nonreflecting and nonconducting chuck must be used in order to avoid the illumination of the nonexposed side. (Stray illumination should be below 3 W/m^2.)

The bifaciality factor φ_x of the solar cell responses is subsequently determined by taking the ratio between the rear and front responses for current J_{SC}, voltage V_{OC}, P_{max} and efficiency η. See also (2.1)–(2.4) in Section 2.3.1. The bifacial efficiency (or power) of the cell/module must then be determined on the front side at equivalent irradiance levels G_{Ei}, that are dependent on the bifaciality factor, and correspond to 1,000 W/m^2 on the front side plus different rear side irradiance levels G_{Ri}. The equivalent irradiance levels are given by:

$$\begin{aligned} G_{E_i} &= 1,000 + \varphi \cdot G_{R_i} \\ \varphi &= \varphi_{\text{Pmax}} \end{aligned} \tag{2.9}$$

where the bifaciality factor φ is equal to the smallest of φ_{Pmax} and φ_{Jsc}. Correcting the irradiance with the bifaciality factor ensures that in the differences that occur in the current due to front or rear illumination are also accounted for.

The measurements should be reported at two different rear irradiance levels of $G_{\text{R}_1} = 100$ W/m^2 and $G_{\text{R}_2} = 200$ W/m^2 as $\eta_{\text{bifi10}}/P_{\text{max}_{\text{bifi10}}}$ or $\eta_{\text{bifi20}}/P_{\text{max}_{\text{bifi20}}}$. To determine this accurately, measurements at least three different equivalent irradiance levels are required ($i = 1, 2, 3, \ldots$). For example, a device with maximum power bifaciality of $\varphi_{\text{Pmax}} = 80\%$ must be irradiated on the front side at $G_{\text{E}_2} = 1{,}160$ W/m^2 to provide the equivalence of $G_{\text{R}_2} = 200$ W/m^2. If the equivalent irradiance levels do not correspond to G_{R_1} and G_{R_2}, $P_{\text{max}_{\text{bifi10}}}$ and $P_{\text{max}_{\text{bifi20}}}$ must be obtained by linear interpolation of P_{max} versus G_E.

The measurements of bifacial modules pose similar challenges, and will be addressed in Chapter 3 (bifacial modules). In Chapter 3, more extensive examples for *IV* characterization in practice, both in laboratory and in industrial environments, are given.

2.5 Different types of bifacial solar cells

In this section, the six most common bifacial solar cells will be described in more detail. Special attention will be given to the processing, metallization and interconnection as well as bifacial and efficiency potential. These main factors are briefly summarized in Table 2.1.

2.5.1 Heterojunction solar cells

SHJ solar cells present the advantages to combine a high efficiency (potential for cells with >25% conversion efficiency) with limited number of production steps (pre-requisite for keeping reduced fabrication costs). SHJ solar cells further exhibit a low temperature coefficient (<−0.3%/°C) as well as a high bifaciality (>92%), triggering high energy yield for bifacial SHJ modules in the field.

2.5.1.1 Cell structure and processing details

SHJ solar cells implement carrier-selective contacts with high surface passivation formed on crystalline silicon wafer with the deposition of stacks of thin layers of hydrogenated amorphous silicon (a-Si:H) and of TCO [44,45]. These hetero-contacts are demonstrations of so-called passivating contacts, enabling for the key advantage of SHJ solar cells: increased operating voltages. Intrinsic a-Si:H layers are deposited by plasma-enhanced chemical vapor deposition (PECVD) at typical temperatures of about 200 °C, directly on the textured and cleaned n-type silicon wafer surfaces, providing excellent chemical passivation properties and yielding

Table 2.1 Bifacial solar cells and their main parameters

	Cell concept	Bifaciality factor (on cell level)	Si base material	Junction and BSF doping method	Contacts	(Front) Efficiency potential
2.5.1	Heterojunction	>92%	n mono	a-Si:H p- and n-type doped	TCO/Ag printed TCO/Cu plated	22%–25%
2.5.2	n-PERT	>90%	n mono	Boron and Phosphorous diffusion	Ag and Ag/Al printed	21%–22%
2.5.3	p-PERT	>90%	p mono	Phosphorous and Boron diffusion	Ag and Ag/Al printed	21%–22%
2.5.4	PERC+	>80%	p-mono	Phosphorous diffusion and local Al BSF	Ag and Al printed	21%–22%
2.5.5	IBC	>70%	n-mono	Boron and Phosphorous diffusion	Ag and Ag/Al printed	22%–23%

minority carrier lifetime potentially on par with theoretical limits [46]. On top of these high band-gap passivating layers, doped- a-Si:H layers are deposited by PECVD at similar temperatures, and allow for the selective collection of one type of carriers while blocking the other type: a p-type doped layer is deposited on one side of the wafer, acting as hole selective contact, while an n-type doped layer is deposited on the other side of the wafer, acting as electron selective contact. Then, TCO layers are deposited, typically by sputtering process, on each side of the wafer, enabling for an efficient contact between the a-Si:H stack and the cell metallization for lateral conduction, as well as for antireflective effect. Finally, the metallization is realized on both sides of the wafers: SHJ cells can be made monofacial with front metallization grid and rear metallization blanket (e.g. sputtered Ag) over the full rear area, or bifacial with front and rear metallization grids.

This SHJ (quasi-)-symmetrical bifacial solar cell architecture is represented in Figure 2.12. It can be noted that SHJ solar cells can be optimized either in a so-called front emitter configuration (with p-type layers/hole selective contact on the sunny-side), or in a so-called rear-emitter configuration (with n-type layer/electron selective contact on the sunny-side). High performance can be achieved in both configurations [46–48], demonstrating the true bifacial nature of this type of solar cell. The symmetrical SHJ architecture and its natural bifaciality therefore enable for high bifaciality factor, typically above 92% and tunable up to 100%. This bifaciality factor will depend on the properties and optimizations of the a-Si:H and TCO materials, respectively, for the hole and electron hetero-contacts, and on the eventual similarity/differences achieved in the transport losses and parasitic absorption properties of these hetero-contacts. On top of these potential differences,

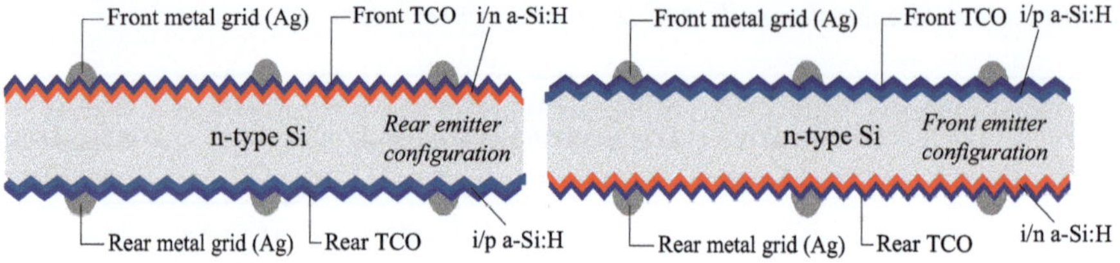

Figure 2.12 Schematic representation of the architecture of bifacial silicon heterojunction solar cells, based on a rear emitter configuration (left) or on a front emitter configuration (right)

bifaciality will be governed by eventual differences between the front and the rear metallization grids applied. For standard bifacial modules applications in tilted configuration, with rear illumination in the range of 10%–30%, best performance will be achieved with a denser rear metallization grid than for the front grid, reducing the bifaciality factor by providing higher shadowing losses for rear illumination, but enhancing the overall performance.

The passivated contacts and the symmetrical structure of SHJ bifacial cell make it also perfectly suited for an efficient use of thin silicon wafers. The excellent surface passivation enables maintaining high performance even for thin wafers <100 μm [49], while the symmetrical layer structure on both sides of the wafer equilibrates mechanical stress, reducing the risk of wafer bending and breakage, rendering the bifacial SHJ technology a technology of choice for pushing forward the industrial implementation of thinner wafers.

2.5.1.2 Metallization and module interconnection

Consecutively to the a-Si:H layers depositions, the further cell processing temperature must be kept typically below 200–250 °C to not degrade the chemical passivation properties. This constraint imposes at the metallization level the use of low-temperature cured silver pastes, which yield silver lines with typical bulk resistivity of 6 to 10×10^{-6} Ωcm, a factor 2 to 3 higher than state-of-the-art metallization based on firing-through of high-temperature silver pastes. The use of standard H-pattern metallization with 2 to 5 busbars for the manufacture of bifacial SHJ solar cells therefore imposes that more silver is required to achieve similar line resistance than for homojunction silicon cells, and/or lines with higher resistivity or larger dimension have to be employed. Still, this cost and/or performance limitation due to the metallization can be completely overcome with two approaches: either by switching to copper plated lines which demonstrate high conductivity without the need for high temperature processing [50–52], or by switching to multiple-wire interconnection [53,54] which enables the implementation of metallic grids with higher resistance (the line resistance required for 5 busbars being typically <0.5 Ω/cm and for 18 wires interconnection for instance <10 Ω/cm). For the latter case of multi-wire interconnection for instance, low silver usage of only 60 mg in total could be demonstrated for bifacial SHJ solar cells while enabling for performance gain, demonstrating that the switch to this advanced module technology enables to completely cancel an initial limitation of SHJ bifacial solar cells [54].

2.5.1.3 Efficiency and bifaciality potential

The passivated contacts of SHJ solar cells enable for its key advantage: increased operating voltages. Open-circuit voltages above 740 mV are for instance achieved with this technology on commercial 180 μm thick (nominal thickness) Cz Silicon wafer [46,49,50,55]. The main performance limitation is then linked to the fact that the full area hetero-contacts exhibit parasitic light absorption in the thin TCO and a-Si:H layers, requiring advanced engineering of these layers to maintain the high operating voltage while maximizing light transmission and therefore photo-generated current in the solar cell. An optimum trade-off can be realized, and at production level, efficiency >22.5% can be achieved [55]. In particular, up to 24.02% efficiency was demonstrated for a busbar-less SHJ cell fabricated in a pilot-line on mass production equipment [56]. The hetero-structure can further be optimized, and in 2015, record conversion efficiency was demonstrated using this SHJ cell structure, with up to 25.1% achieved on 180 cm^2 by the company Kaneka in Japan [50], demonstrating on already large area the potential for >25% efficiency for both sides contacted SHJ solar cells. Using similar hetero-contact structures in a back-contacted cell architecture further allows for reaching record efficiency for a silicon solar cell, with up to 26.7% achieved also by Kaneka in Japan [57]. For bifaciality, the potential of >92% is related to the bifacial nature of the cell architecture with very similar a-Si layers on both sides of the wafer.

2.5.1.4 Ease of implementation in the existing technology

The SHJ cell technology is a disruptive technology for silicon solar cell manufacturing, as it allows achieving high performance and high bifaciality using a simple process flow with the following equipment: wet-bench, PECVD, physical vapor deposition (PVD) and printers. This requires the implementation of equipment (PECVD/PVD) not typically used in standard silicon solar cell production lines, therefore requiring equipment upgrade and investment. The technology is therefore not directly applicable as an upgrade to an existing line of diffused solar cells, but provides a lean manufacturing process flow for the realization of high efficiency bifacial devices based on passivating contacts.

2.5.2 n-PERT solar cells

The passivated emitter and rear totally diffused (PERT) cell concept on n-type base material enables bifacial solar cells with very high bifaciality (in some cases even above 95%) and high front cell efficiencies between 20% and 22% that are relatively easy to manufacture. With only a limited number of additional processing steps the concept is easy to implement in the majority of current (p-type) solar cell production lines. Several manufacturers are already producing n-PERT cells

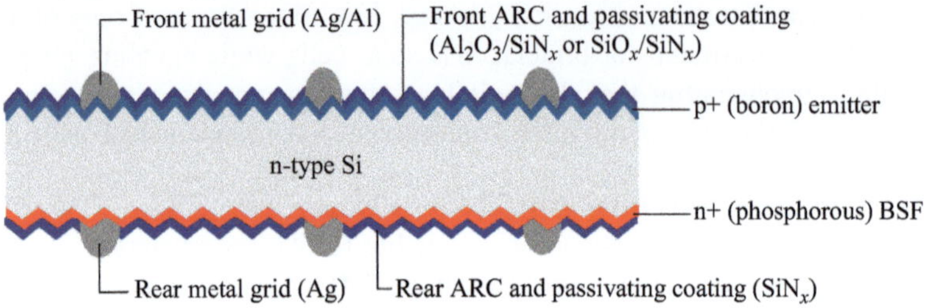

Figure 2.13 Schematic representation of the architecture of bifacial silicon n-PERT solar cell

on a large (>500 MW) scale (Yingli, LG) [58,59], others announced intentions (Jolywood) [60].

2.5.2.1 Cell structure and processing details

The basic configuration of an n-PERT solar cell is shown in Figure 2.13. A p^+ boron (B)-doped emitter is located at the front of the solar cells, while a n^+ phosphorous (P)-doped BSF covers the full rear side. The B-emitter typically has a sheet resistivity between 60 and 100 Ω/□, and is passivated with stack of dielectric layers such as Al_2O_3/SiN_x or SiO_x/SiN_x [61,62]. While the 80 nm thick SiN_x layer provides adequate anti-reflection, the Al_2O_3 or SiO_x interlayer is needed to achieve a low number of interface states and fixed electrical charges of beneficial polarity on p^+ B-emitters [63–65].

On the rear the n^+-BSF provides shielding of minority carriers as well as additional lateral conductivity. Consequently, the BSF is an important element of the cell design, reducing the performance sensitivity to variations in n-type wafer resistivity. Surface passivation of the BSF is typically provided by an 80 nm thick SiN_x dielectric layer. Subsequently, the metallization grid is applied. In the discussions below, the focus will lie on metallization using printing and firing processing, as this is at the moment the prevailing technology for n-PERT solar cells in industry. Alternative metallization using plating and TCO technology, as described in Section 2.5.1 for SHJ can of course be applied to n-PERT cells as well. This typically results in higher efficiencies due to reduced metallized area's and subsequently higher J_{SC} and V_{OC}, but has so far not been adopted by industry for n-PERT production. Researchers at imec have reported n-PERT solar cells with efficiency up to 22.4% using copper and nickel plated metallization [66].

In Figure 2.13, both the front surface and the rear surface of the cell are textured with random pyramids, based on, i.e. the so-called n-Pasha process flow developed by of the Energy Research Centre of the Netherlands (ECN) [61,67]. In contrast, the n-PERT solar cells resulting from the so-called BiSoN process flow developed by ISC Konstanz [62], exhibit a polished rear side and a random pyramid textured front side.

2.5.2.2 Junction formation

There are various methods to realize high-quality doped p^+ and n^+ regions in n-PERT cells:

- High temperature, tube gas diffusions with $POCl_3$ for n^+ and BBr_3 or BCl_3 for p^+ doping are prevalent in industrial production lines as modern tube diffusion furnaces offer high throughput a moderate investment and operational cost. Two subsequent diffusion processes allow careful tuning of junction formation and doping profiles. However, as tube gas diffusion processes typically are not single sided, care has to be taken to avoid "parasitic" doping on the "wrong" (opposite) side. Either a diffusion barrier can be used to prevent parasitic doping, or single-side etching can be used to remove the parasitic doping afterwards.
- A second option is to deposit a diffusion source on the surface and perform the in-diffusion of the dopant species in a subsequent high temperature treatment. Typical source layers are phosphorous silicate glass and boron silicate glass (BSG) which can be deposited by different techniques, such as atmospheric pressure chemical vapor deposition or PECVD [68,69]. The dopant concentration in the layer can be adjusted by the composition of precursor gases. The subsequent temperature treatment is usually performed in a tube furnace. As the source layers can be applied single-sided, so-called co-diffusion processes can be applied in which n^+ and p^+ doping are formed simultaneously during a single high temperature step. By careful tuning of the doping concentration in the source layers as well as temperature profile and atmosphere composition of the subsequent diffusion process, both n^+ and p^+ doping profiles can be optimized. Recently, it has been shown that properly designed doping source layers can also be used as passivation and anti-reflective coating. Cabal *et al.* published a simplified n-PERT process, where the co-diffusion was done by doped passivating and antireflective layers remaining in the cell, effectively combining three processing steps [70].
- A third option to form well controlled doped layers is by ion implantation [71,72]. With this process, shallow, highly doped regions are made by exposing one side of the wafer to a beam or shower of ionized dopant precursors accelerated towards its surface. Subsequently, the wafers need to undergo a high temperature (anneal) step to activate the implanted species, and to repair crystal damage. As this is a single sided process, no additional barrier or etching step is necessary.

These three options for the creation of doped layers can also be combined in different kinds of ways. Especially advantageous can be combinations of a single sided process such as source layer deposition or implantation with a classical tube diffusion process, providing simultaneously temperature treatment of the first side and a gas diffusion of the other side, allowing for short process sequences. Prominent examples are the so-called hybrid implant process combining phosphorous implant with BBr_3 tube diffusion [72,73], and the combination of a deposited BSG with a $POCl_3$ tube diffusion [74,75].

2.5.2.3 Passivation

For long, there was no industrial solution for the passivation of (p^+) boron-doped surfaces that are used as emitters in n-type solar cells. On phosphorous (n^+) emitters in p-type solar cells, SiN_x is commonly used for passivation. When applied on p+-doped surfaces however, the SiN_x results in low or even no passivation at all due to its large positive fixed charge density ($Q_f > 2 \times 10^{12}$ cm^{-2}). This increases the minority carrier (i.e. electron) density near the surface in a p+ emitter.

In 2008, the ECN reported the significant improvements in passivation of p^+-doped silicon by adding a chemical oxide below the SiN_x. This oxide is grown at room temperature by a nitric-acid oxidation of silicon [76] and reduces Q_f and simultaneously results in excellent low density of interface defect density D_{it}. This relatively easy passivation of boron-doped surfaces enabled the industrial development of low-cost n-type Si solar cells [61].

Around the same time, a more superior passivation of p^+ surfaces was developed: Al_2O_3 prepared by atomic layer deposition (ALD) [63,64]. This Al_2O_3 passivating layer is now also widely used to passivate the p^+ rear surface of PERC solar cells. The excellent level of passivation of p^+ surfaces by Al_2O_3 could be attributed to its very low interface defect density and a high *negative* fixed charge density ($Q_f \sim -5 \times 10^{12}$ cm^{-2}). Further it has been demonstrated that BSG layers used as diffusion source, both deposited by PECVD [73] and grown in a BBr_3 tube diffusion process [77] can also passivate the emitter very effectively.

2.5.2.4 Metallization and module interconnection

Industrial n-PERT cells are metallized by printing H-grids of glass-frit containing silver pastes on both front and rear and subsequently applying an IR spike firing process. During this process, the metal contacts are formed by glass-frit etching through the dielectric layers. The similar appearance of n-PERT cells from the front and rear sides due to the symmetric metallization scheme with H-grids for fingers and bus bars from silver paste on both sides is often rated as very aesthetically pleasing in applications were the cells remain pronouncedly visible to the eye such as façades.

As the dominant cell manufacturing technology is still based on p-type material and n^+ phosphorous-doped emitters, relatively little development effort has been spent on p^+ contacting silver pastes so far.

In order to improve contact formation, most p^+ contacting pastes contain aluminum particles. However, the aluminum addition will introduce Al spikes or etching pits during the firing process [78,79], resulting in increased recombination. The effect of this process on the contact recombination parameter (J_{01}) is shown in Figure 2.14.

As can be seen from Figure 2.14 [78], the recombination increases significantly more for larger metallized areas for p^+ emitter areas. Using the existing commercially available Ag/Al pastes, the easiest way to decrease this contact recombination is to reduce the metal contact fraction, for instance by using dual printing, by stencil print or by adopting a metal wrap through cell concept [80].

Alternatively, a deep emitter below the contacts to provide additional shielding by creating a selective emitter can also be applied [81]. This can also be seen in Figure 2.14, where the J_0 values for deep p+ emitters (circles and triangles) are

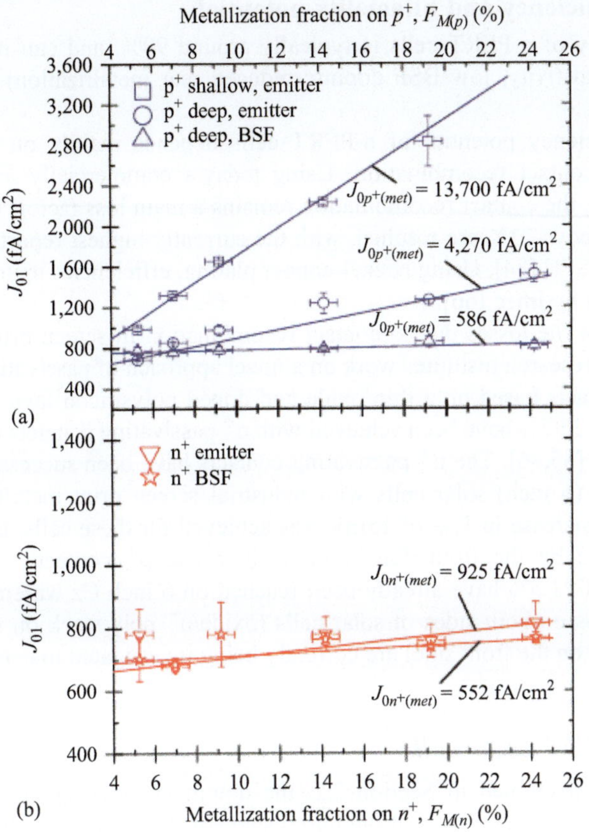

Figure 2.14 *J_{01} extracted using one-diode equation from the experimental IV parameters under illumination (symbols) as a function of metal fraction variation on either p+ (FM(p)) or n+ (FM(n)) diffused regions of the bifacial cells. The J_0 (met) is then extracted from the linear fit (solid lines) to the experimental data*

already much lower than those of shallow p+ emitters (squared symbol). Compared to the p-type case with n^+ phosphorous selective emitters, the selective boron emitters below the contacts need to be quite deep (>800 nm) to realize similar low values for $J_{0_{met}}$ due to more the aggressive etching of the aluminum containing pastes. Recently progress has been reported in contacting p^+ emitters with Al-free Ag-pastes [82,83] to prevent the Al spike formation, but further understanding is required to implement these pastes in industrial cell lines.

The interconnection of n-PERT cells into modules is done very similarly to standard p-type cells. Direct soldering of tabs to both the front and rear busbars is possible as they are both based on Ag-containing metal pastes. Alternatively, also multi-wire interconnections are possible such as described in the heterojunction section. In this case, 18 wires are soldered directly to the front and rear Ag fingers, reducing the silver consumption.

2.5.2.5 Efficiency and bifaciality potential

The bifaciality of n-PERT cells is typically around 90%, and can in some cases (high base resistivity, low BSF doping, reduced rear metallization) reach values above 95%.

The efficiency potential of n-PERT cells depends mainly on the junction, surface and contact recombination. Using today's commercially available contacting pastes, the contact recombination remains a main loss factor. Nevertheless, efficiencies above 21% are reached, with the currently highest reported efficiency of 21.8% by ISFH [84]. Using Nickel-copper plating, efficiencies up to 22.4% have been obtained by imec [66].

To reduce the losses due to contact recombination in screen-printed n-PERT cells, several research institutes work on a novel approach of passivating, or carrier selective contacts based on a thin oxide and doped polysilicon layer stacks. Efficiencies up to 25.7% have been achieved with n^+ passivating contacts on small area lab-type cells [85,86]. The n^+ passivating contacts have been successfully applied on large size (6 inch) solar cells with industrial screen print metallization [83]. Although an increase in V_{OC} of 15mV was achieved for these cells, the efficiency is still limited by the front-side boron emitter and p^+ contacts. Nevertheless, efficiencies of 21.5% have already been reached on 6 inch Cz wafers [87]. Passivating contacts on both sides of solar cells (oxide/n^+ poly stack on rear side and oxide/p^+ poly on the front side) are currently being investigated to exploit their full potential [88].

2.5.3 p-PERT solar cells

As has been mentioned in Section 2.2, the dominant space application of both regular and bifacial solar cells at the initial stage of PV era laid the foundation for the structure and fabrication technology of the terrestrial cells. The better resistance to high-energy space radiation (electrons and protons) of p-type silicon comparing to n-type silicon was the main factor explaining the leading role of p-type in the PV industry for terrestrial application.

Further work on p-PERT solar cells continued in the previous century, mainly focusing on monocrystalline silicon [89–98]. In recent years, p-PERT solar cells were investigated on multi crystalline silicon, also as an alternative to single facial PERC cells [99–101]. However, large scale industrial manufacturing of p-PERT solar cells has not been announced yet.

2.5.3.1 PERT solar cell with n^+-p-p^+ structure

The schematic of the bifacial PERT structure is shown in Figure 2.15. Different modifications of the bifacial cells were fabricated with polished and textured surfaces. The following general factors are crucial for the design of bifacial cells with the n^+-p-p^+ structure:

- Bulk diffusion length L significantly exceeding cell thickness d ($L \gg d$);
- High bulk resistivity, as empirically known factor accompanying long L;
- Low effective recombination at the rear, S_{eff}.

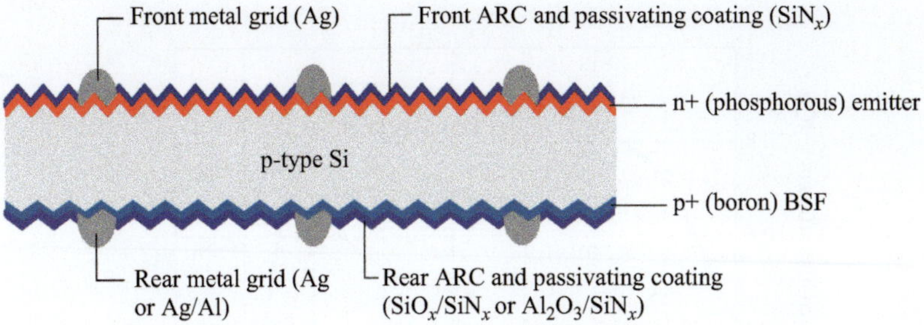

Figure 2.15 Schematic cross section of bifacial solar cell with n^+-p-p^+ structure

The advantage of the bifacial cell with n^+-p -p^+ structure is the relative simplicity of the fabrication technology. The cell has uniform n^+ and p^+ doping layers on textured or flat surfaces, passivating and antireflection coatings, and contacts. However, keeping the long diffusion length (or high lifetime) of the bulk minority carrier during fabrication procedure, which includes high temperature diffusion steps to form the n^+ and p^+ regions, is a challenge. Low back S_{eff} depends on p^+ doping parameters and surface passivation. The relatively simple process of Al alloying does not allow suppressing S_{eff} below ~400 cm/s. Therefore, alternative doping is needed for providing both bifaciality and low S_{eff}.

The typical structure of a p-PERT silicon solar cell is shown in Figure 2.15. The front surface is an n^+ (phosphorous-doped) emitter passivated by a SiN_x layer. Typical sheet resistances are in the range between 90 and 120 Ω/□. Depending on the process flow, the rear surface is either (slightly) polished or textured and has a boron-doped BSF. Sheet resistance values of the BSF can be as high as 250 Ω/□. In some cases, the BSG layer is kept as passivation layer, with SiN_x coated on top to accomplish an optimum ARC layer thickness. Other bottom passivation layer of the passivation stack are AlO_x or SiO_x.

2.5.3.2 Processing details junction formation

Several kinds of fabrication processes for p-PERT structures have been evaluated, two of which are elaborated below.

The combined ion implantation/thermal diffusion fabrication technology

This process is based on boron ion implantation for p^+ layer formation and phosphorous thermal diffusion for n^+-emitter doping. It was first developed by the Russian company KVANT for space cells production and later studied in the Jerusalem College of Technology (JCT) [89–93]. The space cells are unlike the one shown in Figure 2.15, as their surfaces are not textured but polished. Another difference was the front and back Ti-Pd-Ag grid contacts prepared by thermal vacuum evaporation.

This fabrication procedure has distinct advantages: controllability, doping uniformity and reproducibility [89,90]. The starting material is Cz p-Si with resistivity

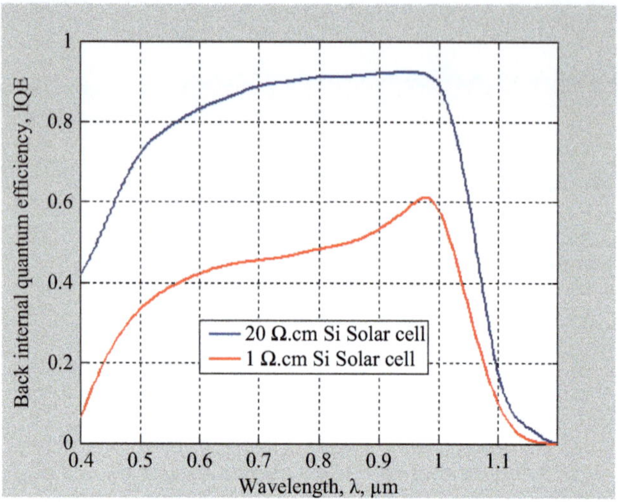

Figure 2.16 Back spectral IQE for solar cells fabricated using FZ p-Si of different resistivity. Implantation dose was 5.6 × 10^{15} cm^{-2}. Measurements under sun light bias

1–25 Ωcm. Solar cell emitter is formed by using P gas phase thermal diffusion at 850 °C with $POCl_3$ as the source, followed by a drive-in stage at ~950 °C.

The drive-in stage is used for simultaneous defect annealing after 30 keV B ion implantation as a doping process for p^+ layer formation. The procedure of ion implantation requires minimizing the impact of imposed structural defects. In addition to defect formation in the doped layer, another effect can appear because of ion implantation—defect creation in the base region just behind the doped p^+ layer [91,92]. The defects can extend in the base region to an effective depth of 0.5–0.6 μm, i.e. significantly deeper than ion range. Damage is less pronounced when Si resistivity is higher [93]. Illustration of the silicon resistivity influence is given in Figure 2.16, where two rear-side internal quantum efficiency (IQE) curves are compared for solar cells fabricated using starting FZ Si of resistivity 1 and 20 Ωcm. Both cells were implanted with the same boron ion dose, 5.6 × 10^{15} cm^{-2} and annealed at the same temperature 950 °C. The main reason for the difference in rear IQE is the influence of defect layers: in the 1 Ωcm it is responsible for $S_{eff} \approx$ 1,100 cm/s, whereas in the 20 Ωcm the effect of defect layer is negligible. The same positive effect of using high resistivity Si was found for Cz Si.

More complete defect annealing can be achieved in a modified fabrication process, in which ion implantation and a higher temperature annealing step are preceded by the phosphorous gas phase diffusion process. Some variations of above design and process flow have been evaluated. One of them includes the modified back BSF and contact structure. A matrix of local Al alloyed ~ 10 μm wide strips under and along the back contact fingers completes the BSF structure [94,95].

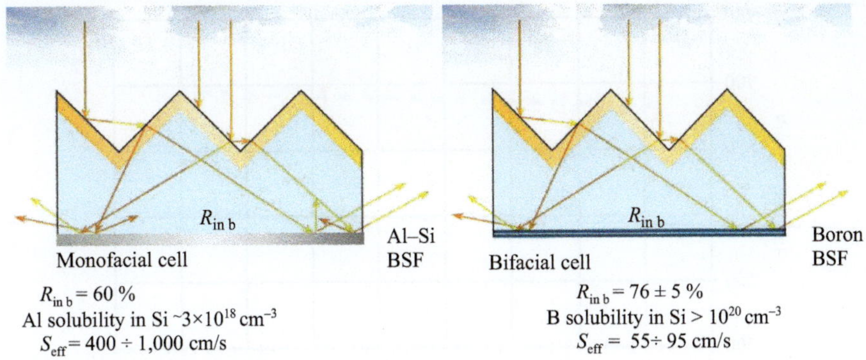

Figure 2.17 Light trapping and recombination improvements in B diffused vs. Al alloyed BSF

Another variation based on the formation of selective BSF was evaluated in experiments with local diffusion of extra B impurity over the uniformly ion-doped p^+ layer [96]. Spun-on B source was treated by laser beam resulting in increasing the maximal B concentration from ~10^{20} to ~3×10^{20} cm^{-3} in local regions. When the area of laser-doped local BSF regions did not extend 1.5% of whole back area their introduction did not deteriorate the cell recombination parameters. However, no real improvement on cell level was observed.

Thermal diffusion only as a doping means for the bifacial cell fabrication. The same cell structure can be fabricated using thermal diffusion only. Analysis of industrially produced terrestrial bifacial Si solar cells with this method was carried out in [97]. The bifacial cells were fabricated using a gas phase diffusion process for uniform phosphorus n^+ layer doping and diffusion from a surface deposited boron source for full p^+ back layer formation. The starting material was 6″ pseudo square wafers of 3 to 6 Ωcm single crystalline Cz Si. The front was texturized, and the back was treated in different ways—textured or smooth. The last one provided a better photo response for front irradiation due to improved rear internal reflection. SiN_x antireflective coatings and screen-printed contacts were applied to both sides.

The schematic of the cell cross section for the cells with a smoothened rear is shown in Figure 2.17, which illustrates the light trapping of front irradiation due to combination of textured front and smooth rear cell surfaces. Substitution of alloyed Al on the back by boron BSF leads to increase of internal cell back reflection $R_{\text{in b}}$ from ~0.6 to 0.76 ± 0.05 and therefore to a better light trapping. Improved light trapping is effective in combination with low recombination losses in the base region.

The high boron doping concentration effects on effective surface recombination in two directions: first—is the increase of recombination rate inside the doped p^+ layer, second is passivation of the surface by BSG layer. Resulting effective surface recombination (75 ± 20 cm/s, as mentioned in the subscription of the Figure 2.17) is significantly lower than S_{eff} on the interface Alloyed Al–Si. More details on measurements and analysis of this cell structure can be found in reference [97].

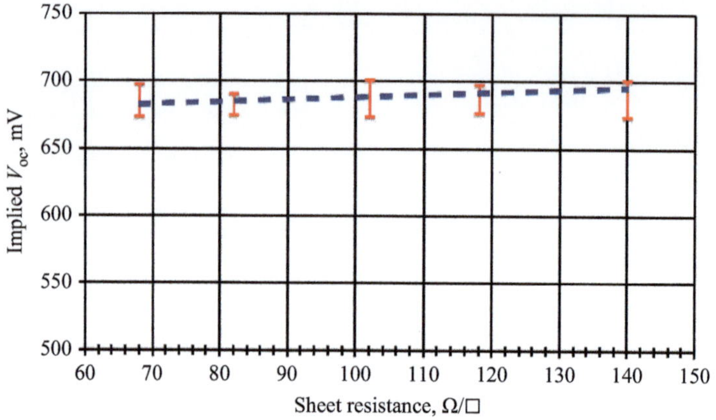

Figure 2.18 Implied open-circuit voltage of n^+-p-p^+ structures vs. doping level of a p^+ layer

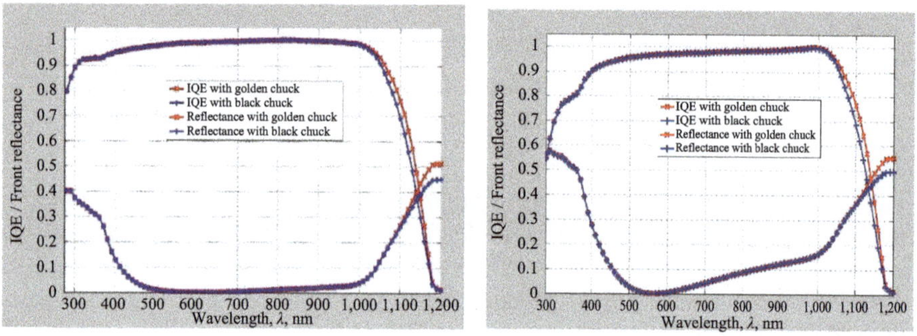

Figure 2.19 Front (a) and back (b) IQE of the bifacial cell placed on reflective and black chucks

The same cell design with an improved p^+ layer structure was realized in [98]. The cell has also the high retained bulk lifetime of minority carriers, which exceeds 0.5 ms, but p-p^+ barrier is improved due to precisely controllable boron doping of p^+ layer. Important data describing the recombination quality of the fabricated cell structures are given in Figure 2.18. The cell structure includes the p-n junction with sheet resistance of phosphorous-doped n^+ layer ~120 Ω/□, passivating and anti-reflective SN coatings on both cell sides. The implied V_{OC} was measured as a function of the p^+ layer doping level (characterized by R_{sh}). Implied V_{OC} of the passivated cell structure (before contact deposition) is quite high: 703 to 678 mV, for p^+ layer sheet resistance varying from above 140 to ~65 Ω/□. Even with some small reduction of iV_{OC} with increased doping, the measurements indicate the high cell open-circuit voltage is promising.

Improved design of the n^+ and p^+ layers are leading to enhancement of both front and rear spectral responses. Examples typical of front and back IQE curves for this cell type are shown in Figure 2.19. The measurements were made using reflective chuck (with long wavelength reflectivity ~0.9) and "black" low reflecting chuck (with reflectivity ≤ 0.06). Front spectral reflectance of the cell installed on the each chuck are shown in the same Figure. Some difference of the two IQE curves due to chuck reflection can be seen at the wavelength longer than ~1,000 nm. The impact of the chuck reflectance on the determination correctness of energetic cell parameters at solar illumination can be evaluated by integration of the cell spectral response over standard solar radiation spectrum and comparison the results for the cases of using reflective and black chucks. Such kind of the calculations based on the front spectral response data for the cell, IQEs of which are shown in Figure 2.19, results in short-circuit current densities, J_{SC}, of 39.47 and 39.28 mA/cm^2 for the reflective and black chucks using, respectively. Therefore, the relative difference in determination of solar cell current when measured using two types of a chuck is ~0.5%. The ~ same small difference will be resulted in the determination of cell efficiency values.

IQE of the back illuminated solar cell, as can be seen in Figure 2.19, right, concedes slightly the front IQE in the short-wavelength region. The depth and profile of doping as well as recombination losses under back contact fingers are explaining this difference. The illuminated surface in this case is not textured, which results in a higher overall reflectance of the cell. A slightly lower short wavelength IQE, ~1% a larger contact shading and mainly higher reflectance result in lower back short-circuit current, $J_{SC\ b}$. The $J_{SC\ b}$ values calculated as above by integration of spectral response over standard solar spectrum are ~92.7% of the respective front J_{SC}. It means that bifaciality factor determined as the ratio of back to front short-circuit currents is above 90% for this precisely B-doped cell.

Efficiency and bifaciality potential of monocrystalline p-PERT cells

An example of the typical I-V curves of a bifacial cells of ~20.2% front efficiency at simulated 1 sun front and rear illumination are shown in Figure 2.20. The V_{OC} value of the cell is significantly lower than iV_{OC} measured for the plain structures (see Figure 2.18). The obvious explanation is the influence of the contact recombination.

A comprehensive analysis of fabricated p-PERT bifacial cells based on approximate statistical data for small experimental batches, demonstrates an intrinsic superiority of the bifacial n^+-p-p^+ design and fabrication technology not only in comparison with regular p-type cells with Al alloyed back, but as well with p-PERT cells of the first generation (with B over doped p^+ layer [97]). According to evaluations of authors of [98], parameters of bifacial solar cells can be improved without any significant change in fabrication processing, and front efficiency is expected to exceed 21% and bifaciality to surpass 92%.

The most imperative factor for attaining the highest cell efficiency is a high τ_b. The fabrication technology provides retaining very high τ_b in the Cz wafers. The measured τ_b values are in the range 0.5–1.2 ms, i.e. in the range typical for the tested starting wafers before thermal processing. Properties of high-low BSF are of the first order of significance in influence on the front and back cell efficiencies. Controllable B doping of the p^+ layer forms a very effective BSF resulting after proper passivation in low S_{eff} [102].

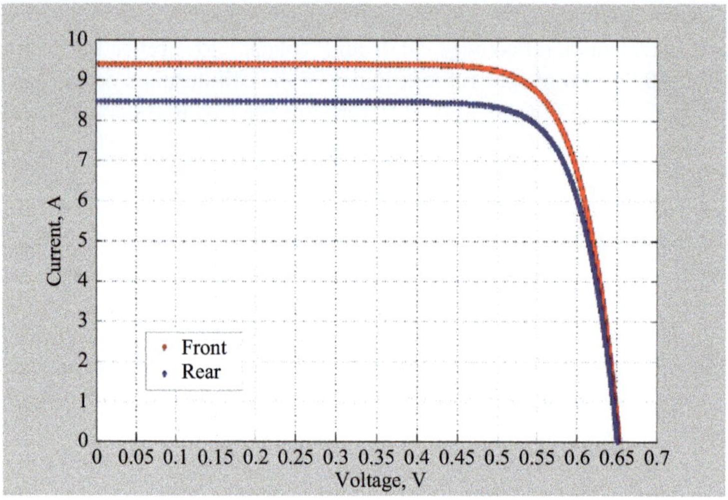

Figure 2.20 The I–V characteristics of a bifacial p-PERT cell at front and back 1 sun illumination

Efficiency potential and bifaciality for multicrystalline-p-PERT

The highest published p-PERT solar cell on mc-Si has an efficiency of 18.9%, measured on a nonreflecting, non-conducting chuck [101], with average efficiencies of about 18.6%. With general advances in solar cell efficiencies of about 0.3%–0.5% abs/year, we expect that mc-p-PERT solar cells will achieve average efficiencies close to 20% in the near future. Besides further optimization of the rear surface (e.g. diffusion and passivation, metallization), also front surface optimization (emitter, screen-printed contacts) as well as material improvements will contribute.

The bifaciality factor of mc-p-PERT solar cells with Ag/Al contact grid are in the range of 85% to 90%. By applying an Al contact grid, the bifaciality factor will be reduced by approx. 5%. Initial tests on mc-p-PERT solar cells and modules did not show a significant LID [101], a very positive effect when compared to mc-p-PERC solar cells and modules. Further investigations are still necessary to understand the LID and LeTID on multicrystalline p-PERT and p-PERC cells and modules.

2.5.4 p-PERC+ solar cells

Bifacial PERC+ solar cells apply the same process sequence as industrial monofacial PERC cells, that are currently introduced into mass production by many leading solar cell manufacturers. Presently, monofacial PERC cells account for 20% of the global production capacity and this share is expected to increase towards 60% within the next years [103]. To convert a monofacial PERC cell production into a bifacial PERC+ cell production, the Al screen-print has to be changed from full-area to an H-pattern layout and the thickness of the rear-side

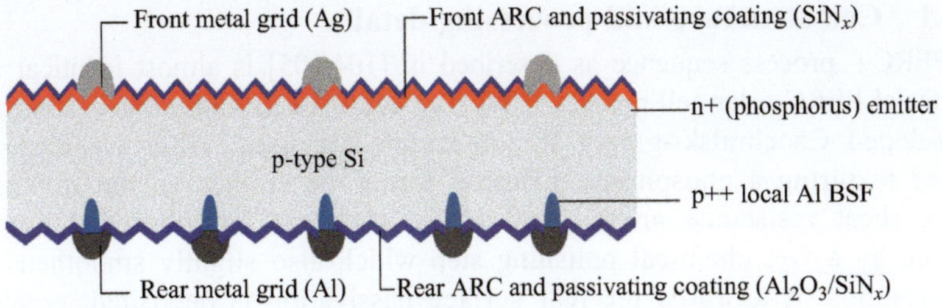

Figure 2.21 Schematic drawing of a PERC+ bifacial solar cell

passivation coating maybe optimized for anti-reflective properties [104]. Hence, for monofacial PERC cell manufactures the PERC+ cell concept offers a very attractive path towards bifacial solar cell production as no tool invest is required to change production from monofacial PERC to bifacial PERC+ solar cells. In Figure 2.21, a cross section of a PERC+ solar cell is shown.

ISFH and SolarWorld first independently and later jointly started the development of a bifacial PERC solar cell design in 2015 by applying a screen-printed rear Al finger grid instead of the conventional full-area aluminum (Al) rear layer while using the same PERC manufacturing sequence with only minimal recipe modifications for rear passivation, LCO and Al screen printing [104]. Hence, a monofacial PERC cell production line can be switched to produce bifacial PERC solar cells without requiring any invest into new or different production tools. The initial publications [104,105] demonstrated several advantages of the PERC+ concept compared to conventional monofacial PERC cells. In particular, the Al finger grid enabled bifacial applications of the PERC+ cells with front-side efficiencies up to 21.2% and rear-side efficiencies up to 16.7% measured with a black chuck [105]. The corresponding bifaciality was up to 80%. When measured with a reflective brass chuck, PERC+ cells demonstrated front-side efficiencies up to 21.5% compared to conventional PERC cells with 21.1% efficiency [105]. The Al paste consumption of the PERC+ cells was drastically reduced to 0.15 g instead of 1.6 g for the conventional PERC cells [104]. Hence, PERC+ solar cells are attractive for both, bifacial and monofacial module applications [104] which is why the naming convention PERC+ was proposed rather than, for example, biPERC or bifiPERC. In 2015, two additional publications addressed the concept of bifacial PERC+ cells. Trina Solar published bifacial glass/glass modules applying bifacial PERC+ solar cells designed for aesthetic optical appearance in building integrated photovoltaic applications [106]. Fraunhofer Institute of Solar Energy Systems assessed the concept of bifacial PERC+ cells mainly by numerical simulations of the potential front and rear conversion efficiencies and corresponding bifacial gains [107]. Since then, several solar cell manufacturers have introduced PERC+ solar cells into pilot production or mass production as will be described in the following sections.

2.5.4.1 Cell structure and processing details

The PERC+ process sequence as described in [104,105] is almost identical to a monofacial PERC solar cell process [108]. Typically, 1 to 2 Ωcm, 156 × 156 mm^2, boron-doped Czochralski-grown silicon wafers are used. After cleaning and alkaline texturing a phosphorus diffusion forms the emitter on the front side with a sheet resistance around 100 Ω/□. The rear phosphorus doping is removed by a wet chemical polishing step which also slightly smoothens the rear pyramids. Afterwards the rear surface passivation is deposited, typically consisting of an AlO_x/SiN_y layer stack. In an alternate processing sequence for PERC cells, a rear-side passivation stack consisting of SiO_x and SiN_y layer is deposited on a flat etched wafer, before applying texturing and phosphorus diffusion on the front side [108]. In case of PERC+ cells, the thickness of the rear-side SiN_y capping layer is set to 80 nm to obtain a low reflection for light incident from the rear side [104]. In contrast, monofacial PERC cells with full-area Al layer typically apply a 100 nm to 150 nm thick SiN_y capping layer, for maximizing internal reflections. The front side is passivated with PECVD SiN_x. Line-shaped LCO locally remove the rear passivation layer. The pitch p of the LCO lines has to be optimized for bifacial PERC+ applications in order to balance shading losses of the Al fingers versus the series resistance contribution of the wafer bulk spreading resistance [104]. Whereas monofacial PERC cells receive a full-area Al screen-print, bifacial PERC+ cells use an Al finger grid screen design. The aluminum screen has a finger pitch p which is identical to the LCO pitch and is printed aligned to the LCOs in order to ensure that the screen-printed Al fingers overlap with the LCO line openings. Since the specific resistance of screen-printed Al is 6 times higher compared with screen-printed Ag [104] the Al screen layout has to be optimized for low series resistance contribution. In particular metal grid designs with 5 or more busbars or multi wire interconnection technologies with short Al finger length are attractive for bifacial PERC+ cells. Whereas monofacial PERC cells consume around 1.0 g of Al paste per wafer, the Al paste consumption of bifacial PERC+ cells is drastically reduced to around 0.2 g per wafer due to the finger grid design [104]. The Ag front grid is typically deposited by screen printing. The front and the rear contacts are fired in a conventional belt furnace during which the Al paste locally alloys with the silicon wafer at areas where the rear passivation has been removed by LCO.

A schematic drawing of the resulting bifacial PERC+ solar cell is shown at the top of this section in Figure 2.21. The front side including the phosphorus-doped emitter, the SiN surface passivation and the Ag fingers is identical to an industrial monofacial PERC cell. The rear side features the rear surface passivation (typically AlO_x/SiN_y) with a thickness optimized for anti-reflection. The Al fingers are printed aligned to the LCOs. During furnace firing, the aluminum alloys with the silicon wafer forming local aluminum BSFs (Al BSF).

When looking at the fast development and industrial implementation of bifacial PERC+ cells as summarized in the next section in Table 2.2, it remains the question

Table 2.2 Published efficiencies of industrial PERC+ solar cells when illuminated from the front or from the rear side. Several leading solar cell manufacturers such as SolarWorld and Trina Solar are producing bifacial PERC+ cells and modules but did not publish any or recent PERC+ cell efficiencies

Year	Efficiency (%) front/rear	Organization	Source/comments
2015	21.5/16.7	ISFH	[105]; industrial process flow. No rear Ag pads
2015	20.3/n.p.	Trina Solar	[106]; optimized for optical appearance in BIPV
2016	20.7/13.9	Big Sun Energy Technology Inc.	[111]
2017	21.5/16.1	JinkoSolar	[112]
2017	21.4/n.p.	Neo Solar Power	[113]
2017	21.6/17.3	LONGi Solar	[114]
2017	21.6*/n.p.	ISFH	[115]; rear side optimized for monofacial use
2017	22.1*/n.p.	ISFH	[116]; Busbar less Ag front grid design

n.p.: not published
*independently confirmed

why it took almost 10 years of industrial monofacial PERC solar cell research and development until the concept of bifacial PERC+ cells was proved and published.

One key issue with the development of bifacial PERC+ cells is the very high specific resistivity of 20 μΩcm [104] of screen-printed Al fingers which is approximately 6 times higher compared to screen-printed Ag fingers. Hence, the rear Al finger grid has to be designed in a way to minimize series resistance losses caused by the Al finger lines. Figure 2.22 shows the calculated series resistance contribution $R_{s,L}$ of the Al finger grid in dependence of the number of busbars/wires and the Al finger width [116]. In order not to significantly reduce the front-side efficiency when changing from PERC to PERC+, as a rule of thumb the series resistance increase caused by the Al finger grid should remain below 0.05 Ωcm^2. As this is not possible with a 3 busbar configuration due to the high Al finger length in-between the busbars, the 5 busbar design can be considered as an enabling technology of bifacial PERC+ cells when applying wide Al fingers around 150 μm. When moving to narrow Al fingers below 100 μm width, multi wire module interconnection technologies with, for example, 20 wires per PERC+ cell drastically minimize resistive losses of the Al fingers to below 0.01 Ωcm^2 [116]. It is, however, challenging to print very narrow Al fingers due to the spreading of Al pastes during screen printing. First Al finger print tests in 2014 with 100 μm screen-opening width and conventional full-area PERC Al pastes resulted in around 200 μm wide Al fingers. Since then, paste vendors have optimized PERC Al pastes for

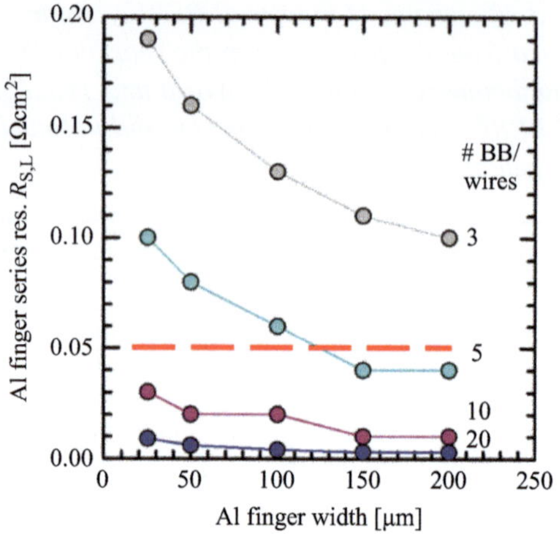

Figure 2.22 Calculated series resistance contribution Rs,L of the Al finger grid in dependence of the number of busbars/wires and the Al finger width [116]. In order not to reduce the front-side efficiency when changing from PERC to PERC+, the series resistance increase caused by the Al finger grid should remain below 0.05 Ωcm^2. As this is not possible with a 3 busbar configuration due to the high Al finger length, the 5 busbar design can be considered as an enabler of bifacial PERC+ cells. When moving to very narrow Al fingers below 100 μm width, smart wire technologies with, for example, 20 wires drastically minimize resistive losses of the Al fingers. Figure reproduced from Ref. [116]. Copyright (2018) The Japan Society of Applied Physics

fine line print capability which now results in about 100 μm to 150 μm wide Al fingers when applying 100 μm screen opening [116]. To further increase rear-side efficiency and bifaciality of PERC+ cells in the future, further developed Al pastes are required with even better fine-line printing capabilities approaching present Ag finger widths around 35 μm.

Another challenge with PERC+ is the precise alignment of the Al finger print on top of the LCO. In case of extreme misalignment when the Al finger does not overlap the LCO area, the open silicon surface of the LCO area leads to very high surface recombination of minority charge carriers and hence strongly decreased open-circuit voltages. Accordingly the alignment tolerances between Al finger print and LCO are in the range of ±30 μm depending on the detailed Al finger and LCO geometries [116]. This requires high-precision laser processes and Al screens as well as camera-based alignment schemes between LCO and Al screen print.

2.5.4.2 PERC+ Al contact formation

The limited Al volume of the Al fingers of PERC+ cells changes the alloying process with the silicon wafer during furnace firing as compared to full-area Al layer in PERC cells resulting in different Al contact properties. Figure 2.23 shows

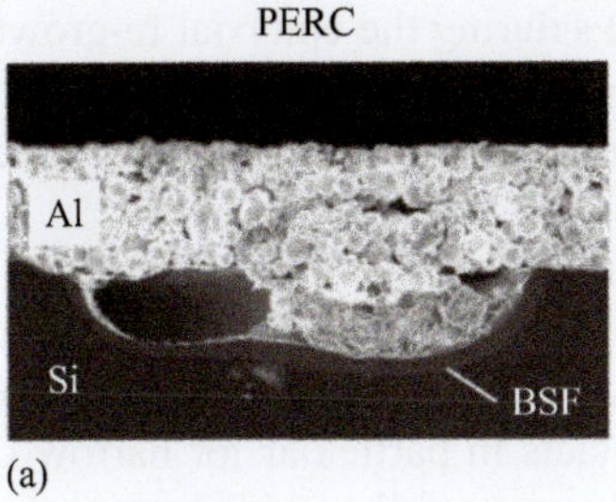

Figure 2.23 Typical scanning electron microscope (SEM) images of locally alloyed aluminum contacts of (a) PERC and (b) PERC+ solar cells. Whereas the local Al contacts of PERC cells often show voids and a thin BSF, PERC+ cells exhibit filled contacts and a deep BSF

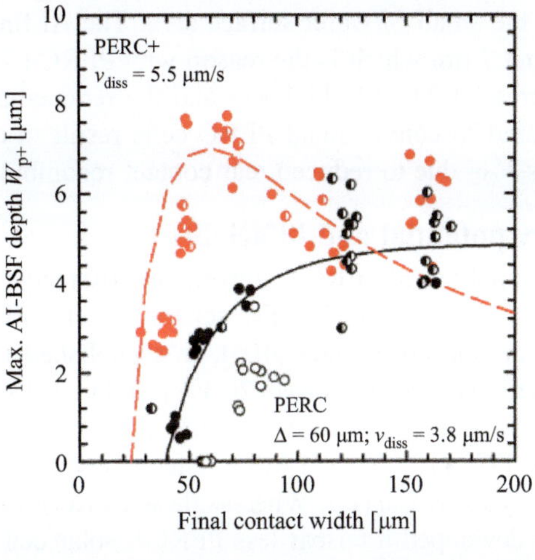

Figure 2.24 Maximum Al-BSF depth of PERC (black) and PERC+ (red) solar cells for different final LCO line widths. The dashed red line shows a fit to the PERC+ data using an extended model that takes the limited Al volume of the Al fingers into account. Figure reproduced from Ref. [110]

typical SEM images of Al contacts of PERC and PERC+ solar cells. Whereas the local Al contacts of PERC cells often show voids and a thin BSF, PERC+ cells exhibit filled contacts and a deep BSF [104,109].

Figure 2.24 shows measured Al-BSF depths demonstrating that PERC+ cells obtain up to 8 μm deep Al-BSFs compared to PERC with up to 6 μm. The black line represents a fit using a published model to the PERC data. The dashed red line shows a fit to the PERC+ data using an extended model that takes the limited Al volume of the Al fingers into account [109]. The limited Al volume of the Al fingers leads to a higher silicon concentration in the screen-printed aluminum during furnace firing causing thicker Al-BSFs during the epitaxial re-growth in the cool-down phase [109,110]. This effect becomes more pronounced for narrow LCO widths around 50 μm, which are industrially preferred in order to increase the throughput of the LCO tool.

Figure 2.24 uses different symbol styles where filled circles indicate a filled local Al contact, half-filled circles a partially voided Al contact and open circles a fully voided Al contact as determined by scanning electron microscopy. Whereas PERC cells exhibit rather many voided contacts in particular for narrow LCOs as displayed in Figure 2.24, PERC+ cells show no fully voided contact. Further analysis reveals that voids in particular occur for Al contacts where the Al–Si eutectic extends more than 20 μm deep into the Si wafer [110]. This finding is explained by an analytical model that calculates the surface energies of the liquid Al–Si melt, the Si wafer surface and the screen-printed Al particle surface [110]. According to this model, voids form for deep contacts since then a sufficient amount of Al–Si melt is available in order to wet the large surface area of Al particles rather than the small Si wafer surface area. The Al fingers reduce the Al contact depth by about 7 μm which is the reason why PERC+ cells do not exhibit voids [110]. The increased Al-BSF thickness and the reduced amount of voids of PERC+ cells compared to conventional PERC cells result in up to 3 mV higher open-circuit voltages V_{OC} due to reduced rear contact recombination [104].

2.5.4.3 Efficiency potential and bifaciality

Table 2.2 summarizes published PERC+ conversion efficiencies when illuminated from the front or rear side. Whereas ISFH set the benchmark in 2015 for high PERC+ front and rear-side efficiencies [105], first published PERC+ conversion efficiencies of solar cell manufacturers were 20.3% [106] in 2015 and 20.7% [111] in 2016 which then continuously improved to 21.6% in 2017 as reported by LONGi Solar [114] and by ISFH [115]. The 21.6% efficiency obtained by ISFH are independently confirmed by ISFH CalTeC. Whereas these PERC+ cells applied a 4 or 5 busbar design, ISFH developed a busbar less PERC+ solar cell by screen printing only the Ag fingers on the front side without printing the Ag busbars [116]. As stated in the last line of Table 2.2, this busbar less PERC+ cell exhibits an independently confirmed conversion efficiency of 22.1% when illuminated from the front side [116], where the missing busbar shadowing contributed around $0.4\%_{abs.}$ efficiency increase [116]. Benefiting from continuous improvements of industrial PERC solar cells with a current record efficiency of 22.8% by JinkoSolar [117], very likely even higher PERC+ front-side efficiencies will be demonstrated soon. At the same time, the conversion efficiency when illuminating the PERC+ rear side published by cell manufacturers improved from 13.9% [111] in 2016 towards 17.3% [114] in 2017. For the 21.6% and 22.1% efficient PERC+ cells fabricated by ISFH in 2017 the rear-side efficiency was not measured as the rear Al finger grid has been optimized for monofacial applications and not for high bifaciality. Accordingly, the bifaciality (rear efficiency divided by the front efficiency) of industrial PERC+ solar cells improved from 67% in 2016 [113] to 80% in 2017 [114]. Unfortunately, several leading solar cell manufacturers such as SolarWorld and Trina Solar which are producing bifacial

PERC+ cells and modules did not publish any or no recent PERC+ cell efficiencies and hence do not appear at all or only with initial results in Table 2.2.

As the PERC technology continuously improved the conversion efficiency over the past 6 years by 0.5%$_{abs}$ per year [108] with actual record values up to 22.8% [117], it can be expected that within the next few years higher PERC+ front-side efficiencies beyond 23% will be achieved. The front-side efficiency improvements may originate from improved Ag pastes and front emitter designs such as selective emitters resulting in reduced contact resistances and emitter saturation current densities [108]. The rear efficiency may be improved by reducing the Al finger widths thereby reducing the Al grid shadowing loss which is currently in the range of 10% to 15% [104]. Recently, major Al paste manufacturers have started to optimize Al pastes for Al finger print resulting in reduced finger width and higher finger aspect ratio which will support higher PERC+ rear-side efficiencies in the future. Also, the anti-reflection properties of the passivated rear surface maybe further improved by optimizing the rear surface roughness in combination with the rear passivation layer thickness. Furthermore, the rear-side efficiency will benefit from reduced carrier recombination, for example, due to emitter improvements as indicated above. In summary, the rear-side efficiency of PERC+ cells may approach 20% in the next few years corresponding to close to 90% bifaciality.

2.5.4.4 PERC+ module interconnection

SolarWorld pioneered mass production of bifacial PERC+ solar cells and fabrication of novel PERC+ glass/glass bifacial modules named Bisun which have been launched at the Intersolar 2015 [118,119]. Since then, Neo Solar Power, Trina Solar and LONGi Solar followed this technology route and are now offering commercial bifacial glass/glass modules applying PERC+ solar cells as well [120–122]. All manufacturers apply 4 or 5 busbar designs and obtain maximum rating powers between 290 W_p and 305 W_p with 60 PERC+ cells per module [119–122]. The rating power is stated for front-side illumination, only. When applying additional rear-side illumination, the output power increases accordingly. For example, with 10% additional rear-side illumination the output power will increase by approximately 8% to more than 320 W_p total output power. A higher number of busbars such as 4 or 5 is preferred for PERC+ modules, since the specific resistivity of Al fingers is 6 times higher compared to Ag front fingers [104]. A higher number of busbars reduces the Al finger length thereby reducing the Al finger line resistance and related resistive power losses. Since the first publication of a PERC cell applying a 5 busbar design [123], the PV industry is migrating from 3 busbars to 4 or 5 busbars which is beneficial to the industrial adoption of PERC+ cells as explained above and shown in Figure 2.22. The interconnection of PERC+ cells to strings is accomplished using conventional tabbing-stringing technology where the Cu ribbons are soldered to the Ag front busbars and to Ag pads on the PERC+ rear side.

As shown in Figure 2.22, one step further in terms of reducing resistive losses caused by PERC+ rear Al fingers is to move from 4 or 5 busbar designs to PERC+ solar cells without any busbars. In this case, the module interconnection is accomplished by, for example, 10 or 20 wires which reduces the Al finger lengths to a few millimeters hence minimizing its grid line resistance. Thereby, it enables a further reduction of the Al finger width to well below 100 μm which reduces shadowing losses of the Al finger grid and increases rear efficiency and bifaciality. ISFH has developed a wire-interconnected PERC+ prototype module [115] where 18 wires are soldered directly to the Ag front and Al rear fingers applying the smart wire connection technology (SWCT) of Meyer Burger [124]. The PERC+ SWCT prototype module applies 18 halved busbar less PERC+ solar cells and InSn-coated Cu wires with a diameter of 200 μm which are embedded in a transparent foil. The wires contact the Ag front and Al rear fingers directly without using Ag busbars or Ag pads. The PERC+ SWCT module exhibits a front-side efficiency of 19.8% and a rear-side module efficiency of 16.4% as confirmed by TÜV Rheinland, Germany. Accordingly, the module bifaciality is 83%.

2.5.5 *Bifacial back contact solar cells*

The IBC silicon solar cells exhibit both polarities of the metal electrodes on the back side. For this reason the IBC cells have several advantages over conventional both sided contacted cells, such as the absence of front-side shading losses, possibility to have large coverage of back side metal grid, potential easy co-planar interconnection designs, and an attractive uniform appearance of the finished modules. Therefore this cell design is thought to be optimized mostly for front-side illumination. There are various fabrication methods of IBC solar cells, from very high efficiency small area laboratory processes to large area commercial cells and modules. The discussion here will focus only on large area industrial applicable IBC processes.

Large area IBC solar cells were commercialized by SunPower Corp. since 1985. Currently SunPower is the only company producing IBC solar cells in mass production, with the highest achieved R&D cell efficiency of 25.2% [125] and module efficiency of 24.1%.

2.5.5.1 Cell structure

The IBC concept refers in literature to the classical *p-n* homojunction structures, where p^+ and n^+ interdigitated structures are formed by boron and phosphorous diffusions or ion implantation processes. On the other hand, Sharp [126], Panasonic [127], and Kaneka [128] combined the technologies of IBC and SHJ to develop large area heterojunction back contact (HJBC) cells with record efficiencies of 25.1%, 25.6%, and 26.7%, respectively. However, the cost per watt of manufacturing IBC or HJBC cells still needs to be reduced in order to be competitive with the conventional both sides contacted cells.

In the last couple of years several research groups and companies have developed large area IBC processes using low-cost conventional industrial fabrication techniques such as tube diffusion or ion implantation, PECVD SiN_x, and screen printing metallization [129–135]. A schematic cross section of such an IBC cell is shown in Figure 2.25. The doped p^+ and n^+ regions are formed mostly by

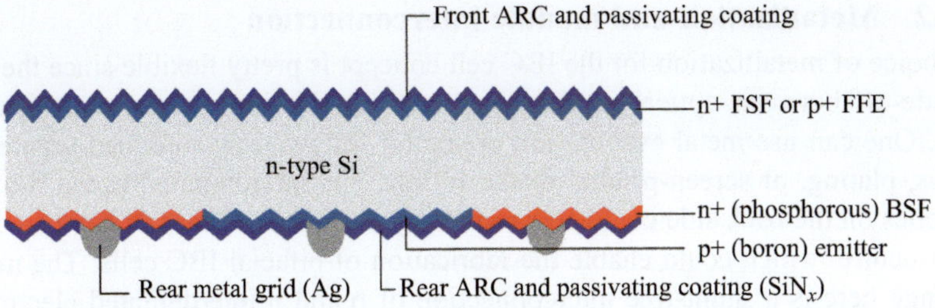

Figure 2.25 Generic cross section of an IBC cell concept with screen-printed metal contacts

diffusion process that is either BBr_3 (p^+) and $POCl_3$ (n^+) or by PECVD-doped layers deposition and a subsequent thermal annealing in order to drive-in the dopant species. Ion implantation has been implemented recently in IBC cell fabrication process due to its possibility for single side doping or even selective doping through a physical mask. Full ion implantation, or a hybrid implantation and diffusion, can be used to fabricate IBC cells. The ion implanted wafers need to undergo a thermal annealing step for electrical activation of the implanted impurities and for healing of implant damage. During this annealing step a thin thermal SiO_2 layer can be grown on Si surface, which can serve as an interface passivation layer in combination with SiN_x. This SiO_2/SiN_x is a widely used passivation stack in silicon solar cell technology and it applies for both, p^+- and n^+-doped regions. For p^+-doped regions alternative passivation stacks, such as ALD Al_2O_3/SiN_x or PECVD AlO_x/SiN_y are also used in industry. The patterning step of the dielectric masking layer, which is required to form an interdigitated grid of p- and n-diffusions on the back side, is done using industrially relevant low-cost techniques, such as screen printing of etching masks or etching paste or by laser ablation. To minimize the number of process steps, a single masking step is used, resulting in p^+-n^+ junctions with low breakdown voltage.

For a low-cost IBC structure, choosing the right device architecture and metallization layout is an important prerequisite to ensure optimum cell performance. Interdigitated p^+ and n^+ regions on the back side are designed to be in the order of hundreds of microns or millimeters to easily facilitate low-cost industrial patterning techniques, such as laser ablation and screen-printed metallization. Thus, the minority carrier collection over the BSF regions can be significantly reduced, as they have to diffuse longer distances to reach the emitter. To improve carrier collection efficiency a floating p^+ emitter can be used on the front side (FFE), instead of front surface field. ISC's "ZEBRA" and ECN's "Mercury" large area IBC cell concepts employ such FFE concept [130–132]. Under operation conditions the FFE, which is covering the entire front surface, efficiently collects the minority carriers generated in the bulk above the base regions and re-injects them back into the bulk above the emitter region.

2.5.5.2 Metallization and module interconnection

The choice of metallization for the IBC cell concept is pretty flexible since there is no trade-off between optical shading on the front-side and the series resistance losses. One can use metal evaporation or sputter followed by a contact separation process, plating, or screen-printed metallization. The interdigitated p and n metal electrodes on the back side could cover virtually the whole surface or have an open grid structure, which could enable the fabrication of bifacial IBC cells. The major challenge here is a shunt-free interconnection of p and n interdigitated electrodes by means of low-cost industrial applicable techniques. For industry standard 6-inch wafers, the extraction of the photocurrent only at wafer edges seems to be unsuited due to high resistive losses. Hence, alternative low-cost metallization concepts have to be used, such as single-layer or multi-layer interconnection of p and n fingers to their respective busbars. This co-planar arrangement of p and n contacts and busbars on the back side allows also for a simplified cell interconnection in the module. Here, various approaches are used in R&D labs and industry, from standard soldering or gluing of ribbons to conductive backsheet or SmartWire® [136].

2.5.5.3 Efficiency and bifaciality potential

The IBC is the most efficient (single junction) solar cell design to date for one-sun applications in large volume manufacturing. This has been demonstrated by SunPower Corp., which showed more than 25% efficiency with cost-effective lithography steps and electroplated metallization. For low-cost IBC cells with screen-printed metallization, the main performance limitation is similar to that of n-PERT cells: the recombination under the metal contacts, which limits the V_{OC} of the cells. The implementation of point contact or selective doping only partially alleviates this limitation. An alternative approach would be to implement high temperature carrier selective junctions, formed by doped polysilicon layers on passivating interfacial oxides, in combination with screen-printed metallization [137].

The IBC cells fabricated with open rear-side gridline metallization are also suitable for bifacial applications. In 1997, SunPower Corp. reported the first IBC bifacial solar cell and mini module, with a front-side efficiency of 21.9% and bifaciality factor at the cell level of $\varphi_{eta} \approx 63\%$ [138]. These cells had 20% metal coverage on the back side and an area of only $\approx$22 cm^2. On smaller area (4 cm^2) and laboratory process Glunz *et al.* demonstrated an IBC cell with a bifacial factor of $\varphi_{eta} \approx 98\%$ [139]. More recently, using screen-printed metallization and a multi-layer interconnection concept, ISC Konstanz has demonstrated 6-inch "ZEBRA" IBC cells with a bifacial factor of $\varphi_{eta} \approx 75\%$ and front-side efficiency of 22% [130,131]. Furthermore, these cells can be interconnected in the module using industry standard ribbon-based soldering techniques and they can be assembled in a bifacial module using glass-glass or glass-transparent backsheet. The bifaciality potential of such an IBC cell is exemplified by *quokka* simulation results in Figure 2.26 (left), where the bifaciality factor φ_{eta} and the front-side efficiency is plotted against metal coverage. All relevant input parameters in the simulation model were extracted from the experimental cell (also shown in Figure 2.26, right).

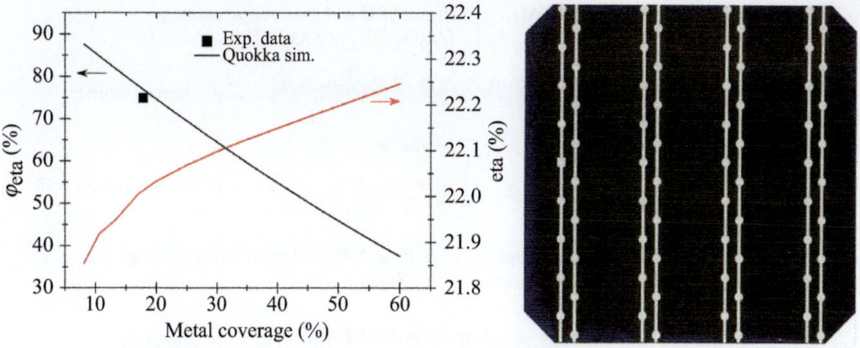

Figure 2.26 (left) Typical bifacial factor, φ_{eta}, and front-side efficiency of IBC cells fabricated by ISC-Konstanz (ZEBRA concept) as a function of metal coverage. In this quokka simulation the metal coverage was varied by extending the width of p and n fingers on top of the passivation layer without changing the contact width. (right) Image of the rear side of a 6-inch bifacial IBC "ZEBRA" cell with screen-printed metallization

In the quokka simulation model the metal coverage was varied by assuming an extension of p and n metal fingers over the passivating layer. Wafer ray tracer from PVlighthouse was used to calculate the generation profiles for the relevant cell regions as input for the *quokka* model. The increase in efficiency with metal coverage is a result of combined improvements in FF, due to reduction in series resistance, and J_{SC}, due to better reflection characteristics on the rear side. With today's screen printing metallization pastes, the simulation shows that a metal coverage (busbars + fingers) of $\leq$ 20% would be sufficient to minimize the grid resistance losses of the cells.

In an outdoor test, the energy yield of bifacial modules fabricated with such IBC cells show an increase of more than 15% as compared with monofacial IBC modules [131]. The bifacial IBC solar cells have not been transferred into the mass-production yet, but the technology and its economics are clearly demonstrated.

Another bifacial back contact solar cell concept on n-type silicon is the metallization wrap through (n-MWT), as shown schematically in Figure 2.27. The MWT solar cells have a fraction of the collecting emitter on the cell rear side and an additional second carrier-collecting junction at the cell front side leading to higher current collection. Because of the open rear-side metallization gridlines and dielectric passivating and anti-reflection coating layers, this cell concept is bifacial. This bifacial n-MWT concept has been introduced in 2014 by ECN [140] and enabled a front-side efficiency of 21% at that time. Till date, only monofacial modules have been made, but in principle the same technology used for bifacial IBC cells can be used to manufacture bifacial MWT modules.

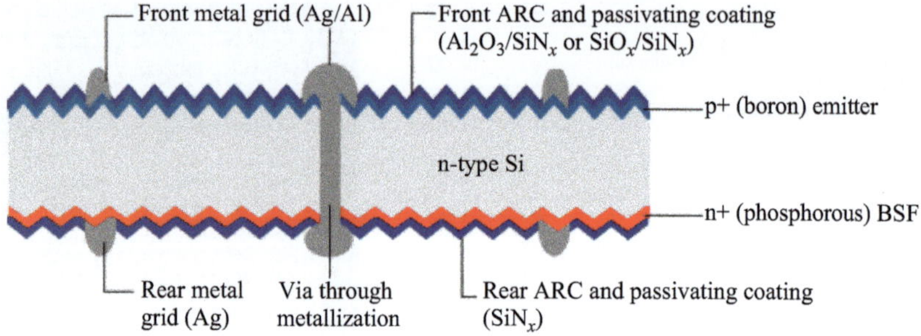

Figure 2.27 Cross section of n-MWT bifacial cell concept

2.6 Industrial solar cell technology roadmap

2.6.1 Industry status in 2017

Since a long time the Al BSF solar cell was dominating the PV market because of its simplicity and high efficiency potential. There were many projections from previous roadmaps, that innovations will enter the market during the feedstock crisis in 2005—but only SoG (Solar Grade) material and partly selective emitters did it with small impact. However, during that times, PERC, HJT, and PERT architectures were developed at many institutes such as INES, ISC Konstanz, ECN, ISFH and others—but only very few producers implemented them; for example, Yingli, PVGS and Sunpreme.

This changed now completely since 2016. At the beginning of 2016, for the first time in PV history, Cz-Si wafers became as cost effective as mc-Si wafers. This happened mostly because of low cost diamond wire sawn mono c-Si wafers, whereas mc-Si wafers still must have been sawn by slurry based technology, as no matching texturization technology was available for mc-Si surfaces at that time. In addition LONGi was pushing down the mono c-Si process costs, so that for first time mono c-Si PERC solar cells became more cost effective as standard Al BSF based mc-Si solar cells. This created something like a "catalytic innovation reaction" and since 2016 p-type mono PERC solar cells were entering the PV market with high impact. Figure 2.28 shows the market share of different cell technologies during that time.

It is clearly visible in that graph, that from 2016 the PERC processes (Mono advanced process) strongly "invaded" the market which also led to the consequence that the high efficiency n-type PERT processes were eaten up by them. Some of at that time established PERT producers like MegaCell and Mission Solar did not manage to compete with the low-cost PERC product. It is predicted that PERC and PERT processes (the so-called "PERX") will be the future of PV which we will see in the following section.

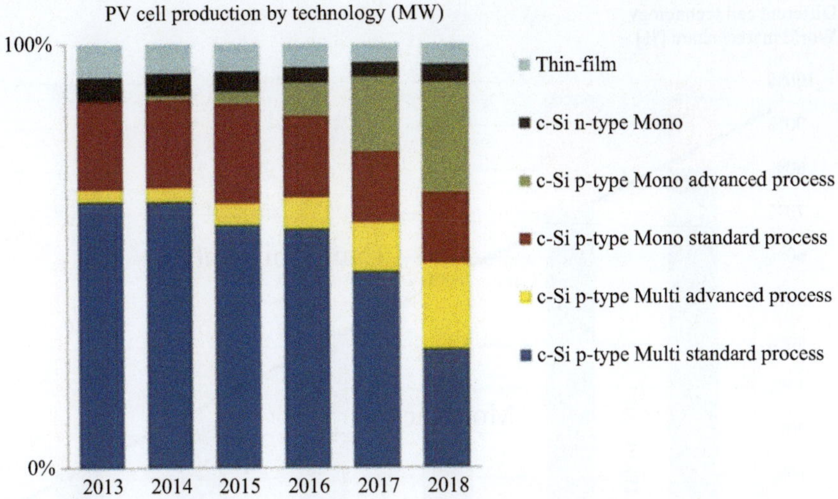

PVTECH CONFERENCE PVCELLTECH

Figure 2.28 PV cell production by technology [141]

2.6.2 Solar cell technology predictions (ITRPV)

The International Technology Roadmap for PV (ITRPV) has forecasted a bright future for bifacial technologies. The reason is quite simple: the Al BSF technology is being pushed out of the PV market as the efficiencies are stuck at about 20%. In order to reach efficiencies above 22% more complex processes and structures are needed which will become bifacial anyhow (PERX). Also in order to save material the rear side will not be covered by metal completely—the bifacial solar cell is a natural consequence of an evolutionary development.

The share of different technologies predicted until 2027 is depicted in Figure 2.29. In 2027 the Al BSF technology will only have a share of 10% and then will be totally gone. All the other technologies can and also will be made bifacial. The most market share will be taken by the so-called "PERX"—PERC, PERL, and PERT technology. Which one of them will take the biggest portion depends weather n-type will come with a strong impact on the market or not. This, for example, also depends how bifaciality will penetrate the market and how carrier selective contacts will be implemented. The SHJ technology will have also a market share of 10%- similar to back contact concepts- we believe IBC technologies and also tandem technologies will be visible. Tandem solar cells will be, for example, Perovskites, GaAs nano wires or CIGS on a c-Si solar cell. These devices can be made bifacial as well reaching front-side efficiencies above 30%.

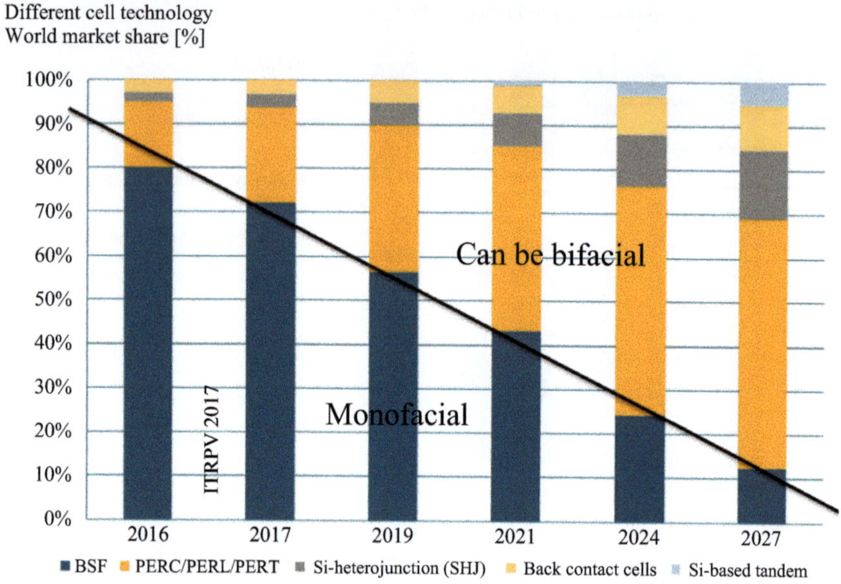

Figure 2.29 Technology share projection by ITRPV [142]

Newest predictions in Q1 2018 show that actually the ITRPV roadmap is much too conservative in terms of PERX market share. Already in 2018, PERC will become the new main technology with a market share of >50%. The total solar cell capacity will be 145 GWp with a PERC capacity of 75GWp. Therefore, bifacial devices will enter much faster the market—depending on the bifacial market. We see therefore that bifacial cell structures have a bright future. In the next chapter we will review the module technologies and summarize the status there—also in terms of standardization.

References

[1] B.B. Van Aken, L.A.G. Okel, J. Liu, S. L. Luxembourg, and J.A.M. van Roosmalen, "White bifacial modules – Improved STC performance combined with bifacial energy yield." Proceedings of the 32nd EUPVSEC, München, Germany, 2016.

[2] Y.S. Khoo, M.H. Saw, J.P. Singh, and Y. Wang, "Enhancing optical performance of glass-glass bifacial PV Modules", 7th International Conference on Silicon Photovoltaics, SiliconPV, Freiburg, Germany, 2017.

[3] S. Guo, T.M. Walsh, and M. Peters, "Vertically mounted bifacial photovoltaic modules: A global analysis", *Energy*, vol. 61, pp. 447–454, 2013.

[4] J.A. Duffie and W.A. Beckman, "Solar Engineering of Thermal Processes", 2nd Ed., Chap.2, (John Wiley & Sons, New York 1991).

[5] A. Cuevas, "The Early History of Bifacial Solar Cells", Proceedings of the 20th European Photovoltaic Solar Energy Conference, Barcelona, Spain, 2005.
[6] H. Mori, "Radiation energy transducing device", U.S. Patent 3.278.811, Oct. 1966 (priority Oct. 1960).
[7] A.K. Zaitseva and O.P. Fedoseeva, "Study of possibility of bifacial silicon solar cell applications", *Teploenergetika*, no. 3, p. 89, 1961.
[8] N.M. Bordina, M.B. Kagan, T.L. Lyubashevskaya, V.A. Letin, T.I. Taurbaev and V.F. Shorin, "High-efficiency GaA1As-GaAs photo converters with two-side sensitivity", *Applied Solar Energy*, vol. 21, no. 1, pp. 1–5, 1985.
[9] I. Chambouleyron and Y. Chevalier, "Silicon double solar cell", Proceedings of the 1st European Conference on Photovoltaic Solar Energy, Luxembourg, Sept. 1977, Reidel, Boston, 1978, pp. 967–976.
[10] A. Luque, J.M. Ruiz, A. Cuevas, J. Eguren, and J.M. Gomez-Agost, "Double-sided solar cells to improve static concentration", Proceedings of the 1st European Conference on Photovoltaic Solar Energy, Luxembourg, Sept. 1977, Reidel, Boston, 1978, pp. 269–277.
[11] K.S. Azimov, N.M. Bordina, G.M. Grigorieva, L.B. Kreinin, and A.P. Landsman, "Some photovoltaic performances of P-I-N solar cells", *Solar Engineering*, 3, pp. 3–6, 1973.
[12] K.S. Azimov, N.M. Bordina, G.M. Grigorieva, L.B. Kreinin, and A.P. Landsman, "About photovoltaic performances of high resistivity silicon photo-cells", *Soviet Physics – Semiconductors*, vol. 7, 11, pp. 1506–1507, 1974.
[13] A. Cuevas, A. Luque, and J. M. Ruiz, "A n+pn+ double-sided solar cell for optimal static concentration", Proceedings of the 14th IEEE Photovoltaic Specialists Conference, San Diego, 1980, pp. 76–81.
[14] H. Ohtsuka, M. Sakamoto, K.Tsutsui, and Y. Yazawa, "Bifacial silicon solar cells with 21.3% front efficiency and 19.8% rear efficiency", *Progress in Photovoltaics*, vol. 8, pp. 385–390, 2000.
[15] N.M. Bordina, V.V. Zadde, A.K. Zaitseva, *et al.*, "Semiconductor Photo-electric Generator". *USSR Certificate of Authorship N* 434872, 1970.
[16] N.M. Bordina, V.V. Zadde, A.K. Zaitseva, *et al.*, "Semiconductor Photo-electric Generator", Patent GFR N 2452263, 1971. U.S. Patent no. 3.948.682, Application 31 Oct. 1974, publication 6 April 1976.
[17] N.M. Bordina, T.M. Golovner, V.V. Zadde, A.K. Zaitseva, A.P. Landsman, and V.I. Streltsova, "Operation of a thin silicon photo converter under illumination on both sides", *Applied Solar Energy*, no. 6, pp. 81–86, 1975.
[18] A. Cuevas, A. Luque, J. Eguren, and J. Del Alamo, "High efficiency bifacial back surface field solar cells", *Solar Cells*, 3, pp. 337–340, 1981.
[19] S.W. Glunz, J. Knobloch, D. Biro, and W. Wetting, "Optimized high-efficiency silicon solar cells", Proceedings of the 14th European Photovoltaic Solar Energy Conference, Barcelona, 1997, pp. 392–395.
[20] A. Meulenberg, J.F. Allison, and R.A. Arndt, "Thin n-i-p silicon solar cell", Proceedings of the 14th IEEE Photovoltaic Specialists Conference San Diego, Jan. 7–10, 1980, pp. 161–165.

[21] M. Giuliano and J. Wohlgemuth, Proceedings of the 15th IEEE Photovoltaic Specialists Conference, Kissimmee, May 12–15, 1981, pp. 111–114.

[22] G. Strobl, C. Kasper, K.D. Rasch, and K. Roy, Bifacial space silicon solar cell, Proceedings of the 18th IEEE Photovoltaic Specialists Conference, 1985, pp. 454–457.

[23] N.M. Bordina, N.A. Borisova, G.S. Daletskii, A.K. Zaitseva, A.P. Landsman, and V.A. Letin, "Using the radiation reflected from the earth for increasing the power of solar batteries," *Cosmic Research*, vol. 14, no. 2, pp. 266–272, 1976, (submitted 24.06.1974).

[24] N.M. Bordina, T.M. Golovner, E.V. Zhidkova, *et al.*, "Optical characteristics of highly efficient silicon photovoltaic cells", *Journal of Applied Spectroscopy*, vol. 32, no. 4, pp. 408–413, 1980.

[25] V.A Letin, M.B. Kagan, V.P. Nadorov, and V.R. Zajavlin, "Bifacial solar arrays of Russian space crafts", *The 28th IEEE Photovoltaic _Specialists Conference*, Anchorage, Alaska, 15–22 September, 2000, pp. 1067–1070.

[26] A. Cuevas, A. Luque, J. Eguren, and J. del Alamo, "50% more output power from an albedo-collecting flat panel using bifacial solar cells", *Solar Energy*, vol. 19, pp. 419–420, 1982.

[27] A. Cuevas, A. Luque, J. Eguren, and J. del Alamo, "Bifacial albedo-collecting photovoltaic module", Spanish Patent no. 493.803, 1980.

[28] Available from https://www.tnc.ch/en/photovoltaics/#pv-noise-barrier [Accessed June 2018].

[29] R. Hezel, "Novel applications of bifacial solar cells", *Progress in Photovoltaics: Research and Applications*, vol. 11, pp. 549–556, 2003.

[30] T. Dullweber and J. Schmid, "Industrial silicon solar cells applying the passivated emitter and rear cell (PERC) concept—A review", *IEEE Journal of Photovoltaics*, vol. 6, pp. 1366–1381, 2016.

[31] L.J.J. Geerligs and D. MacDonald, "Base doping and recombination activity of impurities in crystalline silicon solar cells", *Progress in Photovoltaics: Research and Applications*, vol. 12, pp. 309–316, 2004.

[32] A.R. burgers, L.J. Geerligs, A.J. Carr, *et al.*, "19.5% efficient n-type Si solar cells made in production." Proceedings of the 26th European Photovoltaic Solar Energy Conference, 2011, pp. 1144–1147.

[33] K. Sugibuchi, N. Ishikawa, and S. Obara, "Bifacial-PV power output gain in the field test using "EarthON" high bifaciality solar cells", Proceedings of the 28th European Photovoltaics Solar Energy Conference and Exhibition, 2013, pp. 4312–4317.

[34] T. Dullweber, C. Kranz, R. Peibst, *et al.*, "PERC+ industrial PERC solar cells with rear Al grid enabling bifaciality and reduced Al paste consumption, *Progress in Photovoltaics: Research and Applications*, vol. 24, pp. 1487–1498, 2016.

[35] G. Galbiati, V.D. Mihailetchi, R. Roescu, *et al.*, "Large-area back-contact back-junction solar cell with efficiency exceeding 21%", *IEEE Journal of Photovoltaics*, vol. 3, pp. 560–563, 2013.

[36] R. Kopecek, Y. Veschetti, E. Gerritsen, *et al.*, "Bifaciality: One small step for technology, one giant leap for kWh cost reduction", *Photovoltaics International*, vol. 26, pp. 32–45, 2014.

[37] B.B. Van Aken, C.C.J. Tool, E.J. Kossen, *et al.*, "Bifacial aspects of industrial n-Pasha solar cells", *Journal of Applied Physics*, vol. 56, 08MB03, 2017.

[38] G.J.M. Janssen, C.C.J. Tool, E.J. Kossen, B.B. Van Aken, A,J. Carr, and I.G. Romijn, "Aspects of bifacial cell efficiency", Energy Procedia, Proceedings of the 7th International Conference on Silicon Photovoltaics, SiliconPV, 2017.

[39] J.P. Singh, A.G. Aberle, and T.M. Walsh, "Electrical characterization method for bifacial photovoltaic modules", *Solar Energy Materials and Solar Cells*, vol. 127, pp. 136–142, 2014.

[40] IEC 60904-1: Photovoltaic devices – Part 1: Measurement of photovoltaic current-voltage.

[41] IEC 60904-9: Photovoltaic devices – Part 9: Solar simulator performance requirements.

[42] Available from https://www.astm.org/Standards/G173.htm [Accessed February 2018].

[43] Electric Power Research Institute EPRI, Bifacial Solar Photovoltaic Modules – Program on Technology Innovation, 2016.

[44] M. Tanaka, M. Taguchi, T. Matsuyama, *et al.*, "Development of new a-Si/c-Si heterojunction solar cells: ACJ-HIT (artificially constructed junction-heterojunction with intrinsic thin-layer", *Japanese Journal of Applied Physics*, vol. 31, pp. 3518–3522, 1992.

[45] S. De Wolf, A. Descoeudres, Z.C. Holman, and C. Ballif, "High-efficiency silicon heterojunction solar cells: a review", *Green*, vol. 2, pp. 7–24, 2012.

[46] A. Descoeudres, C. Allebé, and N. Badel, "Silicon heterojunction solar cells: towards low-cost high-efficiency industrial devices and application to low-concentration PV", *Energy Procedia*, vol. 77, pp. 508–514, 2015.

[47] E. Kobayashi, N. Nakamura, K. Hashimoto, and Y. Watabe, "Rear-emitter silicon heterojunction solar cells with efficiencies above 22%", Proceedings of the 28th EU-PVSEC, 2013, pp. 691–694.

[48] A. Danel, S. Harrison, B. Martel, *et al.* "Recent progress on the CEA-INES Heterojunction Solar Cell Pilot Line", Proceedings of the 31st EU-PVSEC, 2015, pp. 279–283.

[49] M. Taguchi, A. Yano, S. Tohoda, *et al.* "24.7% Record efficiency HIT solar cell on thin silicon wafer," *IEEE Journal of Photovoltaics*, vol. 4, pp. 96–99, 2014.

[50] K. Yamamoto, H. Kawasaki, W. Yoshida, *et al.* "Silicon heterojunction solar cell with interdigitated back contacts for a photoconversion efficiency over 26%", *Nature Energy*, vol. 2, 17032, 2017.

[51] J.B. Heng, J. Fu, B. Kong, *et al.* "23% High-efficiency tunnel oxide junction bifacial solar cell with electroplated Cu gridlines," *IEEE Journal of Photovoltaics*, vol. 5, no. 1, pp. 82–86, 2015.

[52] J. Geissbühler, S. De Wolf, A. Faes, *et al.*, "Silicon heterojunction solar cells with copper-plated grid electrodes: status and comparison with silver thick-film techniques," *IEEE Journal of Photovoltaics*, vol. 4, no. 4, pp. 1055–1062, 2014.
[53] A. Schneider, L. Rubin, and G. Rubin, "Solar cell improvement by new metallization techniques – The DAY4™ electrode concept", Proceedings of the 4th WCPEC, Waikoloa, 2006 p. 1095.
[54] A. Faes, M. Despeisse, J. Levrat, *et al.* "Smart wire solar cell interconnection technology", Proceedings of the 29th EU- PVSEC, Amsterdam, 2014.
[55] B. Strahm, D. Lachenal, D Bätzner, *et al*, "The Swiss Inno-HJT project: fully integrated R&D to boost Si-HJT module performance", Proceedings of the EU-PVSEC, Hamburg, 2015.
[56] Available from https://www.meyerburger.com/ch/en/meyer-burger/media/news-detail/article/meyer-burger-technology-day-2017-premiers-heterojunction-smartwire-module-with-335-watt-efficiency/ [Accessed February 2018].
[57] K. Yoshikawa, H. Kawasaki, W. Yoshida, *et al.*, "Silicon heterojunction solar cell with interdigitated back contacts for a photoconversion efficiency over 26%", Nature Energy, vol. 2, p. 17032, 2017.
[58] Available from http://www.yinglisolar.com/en/products/monocrystalline/panda-210/ [Accessed February 2018].
[59] Available from http://www.lg.com/uk/business/solar/neon/lg-neon-2-bifacial [Accessed February 2018].
[60] Available from https://www.pv-magazine.com/2017/02/20/jolywood-breaks-ground-on-2-1-gw-bifacial-cell-factory/ [Accessed February 2018].
[61] A.R. Burgers, L.J. Geerligs, A.J. Carr, *et al.*, "19.5% efficient n-type Si solar cells made in production", Proceedings of the 26th European Photovoltaic Solar Energy Conference and Exhibition, Hamburg, 2011.
[62] V.D. Mihailetchi, J. Jourdan, A. Edler, *et al.*, "Screen printed n-type silicon solar cells for industrial application", Proceedings of the 25th European Photovoltaic Solar Energy Conference and Exhibition/5th World Conference on Photovoltaic Energy Conversion, Valencia, 2010, doi: 10.4229/25thEUPVSEC2010-2DO.2.2
[63] B. Hoex, J. Schmidt, R. Bock, P.P. Altermatt, M.C.M. van der Sanden, and W.M.M. Kessels, "Excellent passivation of highly doped p-type Si surfaces by the negative-chargedielectric Al_2O_3", *Applied Physics Letters*, vol. 91, p. 112107, 2007.
[64] G. Dingemans, N.M. Terlinden, M.A. Verheijen, M.C.M. van de Sanden, and W.M.M. Kessels, "Controlling the fixed charge and passivation properties of Si(100)/Al_2O_3 interfaces using ultrathin SiO_2 interlayers synthesized by atomic layer deposition", *Journal of Applied Physics*, vol. 110, 093715, 2011.
[65] D.S. Saynova, I.G. Romijn, I. Cesar, *et al.*, "Dielectric passivation schemes for high efficiency n-type c-Si solar cells, Proceedings of the 28th European Photovoltaic Solar Energy Conference and Exhibition, Paris, France, 2013.

[66] L. Tous, R. Russel, and Y. Yao, "Progress on bifacial Ni/Ag plated n-PERT cells for module fabrication with SWCT", Proceedings of the 33rd European Photovoltaic Solar Energy Conference, Amsterdam, Netherlands, 2017.

[67] I.G. Romijn, L. Fang, and A. Vlooswijk, "Industrial n-type solar cells: towards 20% efficiency," *Photovoltaics International*, 15th edition, p. 81, 2012.

[68] P. Rothhardt, R. Keding, A. Wolf, and D. Biro, "Co-diffusion from solid sources for bifacial n-type solar cells", *Physica Status Solidi RRL*, pp. 1–4, 2013,doi: 10.1002/pssr.201208055

[69] N. Wehmeier, B. Lim, A. Merkle, *et al.*, "PECVD BSG diffusion sources for simplified high-efficiency n-PERT BJ and BJBC solar cells", *IEEE Journal of Photovoltaics*, vol. 6, no. 1, pp. 119, 2016.

[70] R. Cabal, T. Blevin, R. Monna, Y. Veschetti, and S. Dubois, "Multi-functional dielectric layers for the fabrication of ultra simplified n-PERT c-Si solar cells", *Energy Procedia*, vol. 92, pp. 684–690, 2016.

[71] A. Lanterne, Y. Veschetti, R. Cabal, *et al.,* "Ion implant doping for n-type Cz high efficiency industrial boron emitter solar cells with single activation process", Proceedings of the 27th European Photovoltaic Solar Energy Conference and Exhibition, Frankfurt, Germany, 2012.

[72] F. Kiefer, J. Krügener, F. Heinemeyer, *et al.*, "Bifacial, fully screen-printed n-PERT solar cells with BF2 and B implanted emitters", *Solar Energy Materials and Solar Cells*, vol. 157, pp. 326–330, 2016.

[73] D. Kania, T.S. Böscke, M. Braun, *et al.*, "Pilot line production of industrial high-efficient bifacial n-type silicon solar cells with efficiencies exceeding 20.6%", Proceedings of the 28th European Photovoltaic Solar Energy Conference and Exhibition, 2013, p. 1383.

[74] B. Bazer-Bachi, C. Oliver, B. Semmache, *et al.*, "Co-diffusion from boron doped oxide and $POCl_3$", Proceedings of the 26th European Photovoltaic Solar Energy Conference and Exhibition, 2011, pp. 1155–1159.

[75] R. Keding, P. Rothhardt, C. Roters, *et al.*, "Silicon doping performed by different diffusion sources aiming co-diffusion", Proceedings of the 27th European Photovoltaic Solar Energy Conference and Exhibition, 2012, pp. 1906–1911.

[76] V.D. Mihailetchi, Y. Komatsu, and L.J. Geerligs, "Nitric acid pretreatment for the passivation of boron emitters for n-type base silicon solar cells", *Applied Physics Letters*, vol. 92, p. 63510, 2008.

[77] J. Lossen, F. Buchholz, C. Comparotto, *et al.*, "From lab to fab: bifacial n-type cells entering industrial production", Proceedings of the 31st European Photovoltaic Solar Energy Conference and Exhibition, Hamburg, Germany, 2015, doi: 10.4229/EUPVSEC20152015-2CV.4.22

[78] A. Edler, V.D. Mihailetchi, L.J. Koduvelikulathu, C. Comparotto, R. Kopecek, and R. Harney, "Metallization-induced recombination losses of bifacial silicon solar cells," *Progress in Photovoltaics: Research and Applications*, vol. 23, no. 5, pp. 620–627, 2014.

[79] S. Fritz, S. Riegel, A. Herguth, M. König, M. Hörteis, and G. Hahn, "Preservation of Si surface structure by Ag/Al contact spots – an explanatory model", *Energy Procedia*, vol. 67, pp. 43–48, 2015.

[80] I. Romijn, B. van Aken, J. Anker, *et al.*, "Improvements in advanced industrial n-type solar cells and modules", *Photovoltaics International*, vol. 25, p. 58, 2014.

[81] J. Liu, G.J.M. Janssen, M. Koppes, *et al*, "Selective emitters in n-type c-Si solar cells", Proceedings of the 31st European Photovoltaic Solar Energy Conference and Exhibition, Hamburg, 2015.

[82] S. Fritz, E. Emre, J. Engelhardt, S. Ebert, and N. Nowak, "Contacting BBr3-based boron emitters with Al-free screen-printing Paste", Proceedings of the 6th International Conference on Crystalline Silicon Photovoltaics Silicon PV, Chambery, 2016.

[83] F. Buchholz, P. Preis, H. Chua, J. Lossen, and E. Wefringhaus, "Progress in the development of industrial n-PERT cells", Proceedings of the 7th International Conference on Silicon Photovoltaics, SiliconPV, 2017,

[84] S.W. Glunz, F. Feldmann, A. Richter, *et al.*, "The irresistible charm of a simple current flow pattern – approaching 25% with a solar cell featuring a full-area back contact", Proceedings of the 31st European Photovoltaic Solar Energy Conference and Exhibition, Hamburg, 2015.

[85] A. Richter, J. Benick, F. Feldmann, A. Fell, S.W. Glunz, and M. Hermle, "Silicon solar cells with passivated contacts: influence of wafer resistivity and thickness", Proceedings of the 7th International conference on Crystalline Silicon Photovoltaics Silicon PV, Freiburg, 2017.

[86] M.K. Stodolny, L.J.J. Geerligs, G.J.M. Janssen, *et al.*, "Material properties of LPCVD processed n-type polysilicon passivating contacts and its application in perpoly industrial bifacial solar cells", Proceedings of the 7th International Conference on Silicon Photovoltaics SiliconPV, 2017.

[87] M.K. Stodolny, L.J. Geerligs, J. Anker, *et al.*,, "Material properties of LPCVD processed n-type polysilicon passivating contacts and its application in perpoly industrial bifacial solar cells", Proceedings of the 33rd European Photovoltaic Solar Energy Conference and Exhibition, Amsterdam, The Netherlands, 2017.

[88] R. Peibst, Y. Larionova, S. Reiter, *et al.*, "Implementation of n+ and p+ polo junctions on front and rear side of double-side contacted industrial silicon solar cells", Proceedings of the 32nd European Photovoltaic Solar Energy Conference and Exhibition, Munich, Germany, 2016.

[89] L. Kreinin, N. Bordin, N. Eisenberg, G. Grigorieva, K. Zviagina, and M. Kagan, "Bifacial solar cells fabricated by combined diffusion – implantation technology". Proceedings of the 2nd World Conference on Photovoltaic Solar Energy Conversion, Vienna, Austria, July 6–10, 1998, pp. 1744–1747.

[90] L. Kreinin, N. Bordin, N. Eisenberg, G. Grigorieva, K. Zviagina, and M. Kagan, "Combined thermal diffusion – ion implantation fabrication processing for silicon solar cells". *Opto-Electronic Review*, vol. 8, no. 4, pp. 47–52, 2000.

[91] N. Bordin, L. Kreinin, and N. Eisenberg, "Determination of recombination parameters of bifacial solar cells with a two layer step-like defect distribution in the base region," Proceedings of the 17th European Photovoltaic Solar Energy Conference and Exhibition, Munich, 22–26 October, 2001, pp. 1495–1498.

[92] N. Bordin, L. Kreinin, J. Broder, *et al.*, "Recombination parameters of Si solar cells with back surface field formed by ion implantation," Proceedings of the Fourth World Conference on Photovoltaic Energy Conversion. Waikoloa, Hawaii, May 7–12, 2006.

[93] G. Grigorieva, M. Kagan, V. Unishkov, *et al.*, "Efficiency of bifacial Si solar cells at low irradiance. Effect of design and fabrication technology factors, Proceedings of the 25th European Photovoltaic Solar Energy Conference and Exhibition, 5th World Conference on Photovoltaic Energy conversion, Valencia, Spain, 6–10 September 2010, 2CV.2.26.

[94] L. Kreinin, N. Bordin, J. Broder, *et al.*, "Comparison of two BSF technologies suitable for high efficiency multi-crystalline solar cells", Proceedings of the 21st European Photovoltaic Solar Energy Conference and Exhibition, Dresden, Germany, 4–8 September, 2006, 2CV.3.28, pp. 855–858.

[95] L. Kreinin, N. Bordin, J. Broder, *et al.*, "Comparison of different technologies for bifacial silicon based solar cells," Proceedings of the 22nd European Photovoltaic Solar Energy Conference and Exhibition, Milan, Italy, 3–7 September 2007, 2DV.2.16.

[96] L. Kreinin, N. Bordin, J. Broder, *et al.*, "Silicon solar cells with a back surface field formed by selective doping," Proceedings of the PV in Europe From PV Technology to Energy Solutions Conference, Rome, Italy, 7–11 October 2002.

[97] L. Kreinin, N. Bordin, N. Eisenberg, P. Grabitz, and G. Wahl, "Industrially fabricated bifacial Si solar cells with n+-p-p+ structure", Proceedings of the 28th European Photovoltaic Solar Energy Conference and Exhibition, Paris, France, September 30–October 4, 2013.

[98] Y. Eisenberg, J. Arumughan, L. Kreinin, N. Bordin, and N. Eisenberg "High efficiency p-PERT bifacial solar cell", Proceedings of the 33rd European Photovoltaic Solar Energy Conference and Exhibition, Amsterdam, 25–29 September 2017.

[99] A. Teppe, A. Ramakrishnan, T. Friess, *et al.*, "Novel technology approach based on standard quality MC wafers achieving solar cell efficiencies significantly above 18% in an industrial production environment", Proceedings of the 29th European Photovoltaic Solar Energy Conference and Exhibition, Amsterdam, The Netherlands, p. 1310, 2014.

[100] A. Teppe, C. Gong, O. Voigt, *et al.*, "Bifacial multicrystalline solar cells with efficiencies above 18% processed in an industrial production environment", Proceedings of the 31st European Photovoltaic Solar Energy Conference and Exhibition, Hamburg, Germany, p. 610, 2015.

[101] A. Teppe, E. Wang, W. Guo, *et al.*, "Pilot production of bifacial multi-crystalline PERCT cells achieving 18.5% efficiency and single facial more than 19%," Proceedings of the 32nd European Photovoltaic Solar Energy Conference and Exhibition, Munich, Germany, p. 610, 2016.

[102] Y. Eisenberg, L. Kreinin, N. Bordin, and N. Eisenberg, "Effective surface recombination of p+ layers doped using ion implantation or surface deposited B sources", Proceedings of the 6th International Conference on Silicon Photovoltaics, Silicon PV 2016, Energy Procedia, vol. 92, pp. 16–23, August 2016.

[103] International Technology Roadmap for Photovoltaic (ITRPV.net), 2016 Results, 8th edition, March 2017. Available from http://www.itrpv.net/Reports/Downloads/ [Accessed February 2018].

[104] T. Dullweber, C. Kranz, R. Peibst, *et al.*, "PERC+: industrial PERC solar cells with rear Al grid enabling bifaciality and reduced Al paste consumption", *Progress in Photovoltaics: Research and Applications*, vol. 24, pp. 1487–1498, 2016.

[105] T. Dullweber, C. Kranz, R. Peibst, *et al.*, "The PERC+ cell: a 21%-efficient industrial bifacial PERC solar cell", Proceedings of the 31st European Photovoltaic Solar Energy Conference, 2015, p. 341.

[106] B. Liu, Y. Chen, Y. Yang, *et al.*, "Rear-colored double glass modules fabricated with rear-colored bifacial p-type PERC cells", Proceedings of the 31st European Photovoltaic Solar Energy Conference, 2015, p. 659.

[107] K. Krauß, F. Fertig, J. Greulich, S. Rein, and R. Preu, "biPERC silicon solar cells enabling bifacial applications for industrial solar cells with passivated rear sides", *Physica Status Solidi A*, vol. 213, p. 68, 2016.

[108] T. Dullweber and J. Schmidt, "Industrial silicon solar cells applying the passivated emitter and rear cell (PERC) concept—a review", *IEEE Journal of Photovoltaics*, vol. 6, pp. 1366–1381, 2016.

[109] C. Kranz, B. Wolpensinger, R. Brendel, and T. Dullweber, "Analysis of local aluminum rear contacts of bifacial PERC+ solar cells", *IEEE Journal of Photovoltaics*, vol. 6, pp. 830–836, 2016.

[110] C. Kranz, U. Baumann, B. Wolpensinger, *et al.*, "Void formation in screen-printed local aluminum contacts modeled by surface energy minimization", *Solar Energy Materials and Solar Cells*, vol. 158, pp. 11–18, 2016.

[111] S.-Y. Chen, H.-C. Liu, H.-Y. Chang, *et al.*, "Bifacial solar cells fabricated by PERC process for mass production", Proceedings of the 32nd European Photovoltaic Solar Energy Conference, 2016, p. 772.

[112] Communication with JinkoSolar, October 2017.

[113] Available from https://www.nsp.com/nspsolarcells?lang=en [Accessed September 2017].

[114] Communication with H. Li, LONGi Solar, October 2017.

[115] T. Dullweber, H. Schulte-Huxel, C. Kranz, *et al.*, "Bifacial PERC+ solar cells and modules: an overview", Proceedings of the 33rd European Photovoltaic Solar Energy Conference, 2017, p. 649.

[116] T. Dullweber, H. Schulte-Huxel, S. Blankemeyer, *et al.*, "Present status and future perspectives of bifacial PERC+ solar cells and modules", *Japanese Journal of Applied Physics*, vol. 57, 08RA01, 2018.

[117] Jinko Solar press release, October 2017, Available from https://www.pv-tech.org/news/jinkosolar-pushes-p-type-monocrystalline-perc-solar-cell-to-new-efficiency [Accessed February 2018].

[118] SolarWorld press release, 2015. Available at http://www.pv-tech.org/news/intersolar_europe_solarworld_to_launch_glass_glass_bifacial_modules [Accessed February 2018].

[119] Available from https://www.solarworld.de/en/products/sunmodule-bisun-protect/; [Accessed September 2017].

[120] Available from https://www.nsp.com/nspsolarmodules?lang=en [Accessed September 2017].

[121] Available from http://www.trinasolar.com/de/product/duomax4060/duo-max-twin-deg5c07ii [Accessed September 2017].

[122] Available from http://en.longi-solar.com/Home/Products/module/id/12_.html [Accessed September 2017].

[123] H. Hannebauer, T. Dullweber, U. Baumann, T. Falcon, and R. Brendel, "21.2%-efficient fineline-printed PERC solar cell with 5 busbar front grid", *Physica Status Solidi RRL*, vol. 8, p. 675, 2014.

[124] A. Faes, M. Despeisse, J. Levrat, *et al.*, "SmartWire solar cell interconnection technology", Proceedings of the 29th European Photovoltaic Solar Energy Conference, 2014, p. 2555.

[125] M. A. Green, K. Emery, Y. Hishikawa, W. Warta, and E. D. Dunlop, "Solar cell efficiency tables (version 47)", *Progress in Photovoltaics: Research and Applications*, vol. 24, pp. 3–11, 2016.

[126] J. Nakamura, N. Asano, T. Hieda, C. Okamoto, H. Katayama, and K. Nakamura, "Development of hetero-junction back contact Si solar cells", Proceedings of the 40th IEEE PVSC, Denver, 2014.

[127] K. Masuko, M. Shigematsu, T. Hashiguchi, *et al.*, "Achievement of more than 25% conversion efficiency with crystalline silicon heterojunction solar cell", *IEEE Journal of Photovoltaics*, vol. 4, pp. 1433–1435, 2014.

[128] M. A. Green, K. Emery, Y. Hishikawa, W. Warta, and E. D. Dunlop, "Solar cell efficiency tables (version 48)", *Progress in Photovoltaics: Research and Applications*, vol. 24, pp. 905–913, 2016.

[129] Z. Li, Y. Yang, X. Zhang, *et al.*, "Pilot production of 6″ IBC solar cells yielding a median efficiency of 23% with a low-cost industrial process", Proceedings of the 32nd EU-PVSEC, pp. 571–574, 2016.

[130] G. Galbiati, V.D. Mihailetchi, R. Roescu, *et al.*, "Large-area back-contact back-junction solar cell with efficiency exceeding 21%", *IEEE Journal of Photovoltaics*, vol. 3, pp. 560–563, 2013.

[131] V.D. Mihailetchi, H. Chu, G. Galbiati, C, Comparroto, A. Halm, and R. Kopecek, "A comparison study of n-type PERT and IBC cell concepts with screen printed contacts", *Energy Procedia*, vol. 77, pp. 534–539, 2015.

[132] I. Cesar, N. Guillevin, A.R. Burgers, *et al.*, "Mercury: a back junction back contact front floating emitter cell with novel design for high efficiency and simplified processing", *Energy Procedia*, vol. 55, pp. 633–642, 2014.
[133] C.-L. Lin, C.-C. Chen, Y.-T. Cheng, Y.-H. Huang, C.-P. Tsao, and J.W. Chien, "Large area high efficient back-junction back-contact silicon solar cells with industrial process", Proceedings of the 29th EU-PVSEC, pp. 816–820, 2015.
[134] J. Dong, L. Tao, Y. Zhu, *et al.*, "High efficiency full back contacted cells using industrial processes", *IEEE Journal of Photovoltaics*, vol. 4, pp. 130–133, 2014.
[135] M. Hendrichs, M. Padilla, J. Walter, F. Clement, and B. Rech, "Screen-printed metallization concepts for large-area back-contact back-junction silicon solar cells", *IEEE Journal of Photovoltaics*, vol. 6, pp. 374–383, 2016.
[136] T. Borgers, J. Govaerts, E. Voroshazi, *et al.*, "A woven fabric for interconnecting back-contact solar cells", *Progress in Photovoltaics: Research and Applications*, doi: 10.1002/pip.2851, 2016.
[137] R. Peibst, Y. Larionova, S. Reiter, *et al.*, "Implementation of n+ and p+ POLO junctions on front and rear side of double-side contacted industrial silicon cells", Proceedings of the 32nd EU-PVSEC, pp. 323–327, 2016.
[138] C.Z. Zhou, P.J. Verlinden, R.A. Crane, and R.M. Swanson, "21.9 efficient silicon bifacial solar cells", Proceedings of the 26th PVSC, pp. 287–290, 1997.
[139] S.W. Glunz, J. Knobloch, C. Hebling, and W. Wettling, "The range of high-efficiency silicon solar cells fabricated at Fraunhofer ISE", Proceedings of the 26th PVSC, pp. 231–234, 1997.
[140] N. Guillevin, A. Gutjahr, L.J. Geerligs, *et al.*, "High efficiency n-type metal-wrap-through cells and modules using industrial processes", Proceedings of the 29th European Photovoltaic Solar Energy Conference and Exhibition, Amsterdam, the Netherlands, 2014.
[141] Available from https://www.pv-tech.org/editors-blog/china-and-oem-cell-production-in-2016-delays-shift-to-p-type-mono [Accessed February 2018].
[142] Available from http://www.itrpv.net/Reports/Downloads/ [Accessed February 2018].

Chapter 3

Bifacial modules: design options, characterisation and reliability

Andreas Schneider[1], Bas van Aken[2], Eric Gerritsen[3], Jai Prakash[4], Vahid Fakhfouri[5], Khoo Yong Sheng[4], and Andreas Halm[6]

3.1 Bifacial PV modules: design and characterisation

3.1.1 Design considerations for bifacial modules

Bifacial modules can be applied for large PV plants as well as for residential (flat white roof) and more specific BIPV (façade) applications and can also open up new PV application opportunities like in sound barriers or other vertical installations (fences, balconies).

For bifacial PV plants, the objective is to exploit the main bifacial benefit which is a large reduction of LCOE (due to higher energy yield) with a minimal technical change or investment [1]. The challenge for optimal performance of bifacial plants is mostly in the optimisation of the contribution from the module rear-side. Besides considering the quantity of light reaching the backside (dependent on installation conditions like the ground albedo), the optimisation also concerns the 'quality' of the backside irradiation, that is, its uniformity over the entire module backside which is not straightforward due to the indirect irradiation (diffuse and reflected) of the backside. In general, rear-side uniformity is improved when increasing the module height above ground [2], by reducing self-shading of the module, but the module itself can also be designed to be less sensitive to non-uniform backside irradiation, as will be discussed in Section 3.4. It should be noted that shading, like by mounting structures, on the rear-side, with mostly diffuse irradiation, will be of lesser impact than it would be on the front side, subjected to direct irradiation.

A glass/glass structure is the most obvious choice for bifacial modules with the following benefits. These modules can do without a module frame, thereby offering

[1]Westfälische Hochschule, Germany
[2]ECN part of TNO, The Netherlands
[3]INES – Institut National de l'Energie Solaire, France
[4]Solar Energy Research Institute of Singapore, Singapore
[5]Qualimatest, Switzerland
[6]ISC Konstanz, Germany

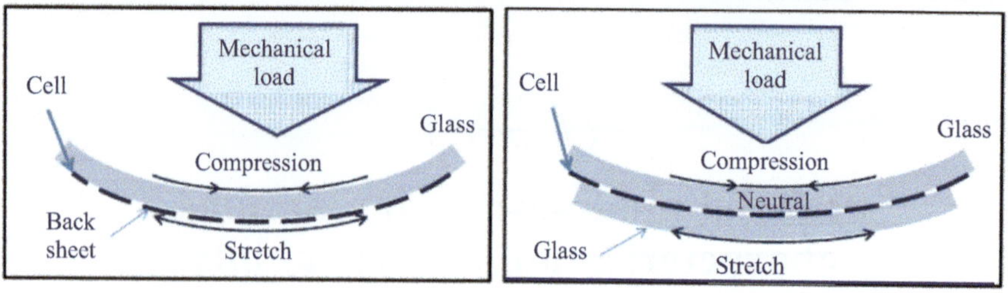

Figure 3.1 Stress distribution in glass/back sheet and glass/glass modules. The cells in the glass/glass module are located in the neutral mechanical fibre of the module and thus not subject to tensile stress

a reduction of the module and BOS costs by using more simple mounting clamps. Moreover, frameless modules will be less sensitive to self-shading of the module edge and to performance degradation due to soiling, which is frequently accumulated near the frame that hinders natural draining of this soiling layer. Furthermore, it is shown that modules without a full edge frame will be less sensitive to degradation by potential induced degradation (PID) [3]. It should be noted that back rail mounting structures can obviously not be used to replace the module frame as they provoke partial shading to the module rear-side.

Apart from the opportunity to construct frameless modules the glass/glass module design offers an inherent benefit to the mechanical integrity of the cells during mechanical load (like by snow or wind). This is due to the fact that the cells are located on the neutral mechanical axis of the module, as illustrated in Figure 3.1, and are thus not subjected to tensile stress that initiates cell cracking.

A glass backside provides a perfect humidity barrier compared to polymer back sheets. Of course the weight of the module is to a certain degree compromised by such a double glass structure but thermally tempered solar glass is now also available at 2 mm thickness. A module with 2 mm front and backside has several advantages in terms of mechanical and chemical durability over a glass/back sheet module with a 4 mm front cover, without increasing the module weight. On the other hand it can be noted that transparent back sheets are available that allow the assembly of lightweight bifacial modules.

The key challenge for bifacial solar modules is the design and placement of the junction box. Since any placement of junction boxes on light-sensitive areas on the module backside leads to undesired shading, the junction box either has to be reduced in size or must be placed in the edge region of the module (see Figure 3.2), if module size is to be kept constant. At the same time, these smaller junction boxes have to handle higher currents because of the extra current generated by the module backside. The latter problem can be solved by cutting the cells in half, thereby reducing the cell current and the associated resistive losses thereby improving the cell-to-module performance. The use of cut cells is discussed in Section 3.4 on electrical design and interconnect options.

Figure 3.2 Edge junction box on a bifacial module

Rooftop and BIPV applications (façade) as well as other types of urban applications can exploit other benefits from bifacial PV modules: for non-standard installation designs, like in vertical installations, these modules offer an energy output independent of the orientation (north, south, east, west). This design freedom can also be exploited by using semi-transparent modules, with larger cell spacing, that can also provide daylighting and reduce self-shading.

3.1.2 Cell-to-module loss analysis in bifacial PV modules

The module fabrication process introduces losses to the solar cells. After encapsulation, the optical performance of the cell changes. Furthermore, the interconnection of solar cells introduces additional losses that affect the electrical performance of the modules. Due to the losses in modularisation process (cell-to-module process), the module power is generally lower than the summation power of all the solar cells used to fabricate the modules. This difference in total cell power and module power is termed as cell-to-module (CTM) power loss (or efficiency loss). Losses in standard monofacial cell have been widely investigated [4–6]. To improve the performance of PV modules, it is important to understand the various mechanisms responsible for cell-to-module losses and minimize them. The losses in cell-to-module process can be broadly categorized into optical, resistive and mismatch components.

While CTM loss is inevitable, there are some issues related to bifacial cell measurements which can lead to overestimation of CTM losses. Bifacial cells transmit radiation in the long wavelength region (900–1,200 nm); hence, extra care must be taken to measure these cells to prevent any measurement artefact in the measurement of current. For example, measuring a bifacial cell on reflective and non-reflective chuck will lead to different values of the cell current and hence the optical losses (see Figure 3.3). Depending on the chuck reflectivity, the long wavelength light that penetrated the cell can be redirected back into the measured cell. Most conventional chucks for monofacial measurement are gold-plated and

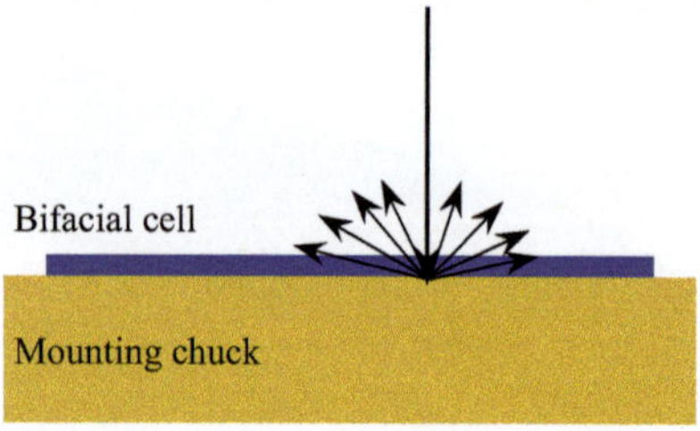

Figure 3.3 Bifacial cell measurements on reflective chuck. Long wavelength light passing through the bifacial cell is reflected back and absorbed by the cell

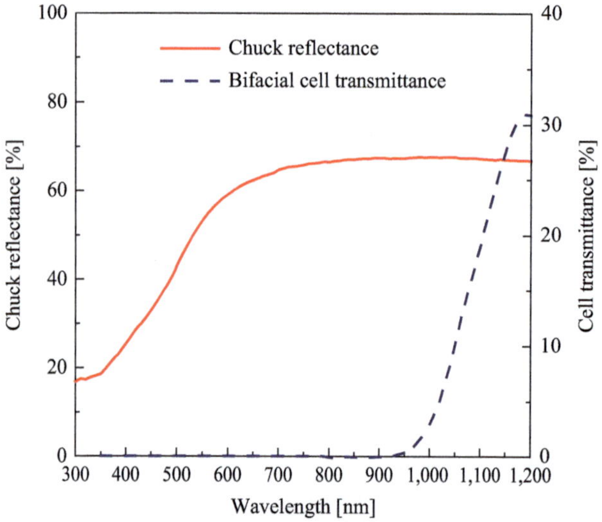

Figure 3.4 Conventional mounting chuck reflectance and transmittance of bifacial cell

are reflective in nature. The reflectance of the conventional chuck is shown in Figure 3.4. The chuck reflectance can contribute to cell current overestimation of up to 1% compared to the measurement on non-reflective chuck [7,8]. As there are no standards on how the bifacial cells should be measured, cell manufacturers are free to use either reflective or non-reflective chuck. In the meantime, bifacial modules are usually measured with non-reflective background. If the cells are measured using reflective chuck and the module is measured using non-reflective background, this can cause a false overestimation of CTM optical losses for bifacial modules. Hence, it is imperative to measure bifacial cells using non-reflective chuck to yield most accurate and fair CTM analysis.

The resistive loss in a wafer-based PV module arises due to various resistive components used to interconnect the solar cells. The additional series resistance in

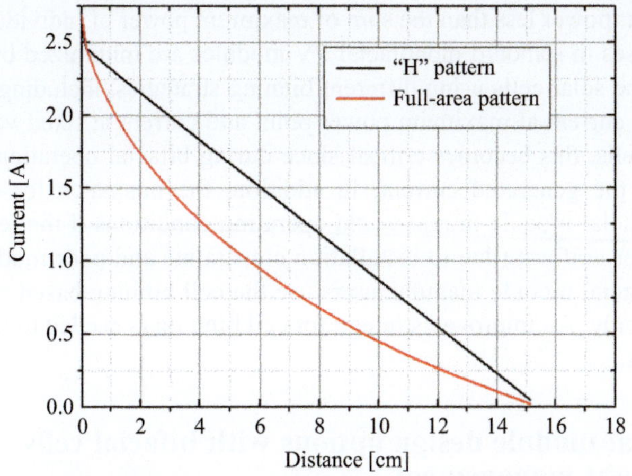

Figure 3.5 Current flow inside a ribbon at the maximum power point for solar cells with different front metallisation patterns and ideal conductor on the rear side [11]

modularisation process is contributed by various components such as soldering-ribbon, bussing-ribbon, contact resistance between cell bus bar and soldering-ribbons, junction box, etc. Bifacial modules typically operate at higher current in the real-world conditions due to albedo from the background [9,10]. Since resistive loss is proportional to the square of the operating current, it is a major concern for bifacial modules. Additionally, the bifacial cells experience different current flow pattern than monofacial cells. For bifacial cells the effective resistance of ribbon is higher compared to monofacial cells with standard Al-BSF; this is mainly due to different metallisation pattern on the back of the cell [11]. Thus, bifacial cells have higher resistive loss compared to the monofacial cell when assembled into modules.

Figure 3.5 shows the current flow pattern in two different types of solar cells, i.e., monofacial and bifacial.

In addition to this, there is certain amount of error introduced by the bifacial cell measurements. If bifacial cells are measured with standard, full area metal contact chuck, the measured fill factor (FF) will be over inflated and does not represent the actual conditions when the cells are interconnected in a module. Studies show that the cell efficiency is overestimated by ~1.4% (~0.3% absolute) for most of the bifacial cells if the bifacial cells are measured on conductive chuck compared to measurements on a non-conductive chuck. Care must be taken, while measuring bifacial cells. Special chuck can be designed which are non-reflective and non-conductive. This will facilitate the user with more precise bifacial cell measurements and provide a fair and accurate CTM loss characterisation for the bifacial modules.

The mismatch loss occurs due to the fact that cells connected in series do not perform at their individual maximum power point simultaneously which results in the total output power less than the sum of maximum power of individual cell [12]. Mismatch losses in standard monofacial PV modules are minimized by measuring and binning the solar cells using different binning strategies, including cell sorting on efficiency, current at maximum power point and current at fixed voltages [13]. For bifacial cells, this becomes critical since during bifacial operation, both sides contribute to the generated current. In addition, the mismatch losses could be severe for bifacial module in real world operating conditions if the rear-side illumination is not uniform (due to installation constraints and poor module design). So far, all bifacial module manufacturers use the cell binning based on front-side performance only. An improved strategy for cell binning is needed to minimise the mismatch loss.

3.2 Optical module design options with bifacial cells and light management

3.2.1 Optical module design options with bifacial cells

In general, bifacial solar cells can be encapsulated into two different module structures, i.e., glass/glass (Figure 3.6) and glass/back sheet, with a back sheet that is, in this case, non-transparent (Figure 3.7).

Due to its bifacial nature, the bifacial cells behave differently in the two different module structures. Both module structures offer certain advantages and disadvantages when used with bifacial solar cells.

With transparent rear glass, the bifacial cells encapsulated in the bifacial structure (Figure 3.6) can absorb albedo (scattered light from ground and surroundings) from the back of the module. This effectively increases the energy yield as compared with a monofacial module when operating in real-world outdoor conditions. However, this advantage is not captured under standard test condition (STC) measurements due to a lack of suitable measurement standards. Most module manufacturers measure bifacial glass/glass modules using only frontside illumination while covering the rear side with a non-reflecting cover.

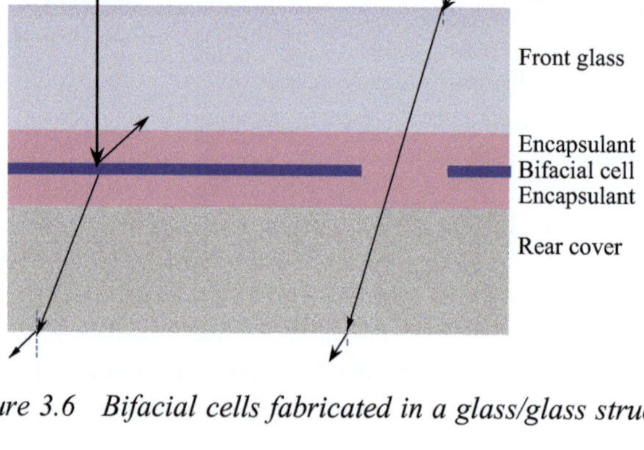

Figure 3.6 Bifacial cells fabricated in a glass/glass structure

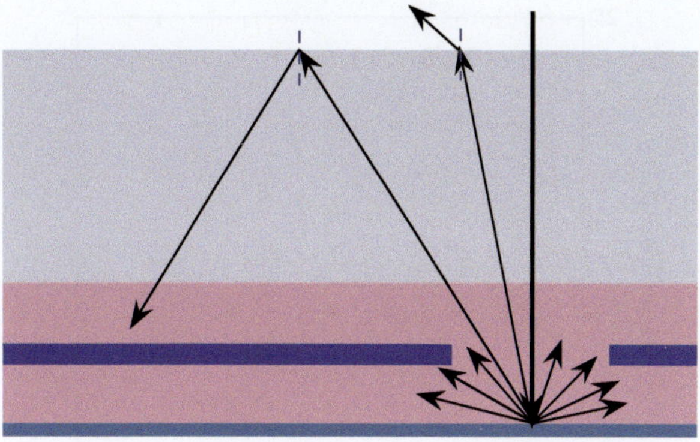

Figure 3.7 Bifacial cells fabricated in a glass/back sheet structure

Compared to the monofacial structure (Figure 3.7), the bifacial cells encapsulated in the bifacial structure (Figure 3.6) have lower power under STCs. This is mainly due to two effects:

1. For a monofacial module structure, the incident light in between the cell-gap region is reflected back by the back sheet (Figure 3.7). Due to the use of transparent rear glass to enable light absorption from the rear of the bifacial module, the incident light from the front is escaping the rear glass through the cell-gap region. This results in power loss of around 3% under STC, depending on the cell-to-cell spacing [14].
2. Due to the solar cell insensitivity to the infrared light, the near infrared light is escaping bifacial cells and not fully absorbed by the cells. For a monofacial module structure, the transmitted infrared light is reflected by the back sheet back to the cells for another absorption opportunity. However, for a bifacial module structure with rear glass, the infrared light will just escape the module. The transmittance loss of various structures is shown in Figure 3.8. Compared to a monofacial module, this leads to around 1.3% loss in current for a bifacial module (Figure 3.9).

Depending on the applications and benefits, some manufacturers use glass/glass, while others use the glass/back sheet structure. Since this is a book on bifacial cell and module applications, the bifacial glass/glass module structure will be predominantly discussed throughout the book (PS: bifacial modules can also be designed with transparent back sheet, to reduce module weight).

3.2.2 Light management in bifacial modules

It was shown in an earlier section that bifacial glass/glass modules could suffer more than 4% lower performance under STC measurements due to light escaping the modules. Since modules are rated under STC and sold in terms of \$/Wp, the

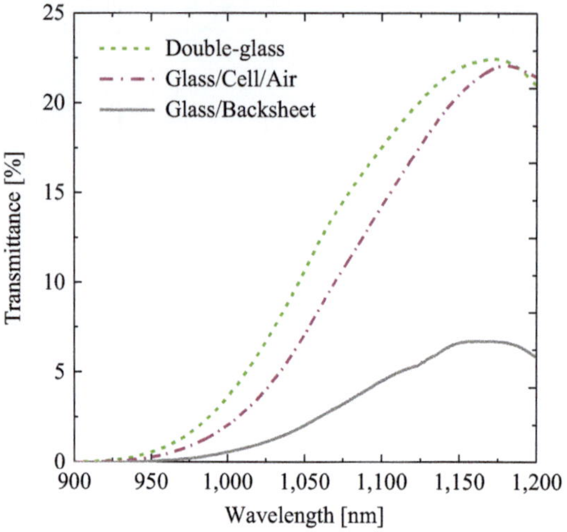

Figure 3.8 Transmittance through different module structures

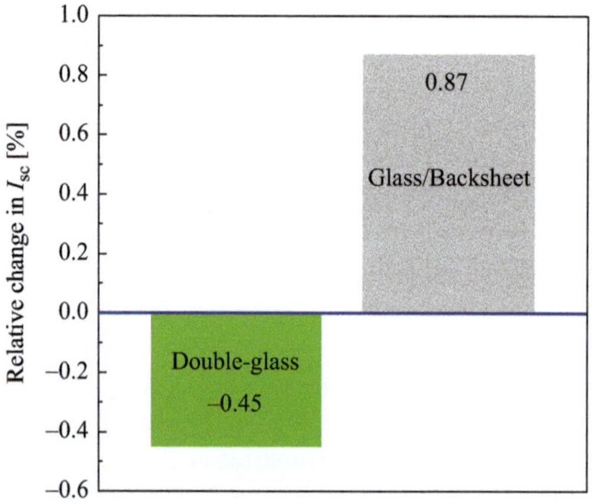

Figure 3.9 Current loss due to bifacial module structure is around 1.3% [14]

current market situation poses a disadvantage towards bifacial modules compared to its monofacial counterpart.

It is possible to increase the STC optical performance of bifacial glass/glass modules while maintaining its bifaciality in the real world. Some research institutes and companies have proposed solutions for this [15–17]. Figure 3.10 and Figure 3.11 show an example to increase the STC optical performance of bifacial modules. By applying reflective coating, the light transmitting through the rear glass in between cell-gap can be reflected back to the module (Figure 3.10). The reflective coating is selectively applied only in between cell-gap region to allow

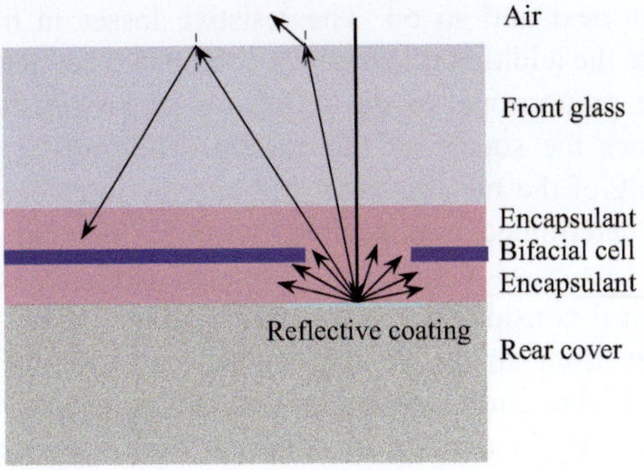

Figure 3.10 Selective coating to prevent light escaping rear glass through the cell-gap region

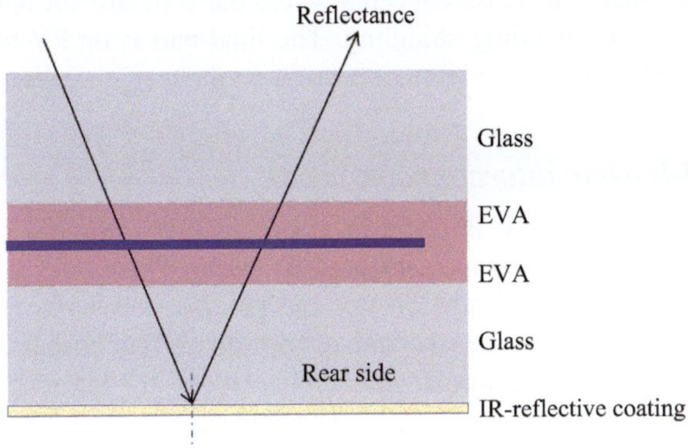

Figure 3.11 Infrared reflective coating to reflect near infrared light for reabsorption by the cells

albedo light from the ground to enter the module from the rear. With specially engineered infrared reflective coating, the near infrared light escaping bifacial cells can be reflected back to the modules (Figure 3.11).

3.3 Electrical design and interconnect options with bifacial cells: half-cut cells, multi-busbar and multi-wire concepts

Most commercial PV modules based on bifacial PERC+ or bifacial n-PERT solar cells apply the same interconnection technology as the vast majority of monofacial PV modules. In these cases, the solar cells are metallised with an H-pattern metal grid on the front and on the rear side. Metal fingers collect the photocurrent from the Si wafer and transport the current to busbars. Cu tabs with solder coating are soldered to the busbars to create a serial connection between the front of a solar cell with the rear of the next and so on. The resistive losses in the cell–cell interconnection dominate the additional losses in a solar panel compared to a bare solar cell. The losses are proportional to the (effective) series resistance of the interconnection tabs times the square of the current. The current depends amongst others on the shading of the metallisation grid and the interconnection parts. The series resistance is determined by the interconnection material, number of interconnecting parts and its cross-section, i.e., the width and height of the tabs.

Apart from optical considerations, we must also take extra care in designing the bifacial modules from an electrical point of view. This is because bifacial modules experience higher current generation due to the double-sided illumination of the modules. Recently, a number of concepts have been explored that offer certain advantages over conventional cell interconnection methods such as multi-busbar, half-cell, shingled, conductive back sheet, etc. In this section, first we cover interconnection technologies beyond 4-busbar soldered tabs. Then we look at the advantages and challenges of smaller cells than the standard 6" Si wafers. The third part covers interconnection concepts where the cells are not placed in a flat, evenly spaced matter, including shingling. The final part is on PV modules made with solar cells that have all the interconnection on the rear, so-called back-contact solar cells.

3.3.1 Multi-busbar interconnection

Over the last decades solar cells have increased in size from 100 mm across to 156 mm and the cell efficiency has increased likewise. Whereas 4″ solar cells were interconnected using soldered tabs on two busbars per cell surface, for 156 mm solar cells three busbars were the standard and nowadays four busbars are the norm. Recently, five and even six busbar cell architectures are introduced, still with interconnection applying rectangular tabs. The next development in this direction is multi-busbar interconnection schemes where the rectangular tabs are replaced by round wires. The wires decrease the contact area between the cell's metallisation and the interconnecting material, thereby even opening up the option of having no busbars at all. The wires are also reported to reduce the reflection losses associated with flat tabs and thus increase the light capturing.

Two companies promote these multi-busbar interconnection schemes. Schmid introduced the multi-wire approach [18] and Meyer-Burger [19] continues the development of the SmartWire technology as pioneered by Day4 Energy [20].

Figure 3.12 shows a schematic view of a single solar cell interconnected by the Day4 technology [20]. The Day4^{TM}Electrode is based on a transparent compound of an adhesive and a polymeric film comprising an embedded copper wire grid. The adhesive on one side acts for the wires as a housing during the electrode production and on the other hand it provides the required adhesion to the solar cell which is established in a standard lamination process at about 150°C. The copper wires are coated with a special low melting point alloy based on the element indium. The alloy composition guarantees for a low contact resistance between the electrode wires and the silver fingers on the front side and to the aluminium layer on the rear side of the solar cell. The distance between the wires varies resulting in about

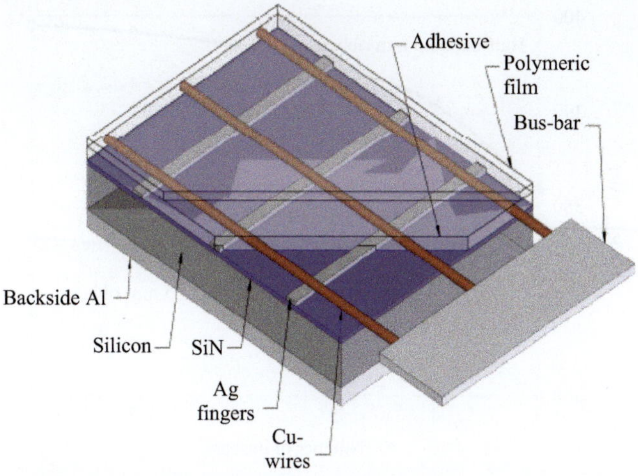

Figure 3.12 Solar cell with Day4^{TM}Electrode contact

36 wires on each electrode. The electrode itself is attached to the screen printed fingers of the cell surface at low temperatures with very low breakage yield. Since this technology provides a way to contact front-side silver fingers and the rear-side aluminium layer without any additional soldering step, there is no need in silver busbars and silver/aluminium pads which significantly reduces the cell processing and material costs. The flexible electrode wire design allows for a reliable cell interconnection for various cell formats. The fill factor thereby remains stable on module level with almost no losses if compared to the fill factor on cell level. The very low series and contact resistance of this interconnection technology allows for lowest CTM fill factor losses and hence the technical application of low concentration panels as shown by Day4Energy.

A further advantage of this interconnection technology is the interconnection redundancy: the Day4 contacting scheme contacts the solar cell by more than 2,000 contacting points on the front side of the solar cell compared to only three, four or five busing lines on standard cells. If the cell integrity is at risk after years of extreme outdoor exposure or in the case of occurring cell breakage, the Day4 technology secures a high redundancy by its unique contacting scheme. This in fact shows only small or no losses in cell power whereas in case of breakage for standard soldered and tabbed solar cells the current mismatch would lead to a strong loss in P_{mpp}. In general, this loss reduces with the amount of busbars existing in the cell metallisation grid.

The GRIDSOL simulation from SERIS was applied to calculate the effect of changing the cell and module metallisation for a module with 60 bifacial n-PERT cells (20% efficiency, 90% bifaciality) under 1,000 W/m^2 front and 350 W/m^2 rear irradiance. The simulation model, labelled GRIDSOL [21], takes into account shading losses due to the metal grid, recombination losses under metallisation lines as well as resistive losses at cell and module level. The simulation in Figure 3.13

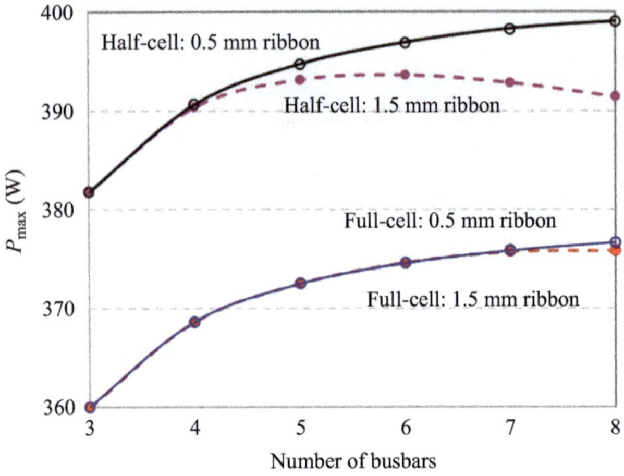

Figure 3.13 GRIDSOL simulation for optimisation of P_{max} of a 60-cell bifacial module with 90% bifaciality for 1,000 W/m^2 at the front side and 350 W/m^2 at the rear-side [21]

varies the number of busbars (from 3 to 8) as well as the ribbon width (0.5 or 1.5 mm) for both full cells and half cells. The simulated module with half cells yields a 20 W higher power output, as half cells give half the current and thus decrease the resistive losses with 75%. Increasing the number of busbars increases the power output both for full and for half cells. As in half cells the current and thus the resistive losses are much smaller than in modules with full cells, an increase in the number of busbars and thus in the number of tabs increases the shading losses especially for wider ribbons, as can be seen in the drop in P_{max} for the dashed pink line (Figure 3.13). In contrast, for full cells, the increase in number of ribbons does increase the shading losses, but also decreases the effective resistivity of the ribbons. These two effects seem to balance each other as can be seen in the more or less constant P_{max} for larger number of busbars in the blue thin line and the dashed red line in Figure 3.13. Experimentally we have found a 6% gain for a 4 cell-module in going from 3 BB full-cell modules to 5 BB half-cell modules, compared to the 9% gain estimated at 60 cells module level by the simulation in Figure 3.13.

3.3.2 Half cells and smaller

The most obvious way to decrease the resistive losses in cell–cell interconnection is by reducing the (total) current to transport. Cutting cells in halves, thirds or even smaller reduces the total current per cell at constant current density, to 50%, 33% and lower, respectively. This reduces the cell-to-module change in FF to a similar percentage. To keep the typical parameters of a 1 m × 1.6 m module with half-cells or quarter cells in the same range as that of a full-size 60-cell module, the cell strings are interconnected partially in parallel so the total current is twice (four times) that of the half (quarter) cells and the module voltage is 60 times the voltage

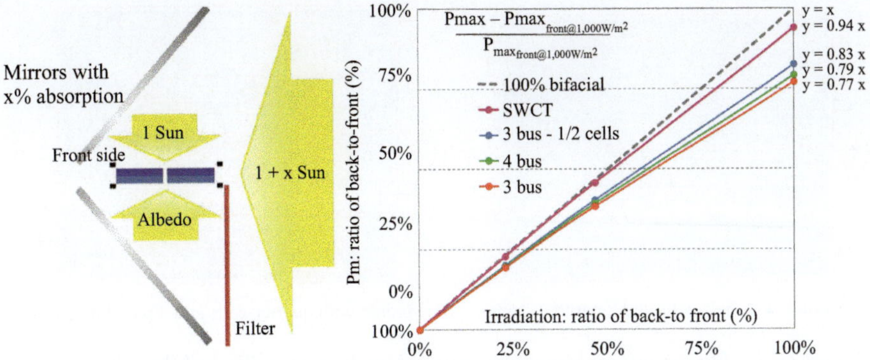

Figure 3.14 Gain in power (Pm) with frontside irradiation of 1,000 W/m² and a backside irradiation varying from 0 to 1,000 W/m², for different interconnection options on 4 cells modules. Measured in a solar simulator with symmetrical mirrors for simultaneous irradiation of front and back sides and a mesh filter to vary the backside irradiation [24]

of a single cell. Note: cutting cells does not change the Voc, assuming that the cutting does not increase the edge recombination losses. Next to the improved output at standard current and voltage levels, the partial parallel interconnections blocks also have a small effect on the sensitivity to partial shading. The effect of half cells and parallel strings on the reduced sensitivity to partial shading of monofacial modules has been reported in [22].

The benefit of using half-cut cells, to reduce the current and the associated resistive losses is illustrated in Figure 3.14 that represents the gain in P_{max} as a function of the backside irradiation, with frontside irradiation fixed at 1,000 W/m^2 for several cell interconnect options. For all these options the maximum power is measured to increase with backside radiation; the curves are not strictly linear due to FF losses with higher currents [23]. The 3 busbar half-cell is found to perform about 3% better than its full-cell equivalent in STC conditions, and even more when backside irradiation increases.

Shadow on part of a module causes inhomogeneous photocurrents in the solar cells. But serial interconnection forces the actual (working point) current to be identical in all cell (strings) in the same loop. Bypass diodes can bypass most of the current around a shaded string, but this will reduce the power output of the system. Parallel interconnection of cell (strings) forces the voltage to be identical, but the output voltage is much less affected by the amount of irradiance. A combination of serial and parallel interconnection can be used to reduce the effect of inhomogeneous irradiance, e.g., due to shading or soiling. Bifacial modules with half-cells also allow to make the module less sensitive to non-uniformity of the rear-side illumination by combining serial and parallel connections. Each substring consists of 60 cut cells connected in series. The substrings are then parallel connected (two

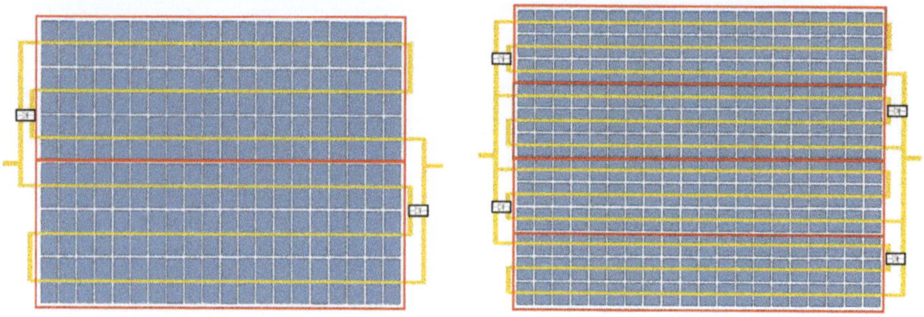

Figure 3.15 Electrical architectures equivalent to a 60 cell module. The cell division factor cell equals the number of blocks in parallel (red rectangles): factor 2 (left) and factor 4 (right) [25]

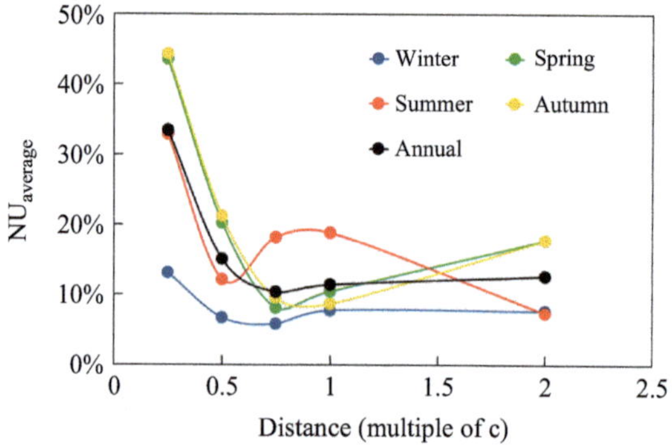

Figure 3.16 Simulation of the non-uniformity on the backside of a bifacial module that is vertically mounted on a building façade. Expressed as a function of the distance to the building wall (distance relative to the module width c) and of seasonal effects [25]

parallel strings when cells are cut in halves, three parallel strings when cut in thirds and so on) as shown in Figure 3.15. A secondary advantage is that the output of such modules, expressed in Voc and Isc, is very similar to that of 60 full-cells connected in a single string.

An example of the advantages of parallel interconnection is for bifacial modules that are mounted vertically on a building with the gap between the module and the building influencing the inhomogeneous rear irradiance. The non-uniformity (NU) on the module backside is defined as: $NU = (G_{max} - G_{min})/(G_{max} + G_{min})$ and Figure 3.16 gives the NU for the four seasons. NU is small in the winter. In spring and autumn NU is largest and more or less the same.

The annual energy yield of this situation has been simulated for modules with full-cells, half-cells and half-cells with parallel interconnection. Figure 3.17 gives

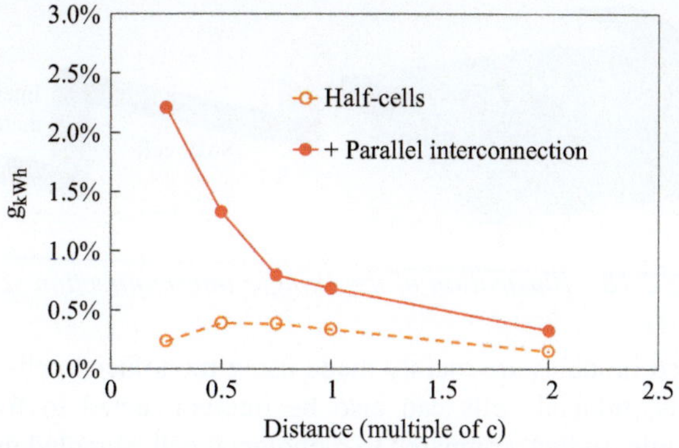

Figure 3.17 Simulation of the annual gain (in kWh) brought by the use of half-cells plus the use of 2 parallel strings for the bifacial module of Figure 3.15 vertically integrated on a building façade at a distance D from the reflective building wall

the relative gain in performance for the half-cell module (open, orange symbols). The combination of vertical mounting and limited rear irradiance, as there is only reflection of transmitted light of the building's wall, limits the occurrence of high current situations. Thus the effect of half-cells by limiting the resistive losses is limited. The relative gain simulated for the module architecture of Figure 3.15 (left) is plotted (red, closed symbols). Clearly this gives a larger gain. The similarity between the annual NU, as function of module-wall distance, and the relative gain is obvious. The larger the non-uniformity, the more advantageous parallel interconnection is.

3.3.3 Shingles and other stacking options

In most Si PV modules, the solar cells are all oriented the same way, with the same side facing the front. In this section, two alternative placings of solar cells are discussed. The first is the so-called shingled interconnection module, where the solar cells are partially overlapping comparable to slate roof shingles. This yields a PV panel with a high packing density and can be done both with monofacial and with bifacial solar cells. The electric interconnection must also change. Advantages and challenges of the shingled module interconnection will be reported. The second alternative gives a sort of checkerboard module where the position of the solar cells is not changed, but their orientation is.

Shingled interconnection of monofacial solar cells is not new and has been reported by a number of researchers [26,27]. Sun Power is one of the few manufacturers that currently produces commercial shingled module. In shingled PV modules, solar cells are cut into small stripes along the busbars. These solar cell stripes are then interconnected in a shingled pattern by connecting the p-busbar of one cell-strip to the n-busbar of the other cell-stripe as shown in Figure 3.18.

Figure 3.18 Illustration of the shingle interconnection scheme

The shingled modules are mainly made using monofacial cells. Similar to the monofacial cells, bifacial cells can also be interconnected to form a shingled bifacial PV module. In fact, compared to monofacial cell, shingled interconnections is more beneficial for bifacial cells [28]. As discussed in Sections 3.1.2 and 3.2.1, there is a significant amount of optical and resistive losses in standard double-glass bifacial modules. One of the key advantages of shingled bifacial interconnection is the reduction of these cell-to-module losses. The shingled interconnection of bifacial cells offers three main advantages over standard bifacial modules: (1) high packaging density and high-efficiency bifacial modules; (2) reduction in optical losses (due to cell-gap and transparent rear side); (3) reduction in resistive losses (due to less current in each cell strip). Thus, the cell-to-module power loss in bifacial shingled modules is less compared to standard double-glass bifacial PV modules when measured under STC.

To achieve the required optimal performance in shingled bifacial PV modules, a number of parameters should be optimised such as front and rear metallisation grids, cell-overlap, number of cell stripes, string connections to form a module, etc. The optimisation of shingled cell interconnection, i.e., number of cell-cuts and cell overlap etc., was performed using "Griddler" (developed by SERIS), is given in Figure 3.19 [28]. The performance of the bifacial shingled module is limited by the number of stripes per cell and cell overlap. Cell overlap is a design constraint due to the lay-up and stringing tools and reaching a minimum is better in terms of performance. The number of cell stripes per cell depends upon a number of factors. As we introduce more cuts, we reduce resistive losses; but the optical losses will increase, depending on the cell overlap. Simulated results in Figure 3.19 show that for more than 1.2 mm cell overlap, 6-cut shingling is no longer offering advantages compared to 4- or 5-cut shingling. In addition to this, the final decision on the number of cuts (stripes) should be made after considering the losses due to cell cutting, throughput and the silver consumption.

Now, as mentioned earlier in this section, the resistive and optical losses in shingled interconnections of solar cells are analysed and compared to the standard cell interconnections using the Griddler simulations tool. In 'Griddler', various resistive, optical shadow and recombination parameters for cell and module are considered to calculate the losses when bifacial cells are cleaved and interconnected in a shingled pattern. First, bifacial cells were optimized for grid metallisation (number of fingers, busbar width, etc.) for standard and shingled interconnection using the same baseline cell parameters. Then, various loss

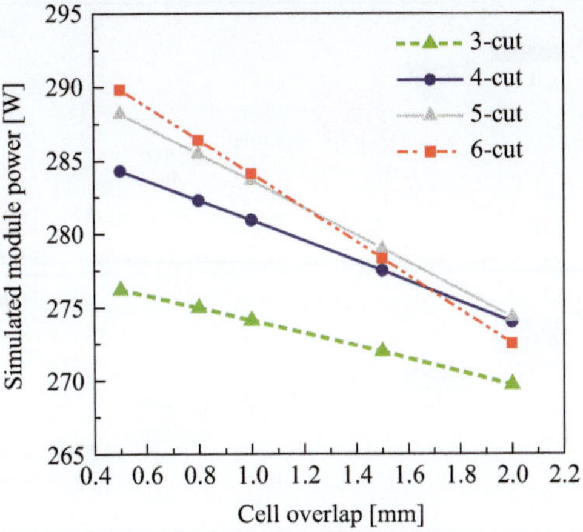

Figure 3.19 Simulated module power of shingled bifacial PV module for varying cell overlap and number of cell cuts

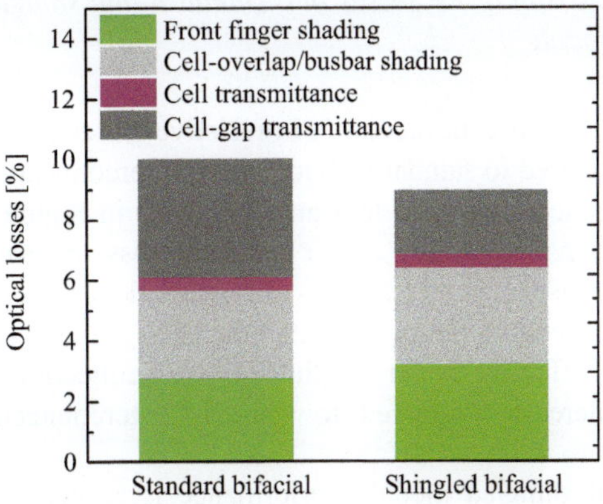

Figure 3.20 Comparison of optical losses in a standard and shingled bifacial PV module

components were estimated in both types of cell interconnections. Figure 3.20 and Figure 3.21 show the detailed optical and resistive loss comparison between standard and shingled bifacial PV modules. To estimate the losses, the design parameters for standard bifacial module are 5-busbar, 0.9 mm busbar width, 3 mm cell-gap and 4 mm string-gap while for shingled bifacial modules, the design parameters, 5-cut, 1.0 mm cell overlap and 4 mm string-gap are chosen. From Figure 3.20, although there is an increase in optical loss due to cell-overlap in shingled module compared to the standard bifacial module, the main advantage of shingled module is reduced optical losses due to reduction in net cell-gap area.

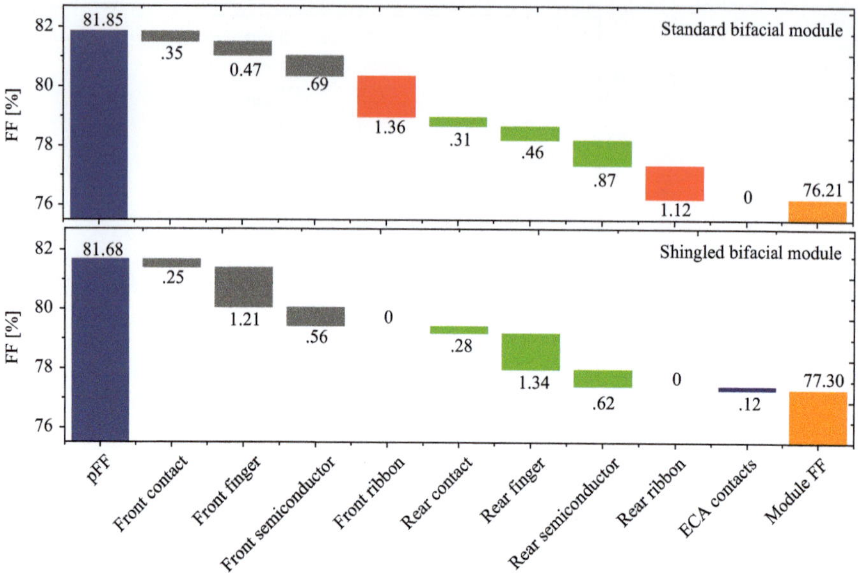

Figure 3.21 Comparison of FF losses in a standard and shingled bifacial PV module

In addition to this, a significant reduction in resistive losses is possible in shingled interconnection compared to standard ribbon based interconnection of bifacial cells as shown in the detailed fill factor loss analysis chart in Figure 3.21. In standard ribbon based interconnection, the major resistive loss mechanism is losses in interconnecting ribbons followed by front- and rear-side semiconductor resistance. For shingled interconnection, there is no interconnecting ribbon involved; however, due to an increase in effective finger length for current collection, the losses in front and rear fingers increase compared to standard interconnection as shown in Figure 3.21.

Overall, bifacial shingled module performance is approximately 2.5% higher compared to standard bifacial modules with similar design parameters as stated above. Moreover, more shingled cells can be accommodated (higher packaging density, e.g., 68 cells) given the same module area. This leads to a further enhancement in module power as shown in Figure 3.22.

Thus, bifacial shingled modules can potentially maximize the energy generated per unit area of installation and can be very useful for high power density applications. Recently, a number of module manufactures have introduced shingled modules [29,30], but commercial deployment of bifacial shingled modules is yet to come. For wide acceptance of bifacial module technology, it is necessary to increase the module front-side power so that the module can be sold at premium. Shingled interconnection could be helpful in increasing the module front-side power by reducing the losses and accommodating more cells on the same glass-size. There are a number of challenges for shingled bifacial technology including

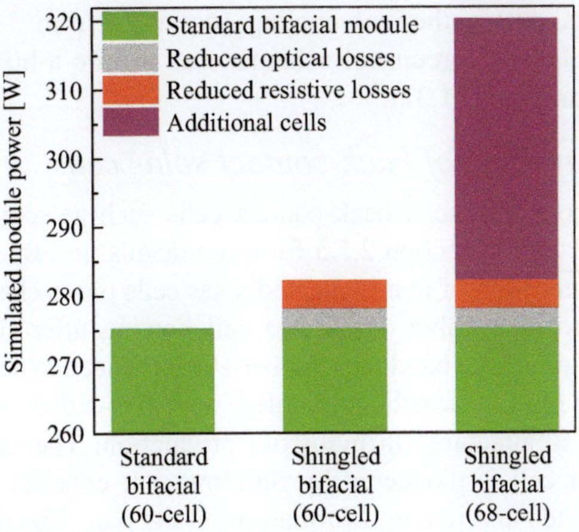

Figure 3.22 Comparison of simulated module power of standard and shingled bifacial PV module. Also shown is the module power for a higher number of cells (same glass-size) in shingled interconnection

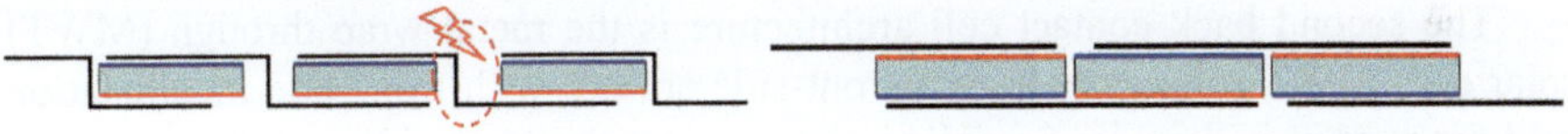

Figure 3.23 (Left) Standard interconnection scheme where all cells are placed sunny-side (blue) up. Tabs connect the front side with the rear side of the neighbouring cell (red). Note that the tab with typical thickness of 200 μm has to go from the front to the rear, a height difference of 180 μm + 200 μm, over a cell–cell distance of 2 to 3 mm, creating a lot of stresses on the Si wafers, as indicated by the red arrow. (Right) Semi-planar interconnect scheme where the bifacial cells are alternatively put sunny-side up and sunny-side down

the cost of electrical conductive adhesive (ECA) or low melting temperature paste interconnection, losses in cell cutting process and throughput. So, the key research focus in coming years will be to overcome these challenges and developing a cost effective bifacial shingled module technology.

Another aspect of bifacial cells is that they can also be exploited in the electrical design of a module (monofacial or bifacial) by the fact that the cells can be interconnected in a planar way (front-to-front and back-to-back), without the need to connect the front side of one cell to the backside of the neighbouring cell. This is achieved by flipping over the bifacial cells during interconnection, as shown in Figure 3.23.

The benefits of this type of 'planar' interconnect are related to increased productivity during the tabbing/stringing process, reduced cell spacing and also increased module reliability due to the reduced thermo-mechanical stress at the front/backside crossing that induces failures like ribbon fatigue and breakage as well as cell cracks during thermal cycling (typically from −40°C to +85°C). Examples of this planar interconnect scheme, that require a bifaciality of at least 95%, are demonstrated in [31,32].

3.3.4 Interconnection of back-contact solar cells

The final section covers bifacial back-contact cells such as screen printed IBC or MWT cells, see Chapter 2, section 2.5.5 for more details. In respect to module integration, the interconnection of rear contacted solar cells poses challenges since classical stringing is not applicable. Excessive cell bowing after single sided ribbon attachment hinders automatic handling. Tabber-stringers need to be specially adapted.

Although, the market is still dominated with two-sided contact solar cells, there are two cell architectures in industrial production. The interdigitated back-contact (IBC) solar cell is pioneered by SunPower. It consists of a front surface field on the sunny side and junction formation on the rear. The rear-side emitter is interspersed with BSF areas, creating an interdigitated pattern. Both rear emitter and BSF areas are contacted, independently, with metal fingers. Cell–cell interconnection can be achieved via tabs contacting the emitter busbar on one cell with the BSF busbar on the next cell or via smart tabs between an emitter contact area on the edge to the BSF contact area on the next cell.

The second back contact cell architecture is the metal wrap through (MWT) solar cell. MWT solar cells have a front-side emitter with front-side metallisation grid for current collection. So-called vias connect the front-side metallisation grid to contact pads at the rear. These emitter contacts are isolated from the BSF [33]. The reduced front-side metallisation coverage limits the shading losses [34] and reduces the overall front-side recombination losses [35]. Like for IBC solar cells, cell–cell interconnection can be achieved via tabs contacting the emitter contacts on the rear of one cell with the BSF contacts on the rear of the next cell. Note that the tabs on the emitter contacts of one solar cell need to be electrically isolated from the BSF rear side of that solar cell to prevent shunt losses.

Whereas for front-to-rear contacting the busbar on the front of the first cell is aligned with the busbar on the rear of the next cell, for back-contact solar cells this is not automatically the case. Symmetric rear metallisation patterns, where rotating the cell 180° in-plane makes no difference in the pattern and polarity, have emitter contacts or the emitter busbar on one line on neighbouring solar cells and likewise for BSF contacts/busbars. For example, ECN's original MWT back contact pattern consists of four rows of four emitter vias with three rows of five base contacts in-between. Interconnecting such symmetric solar cells needs more complicated tabs and cross-connectors to connect the four rows on one cell with the three rows on the next cell.

An alternative interconnection can be achieved with asymmetric solar cells, where rotating the cell 180° in-plane inverts the polarity of each contact row or each busbar. This can be achieved by having a number of pairs of busbars (or pair of contact rows), one contacting the emitter area and one contacting the base. When alternating solar cells are 180° rotated, busbars (rows of contacts) on neighbouring cell will have opposite polarity. In that case, cell–cell interconnection can be

Figure 3.24 Schematic drawing of an MWT module. From bottom to top: patterned Cu layer on polymer back sheet carrier; electrically conductive adhesive (ECA) dots; encapsulant layer with holes at the contact positions for the ECA; back contact cells; front-side encapsulant; glass. The small, solid circles in the metallisation pattern indicate the location of the vias, which have sub-mm diameter [36]

achieved by contacting straight tabs on the emitter busbars on one cell to the BSF busbars on the next cell. This is similar to the right hand side of Figure 3.23, except that now all contacting tabs are on the rear side.

An alternative interconnection technology has been pioneered by Dutch company Eurotron and ECN. Taking advantage of the fact that all contacts are located on the rear side, the cells are interconnected by conductive foil technologies, see Figure 3.24, based on the printed circuit board technology. The conductive back sheet approach overcomes bowing issues since contact between cell and back sheet is made during lamination and thus the thermomechanical stress between cell and back sheet is compensated by the module laminate, especially the rigid front glass plane. As all cell–cell interconnections are underneath the solar cells, no shading losses occur due to this interconnection material. Because the conductors are as wide as the solar cell, the thickness can typically be limited to 35 μm for Cu layers. Due to the combination of very wide and thin conductors, no additional, mechanical stress is applied on the solar cells, in contrast to soldered tabs, with typical dimensions of 1.0–1.5 mm width and 200 μm thickness. The absence of those tabs also allows the cell packaging to be higher, with cell–cell distances of typically 1.25 mm.

The two polarities of the solar cell are separated by an isolating trench forming two continuous circuit tracks for current transport. The circuit tracks are shaped by

Table 3.1 Dependence of the FF loss on the connector dimensions for various interconnection architectures

Interconnection architecture	Connection dimensions	FF loss
4 BB tabbed	4× 1 mm × 200 μm	−3.1%
Multiwire	15× Ø 300 μm	−2.4%
Smartwire	38× Ø 200 μm	−2.0%
Conductive back sheet	4× 20 mm × 35 μm 3× 26 mm × 35 μm	−0.9%

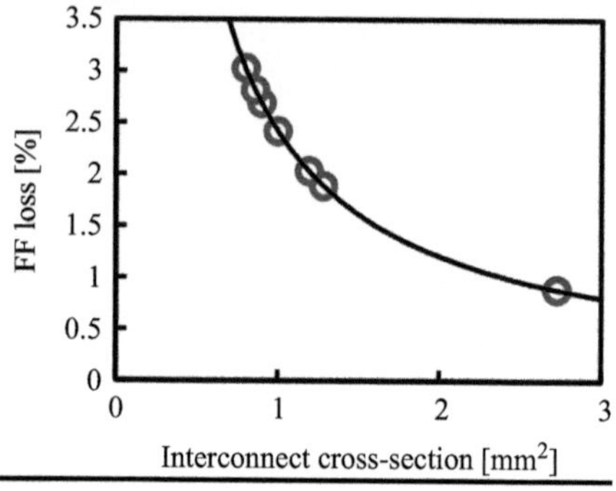

mechanical milling, laser cutting or etching and are isolated against the solar cells by the rear encapsulation layer which is locally opened to enable electrical contact, typically by conductive adhesive gluing. As can be seen in the picture, these separation lines can be designed for optimal conductance and there is freedom of design how to do the cell–cell interconnection. A cell can be connected with any two of its four neighbours, below, above, left or right. In fact, all combinations do occur in a standard 60-cell module. This freedom of design can also be applied to the connection of the bypass diodes by drawing a narrow path between two columns of cells.

The conductive back sheet can be applied to monofacial and bifacial solar cells alike, but obviously conductive back sheet technology results in monofacial modules. The full metal coverage facilitates a low series resistance interconnection. Table 3.1 shows the FF loss for conductive back sheet related to other cell–cell interconnection methods [37]. More, wider and thicker material obviously will increase the conductivity and decreases the FF loss. Note that this analysis does not take into account the effect of the shading of the interconnected solar cells.

These FF losses are in good agreement with observed CTM losses for full-size modules, e.g., Guillevin *et al.* reported a direct comparison between front-to-back contact n-PERT and back-contact MWT solar cells and modules [38]. For the three busbar n-PERT module 3% FF loss and for n-MWT with conductive back sheet 0.8% FF loss was observed.

Following this approach leaves the liberty to realise any circuit track design, even, e.g., ribbon like interconnection track allowing bifacial operation. Figure 3.25 shows a layout for a prototype bifacial conductive back sheet including the cell positions for interconnecting eight half cells. In contrary to both side contacted half-cell modules a more appealing assembly can be realised for back contact cells if cells with asymmetric busbar layout are employed. Connecting the base busbar of

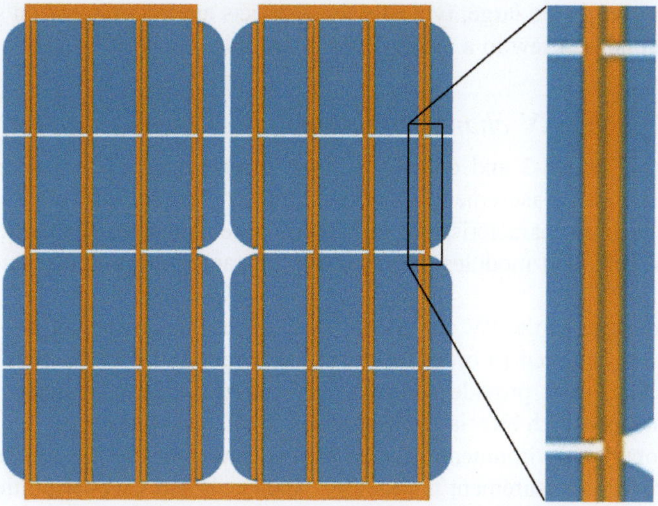

Figure 3.25 *Electrical layout of a bifacial 2×2 mini-module with half cut cells interconnected with a partial back sheet, designed to resemble ribbons. The enlargement shows the asymmetry in the cell orientation. In case of an asymmetric busbar layout, by combining two right or two left half cells, the original cell shape can be preserved*

the left half of a cell to the emitter busbar of the left half of a 180° rotated cell by a straight interconnection pad leaves the original pseudo square shape intact.

On the one hand this approach enables bifacial module assembly for back contact cell without having to consider cell bowing, on the other hand though a trade-off between bifaciality and series resistance of the interconnection tracks has to be made since the metal layer thickness is limited by the production process of the back sheet. For industrial production, instead of local removal, local deposition of metal onto the back sheet would be a more economically viable solution possibly even allowing the increase of the metal layer thickness.

3.4 Characterisation of bifacial devices

As with all other aspects, the characterisation of bifacial devices is affected by their two-sided character. In this section, several characterisation methods are discussed. Section 3.4.1 covers the standardisation of the bifacial *I–V* characterisation, power rating and reporting. Then two imaging techniques are discussed that can be used to spot failures caused by manufacturing or during reliability testing. Finally, an overview is given on the effect of installation parameters on the outdoor performance. As the financial cost of varying parameters like tilt, height, etc. for a utility scale PV system is too large, typically these effects are investigated on small systems consisting of a few to a single full-size module.

3.4.1 Bifacial I–V *characterisation*

IEC standards 60904-3 and 60904-9 [39,40] describe the STCs and appropriate apparatus for the measurement of current-voltage characteristics of PV devices. Accurate electrical characterisation is the basis to set the price. The particularity of bifacial PV cells and modules and their *I–V* characterisation is discussed in this section.

Identification of the PV stakeholder's needs and understanding the technical challenges are required in order to propose coherent and adopted standards. *I–V* characterisation must provide comparability between bifacial modules and must highlight the gain with bifacial compared to monofacial technology.

In laboratory environments, comparable measurement results are required in order to provide measurement traceability. The needs and the possibilities are different in laboratories compared to PV production environments. Laboratories are able to provide additional data, such as the bifaciality, and the quantum efficiencies of the front and the rear sides. Additionally the transmittance of the module is quite different for the areas covered by the bifacial solar cells, which transmit mostly in the near infrared, and the areas not covered by solar cells, which transmit over the whole spectral range.

In production environments, *I–V* characterisation must be well-matched with the production throughputs and the apparatus must be compatible with the production specificities, such as low footprints, automation of the equipment and device handling. Furthermore, *I–V* characterisation of bifacial devices should be available at a reasonable cost.

The data provided by the PV laboratories and productions can be used to predict the power output of bifacial solar power plants or for energy rating purposes. However, it is almost impossible to predict perfectly the real field performances. This fact is understood and accepted in the case of monofacial modules. Standard Test Conditions are defined in order to obtain measurement comparability. In the case of bifacial cells and modules, standard test conditions should also be defined and accepted by the community. These conditions must be easily reproducible in laboratory and production environments and must show the required power gain related to the bifaciality.

3.4.1.1 Bifacial current generation mechanism

It is crucial to first understand the current generation mechanism in a bifacial device and its complexity. Figure 3.26 schematises a bifacial PV device. A bifacial module is, in general, placed in a power plant with a certain distance d_m to the neighbouring modules (see Figure 3.26, part A). The bifacial cells in the module are also placed with a certain distance d_c one to another in the module. The global irradiance G may reach the bifacial device in different manners. Four cases are considered: in case number (1), the light beam does not reach the module directly,

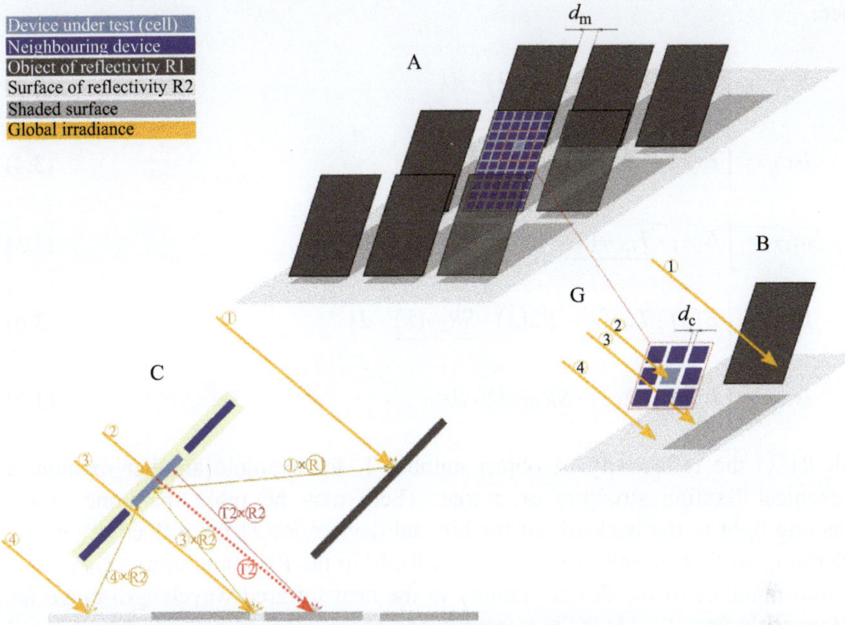

Figure 3.26 Scheme of the bifacial current generation mechanism

but hits an object of reflectivity R1, and is diffused and reflected to the backside of the module ((1) × (R1)). In the second case (2), the ray reaches the front side of the module directly. Part of the photons are absorbed to generate current and heat and a certain quantity corresponding to a certain wavelength range is transmitted through the cell (T2), and is diffused on the underlying surface of reflectivity R2 and reflected to the backside of the module ((T2) × (R2)). Beam numbers 3 and 4 are also reflected in the same way; one goes through the distance d_c between the cells, the other goes through the distance d_m between the modules.

In general, the short circuit current *Isc* is given by the following equation:

$$Isc = \int E(\lambda) \cdot SR(\lambda) \cdot d\lambda \tag{3.1}$$

where $E(\lambda)$ is the irradiance per unit bandwidth at a particular wavelength λ. It refers to the AM1.5g reference solar irradiance and $SR(\lambda)$ is the spectral response of the device. Although not explicitly shown in Figure 3.26, the beams 1–4 and the AM1.5g spectrum also contain diffuse contributions.

For the bifacial device as shown in Figure 3.26, part C, the total short-circuit current is given by the sum of the contributions generated by the different beam paths:

$$Isc = Isc_1 + Isc_2 + Isc_{T2} + Isc_3 + Isc_4 \tag{3.2}$$

where

$$Isc_1 = \int E(\lambda) \cdot R1(\lambda) \cdot \underline{SR_{BS}(\lambda)} \cdot d\lambda \tag{3.3}$$

$$Isc_2 = \int E(\lambda) \cdot \underline{SR_{FS}(\lambda)} \cdot d\lambda \tag{3.4}$$

$$Isc_{T2} = \int E(\lambda) \cdot \underline{T_{DUT}(\lambda)} \cdot R2(\lambda) \cdot \underline{SR_{BS}(\lambda)} \cdot d\lambda \tag{3.5}$$

$$Isc_3 = \int E(\lambda) \cdot \underline{T_{enc}(\lambda)} \cdot R2(\lambda) \cdot \underline{SR_{BS}(\lambda)} \cdot d\lambda \tag{3.6}$$

$$Isc_4 = \int E(\lambda) \cdot R2(\lambda) \cdot \underline{SR_{BS}(\lambda)} \cdot d\lambda \tag{3.7}$$

with $R1(\lambda)$ the reflectivity of object number 1, for example another module, a mechanical fixation structure or a tree. There may be more than one object reflecting light to the backside of the bifacial device; $R2(\lambda)$ the reflectivity of the underlying surface, usually referred to as albedo in the PV community; $T_{DUT}(\lambda)$ is the transmittance of the device, mainly in the near infrared wavelength range for c-Si technologies; $T_{enc}(\lambda)$ is the transmittance of the encapsulant; and $SR_{BS}(\lambda)$ and $SR_{FS}(\lambda)$ are, respectively, the spectral responses of the backside and the front side of the device.

The parameters underlined in the equations are the intrinsic properties of the bifacial module. The final performance of bifacial PV modules in a power plant depends on the spatial distribution of the irradiance incident on the rear surface of the module, which is strongly affected by site-specific conditions such as albedo, reflective surface size, the racking system, the device's elevation and its tilt angle. Due to these dependencies and in order to obtain comparable measurement results, an IEC standard for *I–V* characterisation of bifacial devices is asked for. This proposed standard is based on an extension of the IEC standard for *I–V* characterisation, quantifying the bifaciality of the device and the power generation gain. Bifaciality is an intrinsic property of the module, unlike the site-specific conditions such as albedo and the actual bifacial energy gain. The measurement conditions for bifacial devices should strive to generate extra photocurrents proportional to their bifaciality.

3.4.1.2 Measurement approach

The bifacial *I–V* characterisation procedure, considered as the IEC standard[1] defines two cases: measurements by PV laboratories and measurements in PV production environments [41]. In these two environments, the possibilities and the needs are different and the provided measurement results are complementary. The combination of the measurements from laboratory and production allows getting good knowledge about bifaciality for reasonable cost and complexity.

[1]At the time of writing (Dec 2016), this IEC proposal is at the committee draft stage.

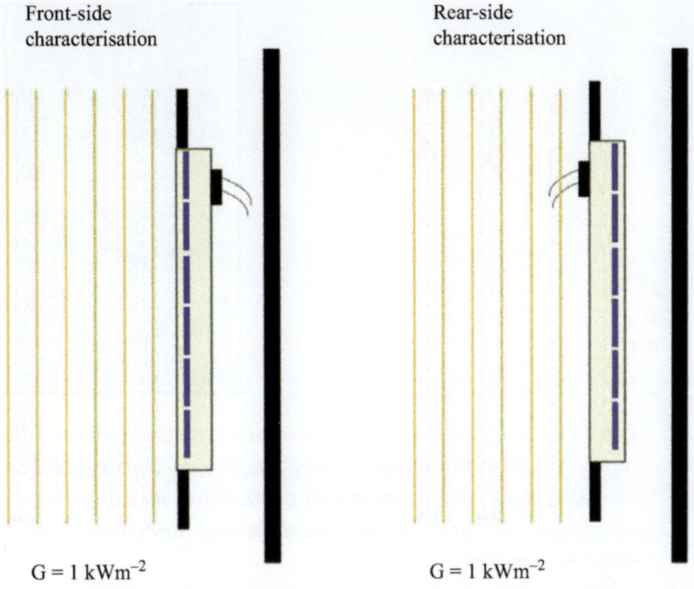

Figure 3.27 Front- and rear-side characterisation for bifaciality. Note the non-reflecting background behind and the aperture around the module to reduce the irradiance on the non-exposed side

Bifaciality coefficients determination

In order to determine the bifaciality coefficients of the test specimen, the main *I–V* characteristics of the front and the rear sides must be measured at STC ($G = 1{,}000$ W/m^2). A non-reflecting background must be used in order to avoid the illumination of the non-exposed side, see Figure 3.27. The background is considered to be non-irradiated if the irradiance is measured to be below 3 W/m^2, on at least two points, on the non-exposed side of the device.

Short-circuit current bifaciality coefficient φ_{Isc} is the ratio between the short-circuit current generated exclusively by the rear side of the bifacial device and the one generated by the front side. Both currents are measured at STC (1,000 W/m^2, 25°C, with the IEC 60904-3 reference solar spectral irradiance distribution):

$$\varphi_{Isc} = \frac{Isc_r}{Isc_f} \tag{3.8}$$

where φ_{Isc} is the short-circuit current bifaciality coefficient. It is usually expressed as a percentage, Isc_x is the short-circuit current at STC under one sided illumination, with index *x*: *f* for front and *r* for rear side.

The spectral mismatch correction shall be applied to the measurement of Isc_f and Isc_r, according to IEC 60904-7, unless it is known that the front and back of the bifacial device have identical spectral responsivity.

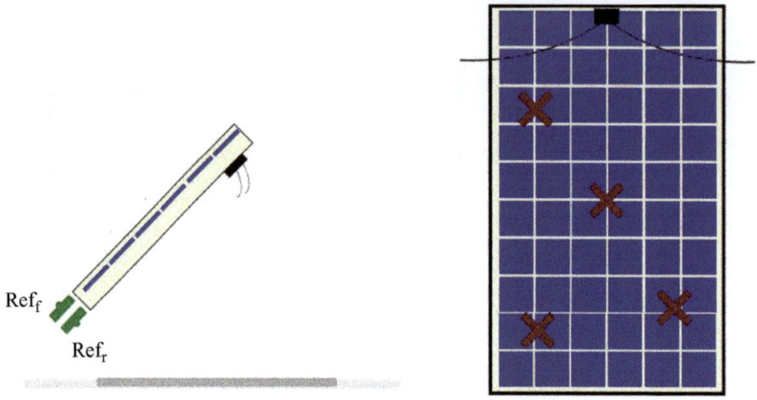

Figure 3.28 Left: Two reference devices are used to measure the irradiance on the front and the rear sides of the device during outdoor measurements. Right: proposed positions to measure the non-uniformity of irradiance outdoor and the way the cables must be handled

Other bifaciality coefficients shall be reported and are calculated by:

$$\varphi_{Voc} = \frac{Voc_r}{Voc_f} \tag{3.9}$$

$$\varphi_{Pmax} = \frac{Pmax_r}{Pmax_f} \tag{3.10}$$

where φ_{Voc} is the open-circuit voltage bifaciality coefficient, φ_{Pmax} is the maximum power bifaciality coefficient, Voc_x is the open-circuit voltage, $Pmax_r$ is the maximum power, both at one sided illumination at STC. Again, the index x indicates front, f, or rear, r, side illumination. The spectral mismatch correction shall be applied, according to IEC 60904-7, for the above-mentioned calculations.

Bifacial power gain estimation

The gain in power generation yielded by the bifaciality of the device under test must be determined as a function of the irradiance on the rear side. To this end, outdoor or indoor measurement procedures must be applied as described in the next section.

Outdoor measurements at STC like conditions

In order to perform outdoor measurements of the power generation gain, the non-uniformity of irradiance on the rear side must be below 5%. To fulfil this requirement and to measure the non-uniformity of irradiance on the rear side, besides the reference device used for the irradiance measurement on the rear side, another reference device must be used to measure the irradiance on the rear side on at least four positions, before the *I–V* characterisation is performed. Figure 3.28 schematises an outdoor measurement set-up. More than one reference device can also be used for non-uniformity measurement.

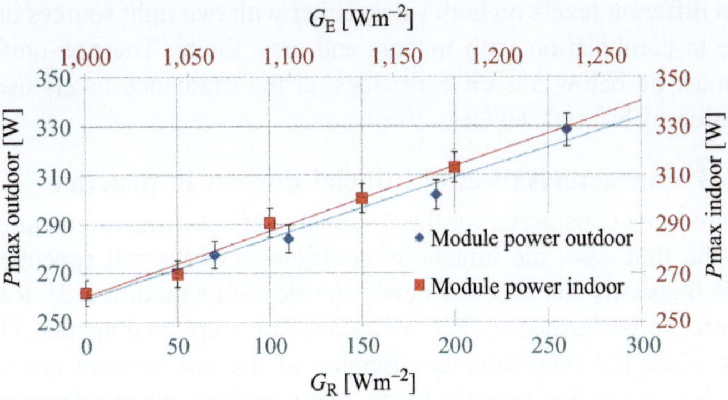

Figure 3.29 Examples of P_{max} as a function of irradiance level on the rear side G_R or its 1-side equivalent irradiance G_E

$P_{\max}$ of the module must be measured at 1,000 W/m^2 on the front side, or corrected to this value, plus different rear-side irradiance levels G_{Ri}. At least three different irradiance levels on the rear side are required. Two specific $P_{\max}$ values, $Pmax_{BiFi100}$ and $Pmax_{BiFi200}$, for $G_{R1} = 100$ W/m^2 and $G_{R2} = 200$ W/m^2, respectively, must be reported. If the irradiance levels on the rear side do not correspond to G_{R1} and G_{R2}, $Pmax_{BiFi100}$ and $Pmax_{BiFi200}$ must be obtained by linear interpolation of the data series $P_{\max}$ versus G_R.

Measurement at equivalent irradiance level

In order to perform indoor measurement of the power generation gain, a standard solar simulator with adjustable irradiance levels for one-side illumination can be used. It is required to restrict the irradiance on the rear side to below 3 W/m^2. $P_{\max}$ of the device must be measured on the front side at equivalent irradiance levels G_{Ei}, which are dependent on the bifaciality coefficient, corresponding to 1,000 W/m^2 on the front side plus different rear-side irradiance levels G_{Ri}. The equivalent irradiance levels are given by:

$$\begin{aligned} G_{E_i} &= 1{,}000 + \varphi_{P\max} \cdot G_{R_i} \\ \varphi &= \varphi_{P\max} \end{aligned} \tag{3.11}$$

where φ is equal to the smallest of $\varphi_{P\max}$ and φ_{Isc}. At least three different equivalent irradiance levels are required ($i = 1, 2, 3, \ldots$). Figure 3.29 is an example of such measurement result. For example: a device with maximum power bifaciality of $\phi_{P\max} = 80\%$ must be irradiated, on the front side at $G_{E2} = 1{,}160$ W/m^2 to provide the equivalence of $G_{R2} = 200$ W/m^2. Two specific $P_{\max}$ values, $Pmax_{BiFi100}$ and $Pmax_{BiFi200}$, for $G_{R1} = 100$ W/m^2 and $G_{R2} = 200$ W/m^2 respectively, must be reported. If the equivalent irradiance levels do not correspond to G_{R1} and G_{R2}, $Pmax_{BiFi100}$ and $Pmax_{BiFi200}$ must be obtained by linear interpolation of $P_{\max}$ versus G_E.

Measurements with double-side illumination

Alternatively, a solar simulator, with the possibility to simultaneously illuminate the bifacial device on both sides can be used. Such simulators are able to provide irradiance at different levels on both sides, either with two light sources or with one light source in combination with mirrors and grey filters. The non-uniformity of irradiance must be below 5% on both sides, at the irradiance levels used for the characterisation of bifacial devices.

3.4.1.3 *I–V* characterisation of bifacial devices in practice

Two cases are to be considered for the *I–V* characteristics measurement of bifacial devices. In the first case, the bifaciality coefficients of the test specimen are not known. This is usually the case for newly developed or modified devices and the measurements are performed by PV laboratories or accredited agents. The second case occurs when the bifaciality coefficients of the test devices are known, as typically is the case in PV manufacturing environments, when reference devices, of the same technology as the devices to be tested are available.

In PV laboratories the procedure is as follows. First *I–V* measurements are performed at STC of both sides of the test device. From these *I–V* curves the bifaciality coefficients of the test device are determined. When the test device is to be used as a reference device, the key data is reported for both sides under STC with monofacial irradiance. To report the bifacial power gain, the $Pmax_{BiFi100}$ and $Pmax_{BiFi200}$ have to be determined, either from measurements or from calculations, both at the appropriate equivalent irradiance levels.

To determine the bifacial power gain in PV production facilities, where reference devices are available, the PV panels are measured at STC, 1,000 W/m^2 only on the front. These monofacial-like values are reported. To report the bifacial power gain, the $Pmax_{BiFi100}$ and $Pmax_{BiFi200}$ are allowed to be calculated, at the appropriate equivalent irradiance levels, applying the bifaciality coefficients of the reference device. The main differences are summarized in Table 3.2.

3.4.2 *Imaging methods*

Imaging methods are used to localise the causes of reduced performance in PV modules directly after manufacturing, during accelerated stress testing or after outdoor exposure. Two methods are widely used, electroluminescence (EL) imaging and methods based on temperature.

Table 3.2 Differences in bifacial I–V characterisation in practice

	PV laboratories	PV production
I–V measurements	STC front STC rear possibly front @ G_E	STC front
Bifaciality coefficients	Calculate ϕ_x	Use ϕ_x (reference device)
Bifacial gain	Measurement or calculation: $P_{max} = f(G_R$ or $G_E)$	Calculation: $Pmax_{BiFi100}$ and $Pmax_{BiFi200}$
Reporting	Key data at STC $P_{max} = f(G_R$ or $G_E)$	$Pmax_{STC}$ $Pmax_{BiFi100}$ and $Pmax_{BiFi200}$

3.4.2.1 Electroluminescence

EL imaging takes a photographic image of the radiative recombination when a module (or solar cell) is put under forward bias in the dark, see [42] for a review on EL imaging. The radiative recombination is proportional to the current density and the logarithm of the local voltage. The EL signal is emitted at the wavelength corresponding to the band gap of the photovoltaic material, e.g., for crystalline Si the EL wavelength is around 1,100 nm. Monofacial solar cells and PV modules can therefore only be imaged from the front side.

EL imaging of bifacial PV modules is very similar to that of monofacial PV modules. The main difference is that both sides are transparent for the relevant wavelength. Figure 3.30 shows two 72-cell modules, fabricated from the same batch of solar cells [43]. The area covered by the solar cells shows no differences caused by the presence or absence of a white back sheet.

Zooming in on the images allows for more detailed analysis, as shown in Figure 3.31. In the left image, the dark grey area between and around the solar cells, is caused by EL light that is scattered off the white back sheet. This scattering process is absent in the bifacial modules (on the right image), only the edge of the glass panel is faintly visible.

3.4.2.2 Infrared and dark lock-in thermography

Thermal imaging techniques show where heat is being generated in the PV module under forward or reverse bias. The simplest method is to look at the actual temperature of PV modules under operation. An alternative method is to expose the module to a

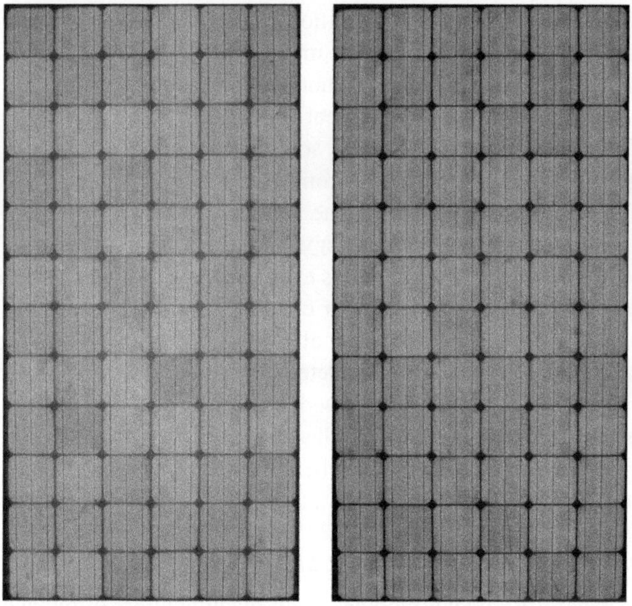

Figure 3.30 EL images for (left) monofacial and (right) bifacial module with transparent back sheet

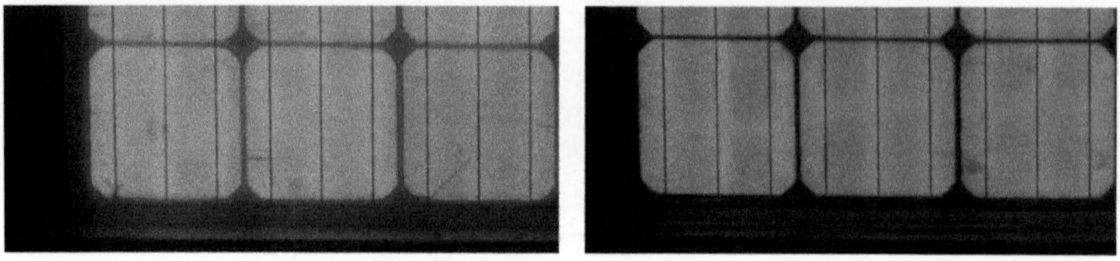

Figure 3.31 Detail of the EL images of Figure 3.30

modulated voltage bias in the dark and to observe the modulated temperature response using the lock-in method. Reference [44] gives a good overview of the theory behind and practical implications of the dark lock-in thermography (DLIT) method.

The drawback of these IR methods is that typical glass is not transparent for the relevant wavelength corresponding to the range between ambient temperature and 200°C. For large features, for example a fully shunted solar cell that is at a much lower temperature than the neighbouring solar cells, this is not an issue. But for temperature differences on a small scale this is not workable. Therefore, the thermography is typically done on the rear side of monofacial modules as the white back sheet has a very good emissivity.

The IR imaging can be used to spot failures like poor interconnection or hot spots under reverse bias. However, the spatial resolution is typically not enough to detect these small features. Using DLIT, it is possible to see failures with a distinctive size of about 1 mm or even smaller. Examples of IR and DLIT images are given in Figure 3.32. The data taken from the four-cells white back sheet module, left hand images, clearly shows the position of the solar cells. Although the three busbars/tabs can be identified from both images, the DLIT image clearly indicates that the heat is generated at the tabs and not in a wider region centred on the tabs. It even shows some inhomogeneities in heat generation along the length of the tab, probably related to variations during the soldering process.

The right hand images are taken from a double glass module. Although the presence of solar cells is visible in the IR image, the edges are blurred and the contrast with the no heat generating environment is low. In the DLIT-amplitude signal image, bottom-right, the contrast is even weaker. Infrared can be used to spot failures in modules that cause a full solar cell to stop generating heat, e.g., a short-circuit. But due to the non-transparency of glass in the relevant wavelength range (about 10 micrometre), it is not easy to detect any small scale details that are easily detectable when a white back sheet is used.

Successful DLIT imaging through glass has been reported in the literature, but these examples are limited to modules where the active, heat-generating, layers are in direct contact with the glass [45].

3.4.3 Outdoor measurements on single modules

To show the benefit of bifacial modules, both manufacturers of bifacial PV panels and research institutes have compared the outdoor power output of bifacial modules

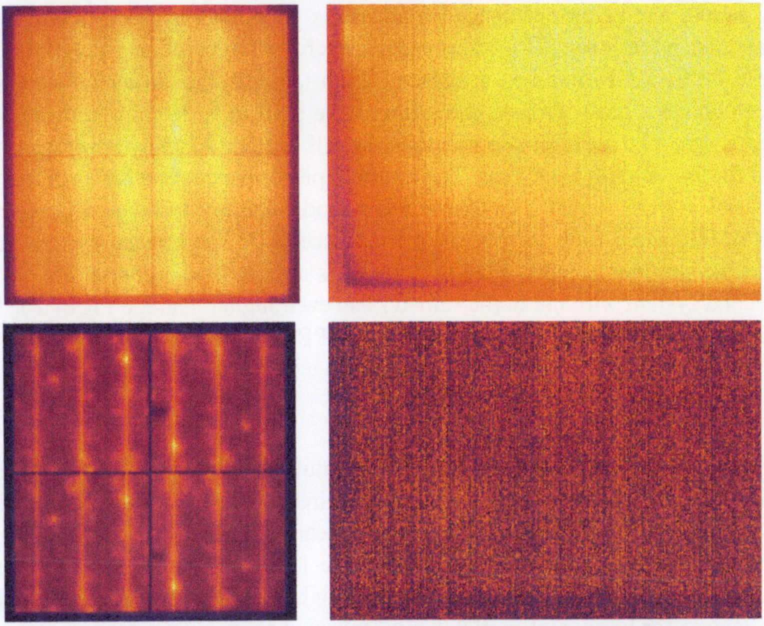

Figure 3.32 (Top) infrared image and (bottom) DLIT image taken from (left) the white back sheet side and (right) the back glass side

to monofacial modules. In the following paragraphs, the effect of external parameters such as the reflectivity of the underground are discussed.

3.4.3.1 Increased albedo

Albedo is the parameter that is used to describe the reflectivity of a surface. For bifacial PV systems it is particularly used to describe the reflectivity of the ground below the PV system. The value of the albedo varies between 0, no reflectivity at all to, and 1, full reflection. A distinction can be made between natural albedo, e.g., sand, stone or grass land, and artificial albedo, like crushed shells, roof surface membranes or solar reflective white roof paint.

For monofacial modules, the albedo has no significant effect on the performance of a PV power plant. For instance, a simple annual energy yield model, shows about 2,150 kWh/kWp for a system in the Californian desert at an albedo value of 0. Each 0.1 increase in albedo increase the yield by an insignificant 10 kWh/kWp, i.e., less than 3% increase at an albedo of 0.5. For bifacial modules, the albedo plays a significant role. The same annual energy yield model shows that for a very low albedo value of 0.05, the rear side receives already an additional 5% light of which about 75% is due to ground reflected light, the other 25% due to diffuse light incident on the rear side. At a high albedo value of 0.5, around 25% of the incident light is ground reflected. This simple model takes only the shade of the device under investigation into account, overestimating the ground reflected light.

A simple test of the influence on albedo is conducted by PVG Solutions on Hokkaido Island in Japan [46]. Two systems, each consisting of an array of three by four bifacial EarthON modules, are placed south-facing at the same tilt. One system is placed above a grass surface, the other above a 10 m × 10 m area made up of crushed shells. The location is prone to snow fall, leading to snow coverage on the ground for five months each year. The monthly performance ratio for a comparable monofacial system at that location in the period February to August is on average 88%, calculated from the observed irradiance [44]. The bifacial system above grass in the summer months, May to August, is almost 10% higher at PR = 95%. The second system with the higher albedo has even higher average PR of 108% in the same period. During the snow season, the PR for both bifacial systems is the same: PR = 110%, a gain of almost 24% relative to the monofacial module.

3.4.3.2 Tilt

The optimal tilt angle for a monofacial PV module is determined by the competition between the tilt angle dependency of the direct light, the diffuse light and the ground-reflected light. Typically, the direct (beam) irradiance is highest at a tilt angle somewhat smaller than or similar to the latitude angle. The diffuse (sky) irradiance is highest for horizontal modules, whereas the ground-reflected irradiance is zero for horizontal modules but highest at vertical orientation.

For bifacial PV modules, the situation is more complicated. As the diffuse, isotropic sky, light will always reach either of the two transparent surfaces, there is only a second-order dependency on the tilt angle due to the angular dependence of the reflectance. A large proportion of the ground-reflected light will be incident on the 'bottom' side of tilted modules. For monofacial modules, the ground-reflected light on the rear has no contribution on the energy yield, but bifacial modules are also transparent on that side and will have a large contribution to the energy yield by ground-reflected light. Another effect to take into account is the (self-)shading.

3.5 Modelling of bifacial modules

In Section 3.4 several methods to determine the power output or *I–V* curves of bifacial solar cells under two-sided illumination have been discussed.

3.5.1 Electrical models

For module or system modelling where the irradiance (on the rear side) is not uniform, the full *I–V* curve is needed to calculate the effect of mismatch between the series connected solar cells that have different maximum power point currents.

A simple method to deal with the non-uniformity in the (rear) irradiance is to determine a mismatch function that depends on the relative standard deviation σ in the irradiance G as presented by Janssen *et al.* [47]. For solar cell in a string the irradiance is calculated as it differs due to position and effective self-shading.

The maximum power of the string, P_{string}, is simulated for the inhomogeneous irradiance using an electronic circuit simulator, e.g., LT-spice. From this set of irradiances also the relative standard deviation is calculated. The mismatch F_m is then given by the ratio of P_{string} and the sum of P_{cell}. Repeating this for a range of irradiance distributions leads to a trend line for the mismatch as a function of the relative standard deviation. For each situation with a given $\sigma(G)$, the mismatch function F_m can then be taken from the trend line. A more elaborate method is to apply an electronic circuit simulator for each time step in the modelling software.

Also the electrical layout could be different for bifacial modules. The most obvious difference is the location of the junction box. Typically, the effect of the electrical layout, cross-connectors and connection to the junction box or boxes is taken into account by the series resistance of the bifacial PV module.

3.5.2 Thermal behaviour

Although the electrical behaviour of (bifacial) PV panels responds directly to both small and large changes in the irradiance, the temperature of the solar cells takes much longer to be in thermal equilibrium with the, ever changing, environment. To accurately simulate or predict the actual operating temperature of (bifacial) PV panels, a detailed dataset, including ambient temperature, irradiance and wind speed and direction is needed with sufficiently short time steps.

An alternative method is the fingerprint method presented by Dekker *et al.* [48]. The method consists of two, simple approximations for the module temperature and the power output, as a function of the irradiance and the ambient temperature. To obtain the module temperature it is assumed that the difference between module and measured ambient temperature is linearly proportional to the irradiance. It also proposes that the P_{max} of the module depends only on its temperature and the irradiance.

These two relationships, the fingerprints (see Figure 3.33), are fitted to observed P_{max} and module temperature data. Plotting the difference between calculated and observed module temperature and between calculated and observed P_{max} shows a large spread, centred around zero difference, see Figure 3.34. Although the maximum deviation was around 6°C, the mean deviation of the module temperature was −0.13°C. The larger deviations are probably due to fast changing conditions and to the cooling effect of wind, which is not taken into account. The difference in modelled energy yield, +0.17%, is slightly larger for bifacial modules than that reported for monofacial modules [46].

If someone wants to study the bifacial module temperature under operating conditions in more details, more factors have to be considered as both the heating by incident sunlight and the thermal properties of bifacial modules are different compared to monofacial, white back sheet modules.

The temperature of bifacial modules is affected by the different optical materials compared to monofacial modules. On the one hand, around 10% of the area of bifacial modules is transparent, which means that the light will pass through the

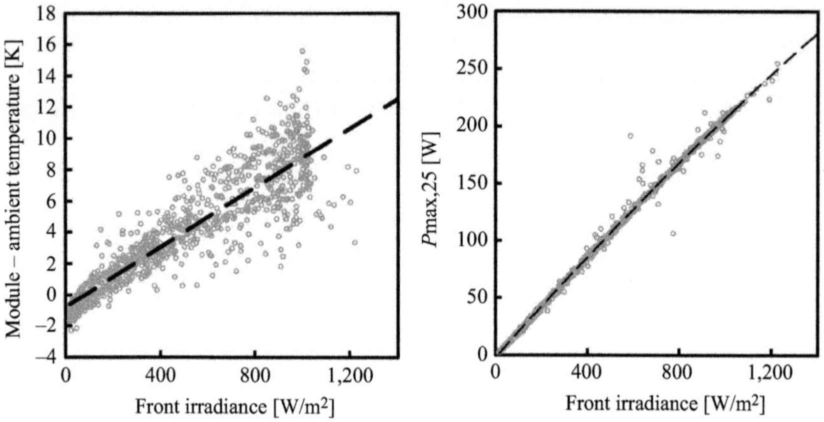

Figure 3.33 *Outdoor data, consisting of 1,000 measurements taken in April/May on a bifacial 60-cell module placed on an open rack on a wind-exposed rooftop. (left) difference between observed module and observed ambient temperature as a function of front irradiance. (right) observed maximum power normalised to 25°C as a function of front irradiance. The dashed lines are the fingerprints derived from the data. Note that tilted roof mounted modules exhibit a much larger temperature increase*

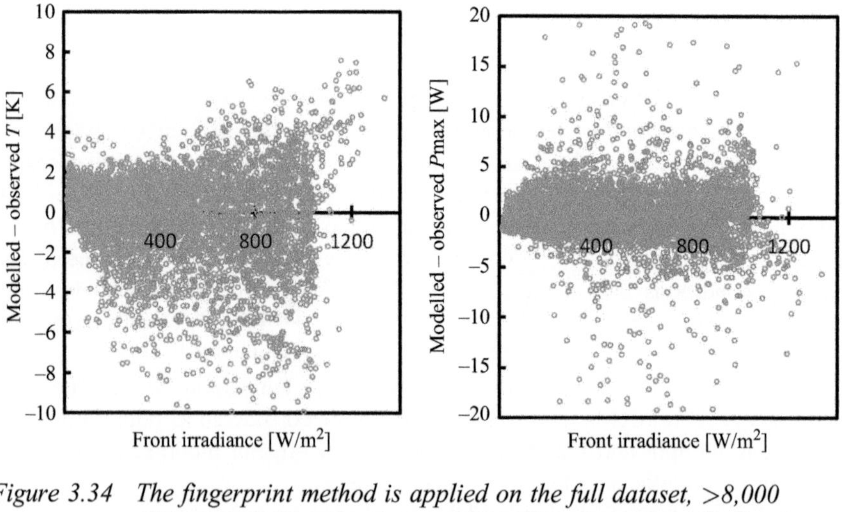

Figure 3.34 *The fingerprint method is applied on the full dataset, >8,000 observations from January to May. Plotted are the calculated differences between the modelled values and the observed module temperature (left) and observed maximum power (right)*

module instead of being scattered or absorbed by the back sheet. This light cannot contribute to the increase of the module temperature above ambient via thermalisation of the excited electrons [49] or other processes such as free carrier absorption.

On the other hand, the increased power output, due to light that is incident on the rear side, means that the heating of the solar cells due to thermalisation and free carrier absorption also increases. Furthermore, the increased power output is mostly due to an increase in photogenerated current and as the resistive losses, both in the solar cells and in the cell–cell interconnection, are quadratically proportional to the current, the heating due to resistive losses will also increase.

The final parameter to consider are the differences in thermal properties. It is well-known that building integrated and even building adapted modules, e.g., regular monofacial modules on slanted roofs, have a much higher operating temperature than modules placed on open frames. Bifacial modules will always be placed on open frames in the field, free flowing air will cool the module more compared to modules placed on a slanted roof. Furthermore, properties like the heat conductivity, the emissivity and the heat transfer to air will be different for glass and polymer rear panels.

3.5.3 Optical modelling

The system modelling chapter will describe the various methods to simulate the two-sided irradiance. The aspect of bifacial modules to be taken into account for, is the partial transparency of such modules. This will change the self-shading correction as discussed, e.g., by Yusufoglu *et al.* [50]. The amount of transparency depends on the dimensions, e.g., the cell–cell spacing and on the optical properties of both glass panels. As can be seen in Figure 3.26, part C, a distinction can be made between light transmitted through the bifacial solar cells in the PV panel (beam 2) and light transmitted through the open area between and around the solar cells (beam 3). Note that the transparency of bifacial PV panels is only about 5%–10%. The deduction on the contribution of the ground reflected light on the rear irradiance due to the self-shading is still 90%–95% of the amount when the partial transparency is ignored.

3.6 Reliability and durability of bifacial modules

The operation of bifacial modules inherently comes with significant changes in terms of the module output current being transported by all electrical components, the heat management of the module design and installation set-up as well as the eligibility of all module materials. On top of this the certification standards currently being used for IEC and UL certification which guarantee long-term and safe operation are definitely not suited in all means for bifacial products specifically not for bifacial modules installed at many locations worldwide. This section first sums up the major differences between the operation of bifacial and monofacial modules. Next, this section gives an overview on the various requirements which have to be implemented to guarantee consistent warranty and safety standards. Moreover, existing individual test procedures currently being applied by certification agencies will be analysed with focus specifically on differences affecting the performance, long-term stability or safety of bifacial products.

3.6.1 Effect of higher output current

At rated irradiation (typically called STC, standard test conditions) the output current at maximum power point of today's monofacial crystalline silicon solar cell modules ranges from 8 to 10 A depending on cell type (mono- or multi-crystalline silicon cell technology), processing technology and module design. For a bifacial module, the output current is generally higher in the real world due to absorption of irradiation from both front and back. Yusufoglu reported [48] that compared to monofacial modules, annual energy output of bifacial modules can be increased by 30%. Some PV manufacturers also came up with a design guide to measure bifacial modules with front and rear irradiation of 1,000 W/m^2 and 300 W/m^2 [51]. Furthermore, bifacial modules deployed such as in the Atacama desert have shown to receive front irradiation values up to 1,200 W/m^2 [52]. Is it evident that bifacial modules in the field can reach very high operating current of up to 15 A. Such high current requires a reconsideration of various details which are discussed in the following.

3.6.1.1 Requirements for the cells architecture

The metallisation grid of standard crystalline solar cells is typically optimized for STC conditions. Ohmic losses beyond STC (in our case the sum of front and rear irradiation acts as a concentration factor of up to 1.5) will lead to a significant power drop. A simple calculation for 3 busbar cells leads to a power loss of 1% on cell level. This power loss directly implies a yield loss in the same order inevitably contributing to other power losses such as stringing and busing related ones. To keep this loss as small as possible the metallisation grid has to be specifically designed for low concentration which includes narrowing the finger distance, increasing the number of fingers and the fingers cross section. Alternatively, the number of busbars can be increased from 3 to 5 busbars or alternative module concepts such as the Smart Wire concept be introduced [53].

3.6.1.2 Requirements for the module architecture

The resistive losses contribute to a majority of the cell-to-module (CTM) losses, as shown in [5]. Resistive losses follow the simple equation $P_{\text{loss}} = I^2 \cdot R$. Based on this equation, the resistive losses can be decreased by reducing the current or the resistance of the module architecture. The module resistance can be for example decreased by increasing the cross section of the ribbon. For typical modules with 3-busbar cells, the interconnecting ribbon has a typical width of 1.6 mm and a thickness of 0.2 mm. By increasing the finger or ribbon width, the resistive loss decreases; however, this will increase the shading losses. Alternatively, the finger or ribbon height can be increased; however, a thicker ribbon will potentially lead to a higher level of electrical faults after soldering such as cracks as reported in [54].

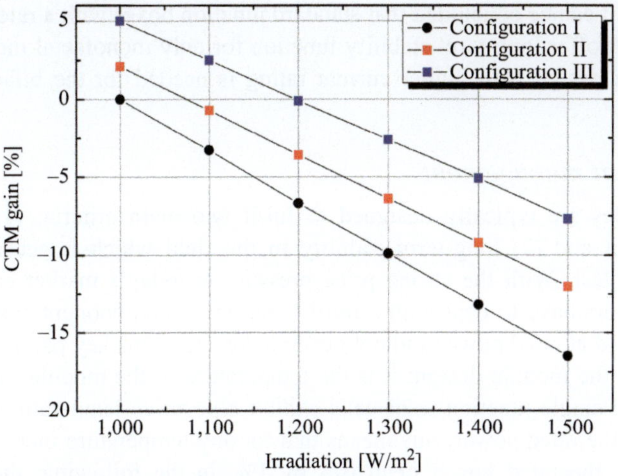

Figure 3.35 CTM gain for various irradiation losses and three configurations

Fortunately, there is another way of decreasing the resistive losses, which is to use solar cells with more number of busbars. As the number of busbars increases, the current that flows through each busbar is reduced.

The CTM gain and losses attributed to various irradiation levels for a typical 60 cells module were calculated for two ribbon types (Configuration I: ribbon width = 1.6 mm; height = 0.2 mm; and Configuration II: ribbon width = 1.8 mm; height = 0.22 mm) and are displayed in Figure 3.35. The losses for a 5-busbar cell module are also included (Configuration III: ribbon width = 1.2 mm; height = 0.22 mm) as well. The starting point was a 20.2% efficient mono-crystalline solar cell.

The results indicate that losses up to 16.5% for ribbon configuration I and 12% for ribbon configuration II occur at irradiation levels of 1,500 W/m^2. Losses can be significantly reduced down to 7.7% if 5-busbar cell technology is used.

One of the key components of solar modules is the junction box which is typically rated according to IP 65. (IP stands for Ingress Protection with a code following the International Protection Rating. 65 stands for No ingress of dust; complete protection against contact and water projected by a nozzle (6.3 mm) against enclosure from any direction shall have no harmful effects.) The maximum tolerated permanent output current on the other hand is specified by the junction box maker. A very typical specification here is a rated current of 12 and 15 A for an operating temperature between −40°C and +85°C. The maximum rated instantaneous forward current of typical Schottky diodes used as bypass diodes by junction box makers is 20 A for a fixed temperature of 25°C. The forward current derating curve of such solar Schottky diodes reveal that even for an application with heat sink the maximum current for such diodes decreases steadily above 25°C. Many junction box makers use the metallic parts for the electrical wiring indirectly as heat sink. However, due to the strongly reduced convection abilities inside the junction box, the metallic parts only partly act as an acceptable heat sink.

In summary it can be concluded that standard junction boxes with a rated current of 12 A or 15 A offer long-term stability function for only monofacial modules. New junction box design with higher current rating is needed for the bifacial module market.

3.6.2 Heat management

Solar modules are typically designed to fulfil two main criteria: (1) optimized power output and (2) long-term stability in the field which is accessed by the certification test. With the strong price pressure in today's market environment, module makers have to deal with a third criterion: the component cost. Since the module is sold at rated power and not generated energy, one key parameter is often neglected in the module design; it is the temperature of the module during operation. Typical single junction solar cells exhibit a negative temperature coefficient of 0.4% for the rated power; this means that for any temperature increase of 2.5 K above 25°C, the rated power decreases by 1%. In the following, the effects of module temperature specifically for bifacial solar modules are discussed in details.

3.6.2.1 Focus on nominal operating cell temperature

Module makers rate their products with a parameter called NOCT (nominal operating cell temperature) to describe the operating module temperature at specified environmental parameters. For crystalline monofacial modules the NOCT typically lies in the range between 40°C and 46°C. In the case of bifacial modules the maximum power current is strongly increased. Furthermore, glass-glass modules come with increased heat isolation due to the glass-sandwich. Besides, the passive heating of solar modules by infrared radiation is partly increased by the impact of the back side illumination. All described effects will impact the modules NOCT once being operated as a bifacial module. As of 2016, a large European module maker is marketing bifacial solar modules (which were already equipped with 5 busbar technology) with rated NOCT values of 48°C where at the same time the NOCT for the standard product is "only" 46°C. The key to interpret this data is not the number itself but it is the NOCT measurement procedure applied which is well defined by standards. In this procedure, no specific rear-side illumination is required and the front irradiation level is only 800 W/m^2 (at a wind velocity of 1 m/s and set temperature of 20°C). This means NOCT will significantly be above 50°C with increasing rear irradiation. Losses of several percentage points are inevitable when compared to monofacial products.

3.6.2.2 Impact on module lifetime

Preventing higher module temperature is the key to reduce the power losses and the chemical reaction rate. In biology and chemistry, the temperature coefficient (Q10) represents the factor by which the rate of a reaction increases for every 10 K rise in temperature. Typically the reaction rate increases by a factor of about 2 when the activation energy of the degradation mechanism is about 0.75 eV as shown in the work of [55]. The main chemical reaction to be prevented in a typical glass-glass bifacial product is related to the degradation of the chemical stability of the encapsulation material. Any degradation will lead to delamination or discoloration over time. Another chemical reaction which will be promoted with a temperature increase is corrosion, e.g., corrosion of the solar cell metallisation or corrosion of the coated copper ribbon. It can be argued that glass-glass modules will see a much lower water ingress as compared to glass-back sheet modules which is indeed true. However, as long as chemical by-products exist inside encapsulation materials, specifically for the most common EVA encapsulants, this degradation will inevitably take place. Various research groups over the past decades have tried to associate a certain minimum real world lifetime for solar modules with certain climatic testing such as damp heat for 1,000 h or 200 cycles of thermal-cycling. However, due to the complexity of the interaction of PV product materials and environmental factors, such approximation can only be given with large temporal error bars. For bifacial modules with higher operating temperature due to significantly increased irradiation levels, the likelihood of failure increases as compared to monofacial solar modules. Only through the careful choice of module materials in combination with a focus on heat management, the module manufacturers can guarantee for a 20–30 years' warranty.

3.6.2.3 Focus on junction box

The junction box is a crucial part of the module in terms of reliability and safety. Overheating of bypass diodes or large contact resistances, caused by for example corrosion or faulty clamp interconnections, may lead to hazardous situations such as fire. For bifacial modules, design of junction boxes becomes even more critical due to the strongly increased module output current. The typical forward voltage of such a diode is about 0.45 V which results in approx. 21 W for three bypass diodes at 15 A. For example, permanent shading will lead to a temperature rise of at least 25°C for each individual diode if a perfect heat sink is attached to such a diode (best case scenario). In case of a module temperature of 65°C the junction box interior partly may rise up to 90°C and higher specifically without heat convection from the interior of the box to the outside. All these examples show that it is challenging and critical for heat management in the junction box. The junction box manufacturers should take extra care when developing junction boxes for bifacial module products.

3.6.3 Selection of module materials for bifacial modules

3.6.3.1 Influence of module materials to lifetime

Bifacial modules consist of a transparent rear side providing a high light transmissivity. This can be glass or a thin transparent plastic material sheet such as ETFE. Besides being impervious to moisture, glass also has the advantage in providing mechanical support. Furthermore, glass is very stable even at high temperatures and UV concentration. Besides, glass does not lose its sealing ability in a harsh and abrasive environment. With all the advantages, glass is an excellent choice for front- and rear-side superstrates. The next important module component is the encapsulant. It has to withstand high temperature and irradiation. Degradation of encapsulant will: (1) introduce base or acidic environment that can cause corrosion to cell or module components and (2) reduce adhesion strength that will eventually lead to delamination.

Due to the more stringent operating conditions of bifacial modules, careful material selection has to be performed.

3.6.3.2 Impact of material selection to module power

As described earlier, the key to longevity of the module is the right material selection which is even more important for bifacial products. The focus has to be set on the encapsulation materials to provide the long term functionality requirements. This will directly affect the generated module power on front and for bifacial modules as well on rear side. Several materials are provided to the market such as EVA with low UV cut-off, POE and silicone-based products or TPU/TPO sheets. In terms of energy production, most of materials with cut-off wavelengths of around 320 nm will perform alike, at least in the beginning. Since degrading effects are inevitable causing a loss in transmissivity the key performance indicator is the stability of the encapsulant itself and the combination of all materials. Hence, maximizing output power is only one consideration factor; the module manufacturers must carefully evaluate the material for overall long term durability.

3.6.4 Discussion on current IEC 61215 testing and its suitability for bifacial modules

The international standard IEC 61215 "Crystalline silicon terrestrial photovoltaic (PV) Modules – Design qualification and type approval" is designed to examine the parameters responsible for the aging of PV modules. The ultimate goal of the testing is to early detect design-, materials- and process-related defects which may lead to premature field failures. In many countries, solar modules must pass IEC61215 before getting approval to be connected to the grid.

The IEC 61215 standard needs to be adapted on several points to account for the specific properties and benefits of bifacial solar modules:

- Bypass diode thermal test (IEC 61215 - 10.18)
- Maximum power determination (IEC 61215 - 10.2)
- Determination of NOCT (IEC 61215 - 10.5)
- Performance at NOCT (IEC 61215 - 10.6.3.2)
- Performance at low irradiance (IEC 61215 - 10.7)
- Outdoor exposure (IEC 61215 - 10.8)
- Hot-spot test (IEC 61215 - 10.9)
- UV preconditioning (IEC 61215 - 10.10)

It is beyond the scope of this work to describe each point in detail individually. Exemplary procedures 10.5 and 10.8 are taken:

In 10.5 the NOCT is determined at an irradiation of 800 W/m^2, ambient temperature of 20°C and wind velocity of 1 m/s. The standard did not mention about the ground reflectance nor the rear illumination. As rear irradiation will increase NOCT, it should be regarded in 10.5 as well.

The hot spot test in 10.9 is performed during 5 h at an irradiation of 1,000 W/ m^2 taking the worst scenario for hot spots into account. Again, this test does not take into account the additional irradiation from the rear of bifacial modules. Since bifacial modules output currents are typically above 10 A reaching values of up to 15 A, the bifacial modules are more susceptible to hotspot failure.

3.6.5 General discussion on safety aspects

IEC 61730-1 and IEC 61730-2 are the international standards for Photovoltaic (PV) module safety qualification and are a requirement in many countries for solar modules to be certified prior to getting permission to be connected to the grid. The main idea of this qualification is preventing any hazardous situation from occurring during operation. The question which arises is if the same design rules and testing specifications in a monofacial module can be applied to a bifacial module. Taking a glance at the standards, the following items are not well defined for bifacial solar modules:

- Temperature test (IEC 61730 - MST 21)
- Hot-spot test (IEC 61730 - MST22)
- Bypass diode thermal test (IEC 61730 - MST25)
- Reverse current overload test (IEC 61730 - MST 26)
- UV preconditioning (IEC 61730 - MST 54)

Again for most of the tests the specific conditions for bifacial modules are not taken into account. Two measures can be introduced to deal with this situation: (1) The standards are modified to account for alterable rear-side illumination but even in this case if the installer adapts the installation site for extremely high ground reflection the standards will not accommodate for such a situation. (2) The other possibility is the module maker to label the product for a maximum total irradiation hence current, meaning the installer should refrain from such high level of irradiation. This brings the advantage that certification agency and installation site owner know the specific conditions for the warranty terms. Another item to consider is the UV pre-conditioning test. The UV energy dose exposed to a solar module has to be accommodated to meet the scenario of a solar module being exposed from front and rear side by UV. For the standard, this means the UV energy exposure time has to be significantly increased to warranty that no UV related defects occur during its lifetime since the UV energy from front and rear side reaching the inside of the module is significantly higher.

The discussion led in this section shall remind the certification bodies and the module makers that bifacial products entering the market currently are not fully rated and tested for bifacial module configuration; more work has to be done to ensure the reliability and safety of such products.

References

[1] Kopecek R., Veschetti Y., Gerritsen E., *et al.* 'One small step for technology, one giant leap for kWh cost reduction'. Photovoltaics International. 2015; 26: 32–45.

[2] Kreinin L., Karsent A., Grobgeld D., and Eisenberg N., 'PV systems based on bifacial modules: performance simulation vs. design factors'. Proc. 43rd Photovoltaic Specialists Conference (PVSC), 2016.

[3] Hacke P., Smith R., Terwilliger K., Perrin K., Sekulic B., and Kurtz S., 'Development of an IEC test for crystalline silicon modules to qualify their resistance to system voltage stress'. Proc. 26th European Photovoltaic Solar Energy Conference, Frankfurt, D, 2012.

[4] Peters I. M., Yong Sheng K. and Walsh T. M., 'Detailed current loss analysis for a PV module made with textured multicrystalline silicon wafer solar cells'. IEEE Journal of Photovoltaics. 2014; 4: 585–93.

[5] Singh J. P., Khoo Y.S., Chai J, Liu Z and Wang Y., 'Cell-to-module power loss/gain analysis of silicon wafer-based PV modules'. Photovoltaics International 2016; 31.

[6] Dasari S. M., Srivastav P., Shaw R., Saravanan S. and Suratkar P., 'Optimization of cell to module conversion loss by reducing the resistive losses'. Renewable Energy. 2013; 50: 82–5.

[7] Hohl-Ebinger J. and Warta W., 'Bifacial solar cells in STC measurement'. Proc. 25th European Photovoltaic Solar Energy Conference; Valencia, SP, 2010.

[8] Singh J. P., Khoo Y. S., Chai J. and Wong J., 'Bifacial solar cell measurements under STC and its impact on bifacial module measurements'. Japanese Journal of Applied Physics. 2017; 56: 08MD04.

[9] Chieng Y. K. and Green M. A., 'Computer simulation of enhanced output from bifacial photovoltaic modules'. Progress in Photovoltaics: Research and Applications. 1993; 1: 293–99.

[10] Cuevas A., Luque A., Eguren J., and del Alamo J., '50 per cent more output power from an albedo-collecting flat panel using bifacial solar cells'. Solar Energy. 1982; 29: 419–20.

[11] Guo S., Singh J. P., Peters I. M., Aberle A. G. and Wong J., 'Two-dimensional current flow for stringed PV cells and its influence on the cell-to-module resistive losses'. Solar Energy. 2016; 130: 224–31.

[12] Wilson K., De Ceuster D. and Sinton R. A., 'Measuring the effect of cell mismatch on module output'. Conference Record of the 2006 IEEE 4th World Conference on Photovoltaic Energy Conversion 2006; 916–19.

[13] Field H. and Gabor A. M., 'Cell binning method analysis to minimize mismatch losses and performance variation in Si-based modules'. Photovoltaic Specialists Conference, 2002. Conference Record of the 29th IEEE. 2002: 418–21.

[14] Singh J. P., Guo S., Peters I. M., Aberle A. G., and T. M. Walsh, 'Comparison of glass/glass and glass/backsheet PV modules using bifacial silicon solar cells'. IEEE Journal of Photovoltaics; 5(3): 783–91.

[15] Khoo Y. S., Singh J. P and Saw M. H. 'High performance double-glass bifacial PV modules through detailed characterization'. Proc. 3rd bifacial PV workshop, Miyazaki, 2016. [Online]. Available: http://bifipv-workshop.com/fileadmin/images/bifi/miyazaki/presentations/4_1_5-SHENG-High_performance_double-glass.pdf, last accessed on April 24, 2018.

[16] Van Aken B., 'Innovative white layer improves efficiency of bifacial solar panels'. Proc. 3rd bifacial PV workshop, Miyazaki, 2016. [Online]. Available: https://www.ecn.nl/news/item/innovative-white-layer-improves-efficiency-of-bifacial-solar-panels/, last accessed on April 24, 2018.

[17] Mrcarica M., Panofen C., Pasmans P., Schneider A. and Rabanal-Arabach J. 'DSM materials application in bifacial modules light management,' Proc. 3rd bifacial PV Workshop 2016. Available from http://bifipv-workshop.com/fileadmin/images/bifi/miyazaki/presentations/4_2_6_-_MRCARICA_-_DSM_materials_application_in__Bifacial_modules_light_management.pdf [Accessed 24 Apr 2018].

[18] Walter J., Tranitz M., Volk M., Ebert C. and Eitner U. 'Multi-wire Interconnection of Busbar-free Solar Cells'. Energy Procedia. 2014; 55: 380–88.

[19] Yao Y., Papet P., Hermans J., *et al.* 'Module integration of solar cells with diverse metallization schemes enabled by SmartWire Connection Technology'. Proc. IEEE 42nd Photovoltaic Specialist Conference. 2015.

[20] Schneider A., Rubin L. and Rubin G. 'Solar cell efficiency improvement by new metallization techniques – the Day4 Electrode Concept'. Conference Record of the 2006 IEEE 4th World Conference on Photovoltaic Energy Conversion, 2006, pp. 1095–1098.

[21] Joanny M., Bertrand D., Veschetti Y., *et al.* 'Module architecture optimization for bifacial applications'. Proc. 3rd bifacial PV workshop, Miyazaki, JP. 2016 http://bifipv-workshop.com/fileadmin/images/bifi/miyazaki/presentations/7_1_3_-_JOANNY_-_bifi_Module-FINAL.pdf [Accessed 24 Apr 2018].

[22] Hanifi H., Schneider J. and Bagdahn J., 'Reduced shading effect on half-cell modules – measurement and simulation'. Proc. 31st EUPVSEC conference, Hamburg, Germany, 2015.

[23] Joanny M., Razongles G., Sicot L., *et al.* 'Bifacial PV modules: measurement challenges'. Proceedings of the Workshop PV Modultechnik, November 13, 2015, Cologne (Germany). Available from http://www.tuv.com/media/germany/10_industrialservices/download_landingpage_solar/12_workshop_photovoltaik_modul/42_Joanny_Bifaciales_PV_Modules.pdf [Accessed 24 Apr 2018].

[24] Joanny M., Soria B., Bettinelli A., *et al.* 'Cell interconnection challenges for bifacial modules'. Proc. 2nd bifacial PV workshop, Chambery, 2014 (Online). Available from http://www.slideshare.net/sandiaecis/16-m-joanny-ok [Accessed 24 Apr 2018].

[25] Soria B., Gerritsen E., Lefillastre P. and Broquin J.E., 'A study of the annual performance of bifacial photovoltaic modules in the case of vertical facade integration'. Energy Science and Engineering. 2016; 4 (1): 52–68.

[26] Zhao J., Wang A., Yun F., *et al.*, '20000 PERL silicon cells for the "1996 WorldSolar Challenge" solar car race'. Progress in Photovoltaics: Research and Applications. 1997; 5 (4): 269–76.

[27] Glunz S. W., Kamerewerd F.J., Knobloch J., *et al.* 'High efficiency silicon solar cells for low-illumination applications'. Proc. 29th IEEE PVSC, New Orleans, Louisiana, USA, 2002: 450–53.

[28] Available from http://bifipv-workshop.com/fileadmin/layout/images/Konstanz-2017/4__J._Prakash_SERIS_Shingled_Bifacial_PV_modules.pdf [Accessed 24 Apr 2018].
[29] Available from https://us.sunpower.com/blog/2015/11/12/sunpower-introduces-performance-series-solar-panel/ [Accessed 24 Apr 2018].
[30] Available from http://www.apvia.org/upload/file/TaiyangNews_Report_Advanced_Solar_Module%20Technology_2017_download_version.pdf [Accessed 24 Apr 2018].
[31] Schulte-Huxel H., Kiefer F., Blankemeyer S., *et al.* 'Flip-flop cell interconnection enabled by an extremely high bifacial factor of screen-printed ion implanted N-PERT Si solar cells'. Proc. 32nd European Photovoltaic Solar Energy Conference, Munich, Germany, 2016.
[32] Kopecek R., Buck T., Kränzl A., *et al.*, 'Module interconnection with alternate P- and N-type Si solar cells'. Proc. 21st European Photovoltaic Solar Energy Conference, Dresden, Germany, 2006: 111–15.
[33] Newman B. K., Bende E. E., Aken B. Van, *et al.* 'Manipulating reverse current in 21% n-MWT cells'. Proc. 31st EU PVSEC, 2015: 829–34.
[34] Romijn I., Mewe A. A., Lamers M. W. P. E., Kossen E., Bende E. and Weeber A. W. 'An overview of MWT cells and evolution to the ASPIRe concept: A new integrated mc-Si cell and module design for high-efficiencies'. Proc. 23rd EU PVSEC, 2008: 1000–5.
[35] Stassen A. and Zhang W. *New Ag metallization pastes for solar energy cost reduction*, Heraeus white paper. Available from https://www.heraeus.com/media/media/hpt/doc_hpt/whitepapers/New_Ag_metallization_pastes_for_solar_energy_cost_reduction.pdf. Accessible via www.heraeus.com [Accessed 24 Apr 2018].
[36] Bennett I. J., Bende E. E., Goris M. J. A. A. and Eerenstein W. 'An overview of developments in foil-based back-contact-modules'. Proc. 29th European Photovoltaic Solar Energy Conference, Amsterdam, NL. 2014, 5–8.
[37] Van Aken B. B. and Slooff-Hoek L.H. 'Positive cell-to-module change: getting more power out of back-contact modules', Photovoltaics International **33** (2016).
[38] Guillevin N., Heurtaulta B. J. B., Aken B. Van, *et al.*, 'High efficiency n-type metal wrap through cells and modules', Energy Procedia. 2012; 27: 610–16.
[39] IEC 60904-1: Photovoltaic devices – Part 1: Measurement of photovoltaic current-voltage characteristics.
[40] IEC 60904-9: Photovoltaic devices – Part 9: Solar simulator performance requirements.
[41] Fakhfouri V., IEC project team, 'IEC 60904-1-2: Measurement of current-voltage characteristics of bifacial photovoltaic devices'. Presented at third Bifi PV workshop; Miyazaki, Japan, 2016.
[42] Kontges M., Kurtz S., Packard C., *et al.* 'Review of failures of photovoltaic modules'. Technical Report IEA-PVPS T13- 01:2014, International Energy Agency, 2014.

[43] Romijn I. G., Van Aken B. B., Bennett I. J., *et al.* 'Improvements in Advanced Industrial n-type Solar Cells and Modules'. Photovoltaics International (PV-Tech). 2014; 25: 58–68.

[44] Breitenstein O., Warta W. and Langenkamp M., Lock-in Thermography – Basics and Use for Evaluating Electronic Devices and Materials. 2nd ed. Springer, Berlin/Heidelberg, 2010.

[45] Straube H. and Breitenstein O., 'Infrared lock-in thermography through glass substrates'. Solar Energy Materials and Solar Cells. 2011; 95: 2768–71.

[46] Sugibushi K., Ishikawa N. and Obara S. 'Bifacial-PV power output gain in the field test using "EarthON" high bifaciality solar cells'. Proceedings of the 28th European Photovoltaic Solar Energy Conference; Hamburg, Germany, 2013: 4312–17.

[47] Janssen G. J. M., Carr A. J., de Groot K. M., Van Aken B. B. and Romijn I.G., 'An energy yield model for bifacial photovoltaic systems'. Proceedings of the 26th Photovoltaic Science and Engineering Conference; Singapore, 2016: to be published.

[48] Dekker N. J. J., Kroon J. M., Jansen M. J. and Eerenstein W., 'Accurate yearly yield calculation using PV module fingerprint method – applied for MWT, H-Pattern and thin film modules'. Proceedings of the 31st European Photovoltaic Solar Energy Conference; Hamburg, Germany, 2015: 2039–42.

[49] Hirst L. C. and Ekins-Daukes N. J., 'Fundamental losses in solar cells'. Proceedings of the 24th European Photovoltaic Solar Energy Conference; Hamburg, Germany, 2009: 457–61.

[50] Yusufoglu U., Hun Lee T., Pletzer T, *et al.* 'Simulation of energy production by bifacial modules with revision of ground reflection'. Energy Procedia. 2014; 55: 389–95.

[51] Castillo-Aguilella J., Annual Bifacial Energy Yield, Best Fit Model, https://www.slideshare.net/sandiaecis/4-2-castillo-aguilella-annual-bifacial-energy-yield-bestfit-model [Accessed 24 April 2018].

[52] Rabanal-Arabach J., Schneider A. and Cabrera E. 'Minimization of electrical losses of PV modules located in places with high solar irradiance'. Energy Procedia. 2015; 77: 402–6.

[53] Faes A., Despeisse M., Levrat J., *et al.* ' Smartwire solar cell interconnection technology'. Proceedings of the 29th European Photovoltaic Solar Energy Conference; Amsterdam, NL, 2014: 2555–61.

[54] Schneider A., Fernada R., Link T., *et al.* 'Impact of ribbon specification and handling during PV module manufacturing to module reliability'. Proceedings of the 32nd European Photovoltaic Solar Energy Conference; Munich, Germany, 2016: 1583–8.

[55] Koehl M., Hoffmann S. and Wiesmeier S. 'Evaluation of damp-heat testing of photovoltaic modules'. Progress in Photovoltaics. 2017; 25 (2): 175–83.

Chapter 4

Simulation models for energy yield prediction of bifacial systems

Ismail Shoukry[1], *Djaber Berrian*[2], *Joris Libal*[2], *and Florent Haffner*[3]

4.1 Introduction/motivation

Despite the early work that demonstrates power gain exploiting bifacial modules goes back to the late 20th century [1–3]. Recently, a significant endeavour has been put to quantify and predict the energy yield of bifacial PV modules for different installation configurations: stand alone, PV plants, east and west orientation [4–9]. The main reason behind these relevant investigations is to determine the levelized cost of electricity (LCOE) of bifacial modules and therewith their profitability. Hence, it is necessary to determine the annual energy yield of the bifacial modules [9].

Compared to standard modules, the calculation of the energy of bifacial modules is more complex, introducing further dependencies, such as the module installation height, the ground reflection coefficient, the distance between module rows and between neighbouring modules of the same row, and the self-shadowing of the modules on the ground.

To solve this question, mathematical modelling and simulation were found to be a key solution; especially bifacial modules are poorly installed worldwide compared to standard solar panels in present time. However, at the time of writing this book, commercial software for the energy yield simulation of bifacial modules, that delivers reliable results, did not exist. The development of such a tool would bring on several advantages for bifacial photovoltaic, both academically and commercially. It would allow for better understanding of the performance of bifacial modules and would advance the bankability of projects utilizing bifacial PV modules.

On the other hand, using simulation it turns out that there several challenges to overcome, particularly when it comes to quantifying the rear irradiance perceived by bifacial modules, where the degree of complexity increases from standalone

[1]Adaptricity AG, Switzerland
[2]ISC Konstanz, Germany
[3]INES – Institut National de l'Energie Solaire, France

bifacial panel to adjacent modules and ultimately at field level. Moreover, existing simulation tools only partially tackle such issues, most of which are limited to modelling single module setups.

This chapter is therefore dedicated to outlining existing know how on energy yield simulations of bifacial modules and to introducing in more detail one simulation tool developed for that purpose. The models used for this tool are introduced and the accuracy of the results of the simulations are then experimentally validated.

4.2 Critical review of current status of bifacial simulations

Two existing commercial software propose bifacial simulations. First, widely used PVSyst introduced a 'complete model [ization] for shed-like PV systems pedagogical tools for a deep understanding of the issues' with the version 6.6.0, in March 2017 [10,11]. 'Other configurations (vertical, tracking, acc. to the 3D scene)' are not implemented yet. Secondly, Polysun, developed by Vela Polaris, can simulate bifacial modules since February 2016 [12]. Nevertheless, many publications were made before the release of this two software. Indeed, the lack of bifacial extensions in commercial software encourages many research organizations and companies to develop their own tool. Thus, many players in the field of bifacial have already proposed simulations of the production of a single cell or a single module. These studies on single module propose first assessments about the effect of albedo, the optimal module elevation and optimum tilt angle. We can cite for example Yusufoglu who numerically studied and modelled the rear side irradiance of a bifacial module in 2013. Johnson *et al.* presented a simulation of a vertical row of five modules in an article in 2015 [4]; most of these tools use generally the view factor approach in 2D approximation.

We summarize below different simulation approaches that have been explored so far in order to quantify the rear side irradiance perceived by bifacial modules and discuss their advantages and disadvantages.

1. View factor and configuration factors
   The view factor concept was the first approach adopted to model the irradiance seen by the front and rear side of bifacial modules [2]. Three examples for complete tools are those developed by EDF [13], ISC Konstanz [8], and CEA-INES – many more are under development at time of writing of this book. They allow simulating bifacial systems using view factors in 3D. These tools propose each different option as seasonal changes of albedo [13], the influence of white sheets on the ground [8] and tracking [8,13]. Similar to the view factor concept, the so-called configuration factors, which are used to model front irradiance, have been adopted by the National Renewable Energy Laboratory (NREL) recently to model the backside irradiance of bifacial modules, taking into account the impact of shadows on the ground reflection and of the restricted view of the sky by neighbouring modules and neighbouring module rows. Both reduce the amount of diffuse light accepted by the rear side of the bifacial modules. In the same way, as later explained in detail in this chapter for the

view factor concept, in order to calculate the ground-reflected irradiance using the configuration factor, the ground surface has to be divided in a shaded and a non-shaded area and for both areas, and separate configuration factors have to be calculated, respectively. A related study has been published by Marion *et al.* [14]. The model presented there implies a simplification that consists in the fact that the rear side irradiance model takes into account only for variations along the slant height direction of the module but not of variations in the direction of the cell and module rows. This simplification is suitable for PV arrays consisting of at least several rows where each row contains at least 12 modules, an assumption that easily applies to typical application fields of bifacial PV such as commercial or utility scale ground mounted PV systems. For suitable test set-ups, a mean bias deviation between the simulated and the measured rear side irradiance has been found. 'Suitable test set-up' means in this case a set-up that is not disturbed by factors that are not included in the input parameters of the model and that cannot be simulated by the model based on configuration factors. Examples of such factors are listed in the article and consist, e.g. in light reflection by a nearby, vertical wall as well as by concrete foundations that show an albedo that is different from the remaining ground surface and that create additional shadows. The authors of [14] show that such features can be modelled by using ray tracing for the calculation of the rear side irradiance.

- Advantages
  - The view factor concept can be implemented fast and easily for rather regular mounting geometries.
  - The view factor concept is able to reproduce the inhomogeneity of the rear side irradiance of bifacial modules.
  - The use of the configuration factor concept allows for very short computation times even for simulation of the rear side irradiance for a complete year with an hourly timestamp
- Disadvantages
  - The accuracy depends on meshing of the module rear side (i.e. on the size of the area elements of the module rear side over which the irradiance is averaged).
  - The view factor model requires a large amount of computation power, and consequently the respective simulations can be time-consuming when using standard personal computers.
  - Without simplifying assumption, the simulation time explodes when simulating large bifacial PV system
  - Using the view factor concept, it is difficult to take into account for irregular geometries of mounting structures or of non-planar ground surfaces

2. Ray tracing

   Ray tracing is another alternative for the calculation of incident irradiances on both faces of a bifacial PV module. An algorithm determines the path of many sunlight's beams, using a Monte Carlo approach. Indeed, the light beams are propagated from the sky onto the panels (forward ray tracing) or from the panels to the sky (reverse ray tracing) using optical physics. Taking into account for anisotropic diffuse irradiance with the ray tracing approach is easier than when using view factors. Since the sky is the source of this radiation, random beams' paths have to be traced from the sky, the intensity of the beam is well-described by anisotropic models according to the position of the emission. By summing the contribution of each traced ray and by supposing a certain uniformity, the incident irradiance on each side of the bifacial panels can be determined. This method seems to be more appropriate for uneven shadows but the computation time can be higher too.

Radiance software was used jointly with other open source software (SMART and PC1D) to estimate the power gain of bifacial modules in [6,15–17]. Besides, NREL, Sandia National Laboratory and the University of Iowa are developing jointly a tool under the Bifacial PV Project using ray tracing as well [18]. However, the concept of ray tracing applied to rear irradiance simulation of bifacial modules demonstrates some advantages and disadvantages which are listed below:

- Advantages
  - Ray tracing is more suitable to model the inhomogeneity of the rear side irradiance of bifacial modules.
  - The impact of structures, module frames on the rear side irradiance can be taken into account.
- Disadvantages
  - The implementation of ray tracing in a modelling tool is more complex compared to the view factor concept.
  - The computation power required by ray tracing simulations is rather high, leading to long computation times.
  - *The Radiance* software can simulate only three different wavelengths of light at a time, i.e. red, green and blue colours. Thus, using the Radiance software, the simulation has to be repeated many times in order to simulate the full sun light spectrum.

3. Empirical modelling
   Currently, there are very few attempts to predict the energy yield of bifacial plants using empirical modelling. The clear reason is the lack of experimental data to build good and reliable empirical models. In fact, with this approach, it is only possible to predict the cumulative percentage gain (over a given time period) in energy yield of bifacial modules or systems compared to monofacial ones and not the absolute energy yield of bifacial systems [19,20].
   - Advantages
     - The computation power required for such models is much lower than for the view factor concept and for ray tracing based simulations.
     - There is a good agreement between experimental and modelled results for a certain range of applications (system configurations).
   - Disadvantages
     - The application of empirical modelling is only restricted to some geographical locations and certain system configurations.

- Only an estimation of bifacial gain over a given time period can be predicted and no absolute values of power for the bifacial module or system for a given timestamp.

A comprehensive collection of publications in the field of modelling of bifacial modules and arrays as well as of open source software tools for modelling of irradiance for bifacial modules can be found in [21]. Other topics that are undergoing further investigations in order to increase the accuracy as well as the range of application of the existing simulations tools for predicting the energy yield of bifacial modules and systems are a standardized electrical characterization for bifacial modules (see also Chapter 3) and an improvement of the existing electrical models to simulate bifacial modules correctly [22], a more accurate modelling of the inhomogeneity of rear irradiance and – related to this – the modelling of the operation of bypass diodes in bifacial systems under operating conditions [15] and finally, the temperature behaviour of bifacial module which will be different than for standard monofacial solar modules even when comparing the same module bill of materials (front cover, encapsulants and rear side sheet) for both, since bifacial solar cells have an open rear side metallization while monofacial solar cells feature a fully metallized rear side [23,24]; in addition, the higher total irradiance absorbed by the bifacial cell leading to a higher current generation within the cell will lead to an increased heat production under operating conditions (i.e. when the module is operating at its maximum power point). There are many studies that aim to improve the accuracy of the modelling of the temperature behaviour of monofacial and bifacial modules under operating conditions (see e.g. [25–27]).

4.3 Bifacial gain simulation model

If used correctly, simulations can be powerful tools with limitless applications including commonly used ones, such as weather predictions. Imitating the characteristics and key functions of a system or a process, simulations can be used in various contexts, including performance optimization, safety testing, visual effects, and the functioning of natural or human systems. Simulations are often used, when the real system or process is not accessible, whether because the process is dangerous, or it is in the design phase, or it does not exist and can therefore not be experimentally tested.

As an example, in the following sections, the main characteristics of a simulation tool for the energy yield prediction of bifacial modules and systems developed by Shoukry [8,28] are presented. This simulation tool has been developed to model the behaviour of different configurations of bifacial solar systems which are not yet installed or built. The model allows for an optimization of the electrical performance of bifacial systems and the influence of the various installation parameters as well as of climatic conditions (irradiance, ambient temperature, etc.) on the energy yield can be determined. The developed optical and electrical models and the functions used in the simulation will be explained in detail in addition some results of the performed simulations will be shown.

4.3.1 Optical model

The optical model estimates the irradiation reaching the front and rear sides of bifacial PV modules. This is possible through a number of steps, which will be detailed in the following sections.

4.3.1.1 Definitions

To avoid confusion about the geometric quantities of a solar module set-up, a unified definition of such quantities will be introduced. The geographic coordinate system is defined in a way that North is set at 0°, East at 90° and South at 180°. Figure 4.1 shows a single solar module with the width w_M and the length l_M installed at a certain elevation of the lower edge of the module h_M. The tilt angle of the module is given by γ_M, whereas the orientation of the module is given by α_M. The position of the sun is described using two angles, the sun elevation angle γ_S and the sun azimuth angle α_S. The angle of incidence θ_{SM} is defined as the angle formed between the two normal vectors of the sun and the module, n_S and n_M, respectively.

Further quantities are visualized in Figure 4.2. These include the distance d_M between the modules in the same row and the distance d_R between module rows.

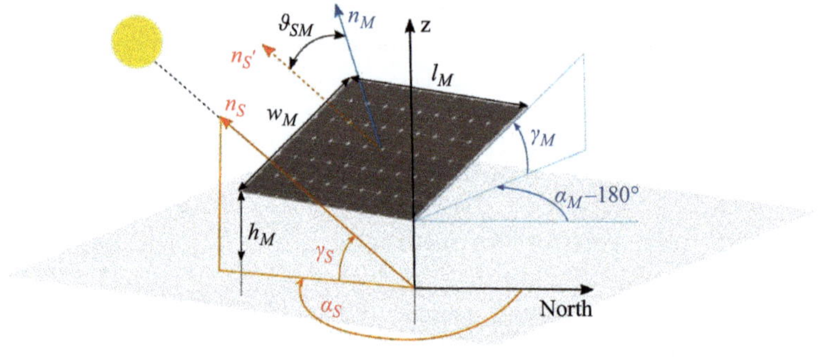

Figure 4.1 Stand-alone module set-up and definition of the module installation parameters and the position of the sun [8]

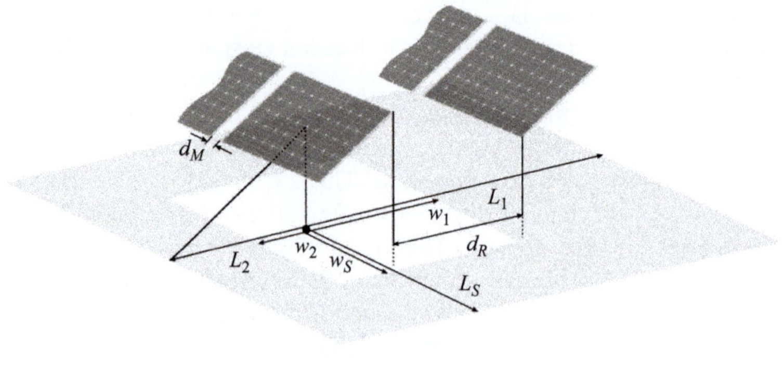

Figure 4.2 Definition of the in-field installation parameters and other input parameters of the simulation [8]

L_S, L_1, and L_2 describe the size of the surface reflecting solar irradiation onto the module rear side. In the case of using white reflective plates or sheets beneath the modules to increase the albedo coefficient, the size of the sheet can be given by w_S, w_1, and w_2.

4.3.1.2 Sun's position

The sun's position is dependent on the date, time and location and can be described using the elevation and azimuth angles, α_S and γ_S. The two angles are calculated according to the DIN 5034 algorithm. As an example, simulation results for the locations El Gouna, Egypt, and Constance, Germany are depicted for an entire year in Figure 4.3(a) and (b).

4.3.1.3 Irradiances

Direct irradiance

Using the position of the Sun and data from measurements of the global, direct (beam) and diffuse horizontal irradiances, *GHI*, *BHI* and *DHI*, the total irradiances on the module front and rear side can be estimated. In the case of the direct irradiance on the front surface of a module $I_{dir,f}$ this is a straightforward geometrical relationship [29–33]. The normal vector of the module front side $n_{M,f}$ is used in the

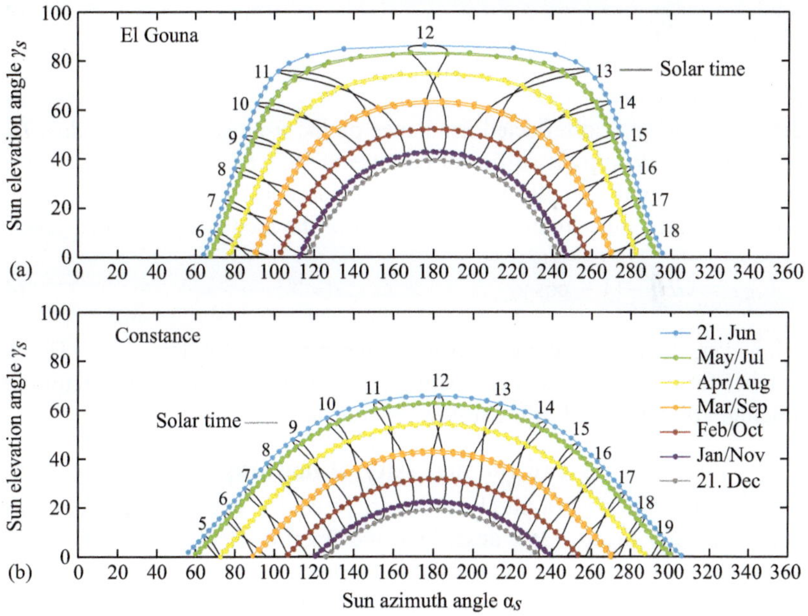

Figure 4.3 The position of the Sun given by the azimuth and elevation angles for (a) El Gouna and (b) Constance. Notice that Constance has longer summer days, shorter winter days and overall lower γ_S than El Gouna [8]

methods described in the aforementioned sources. To calculate the direct irradiance reaching the rear side of the module $I_{dir,r}$, the same equation is used, albeit with one difference, namely that the normal vector of the module is reversed, so that it is facing backwards. This can be mathematically described by inverting the sign of the normal vector using

$$n_{M,r} = -n_{M,f}. \tag{4.1}$$

Diffuse irradiance

Several approaches exist for the calculation of the diffuse irradiance on a tilted surface I_{diff}, which can be categorized under isotropic and anisotropic approaches. A thorough comparison of the different models is given by Noorian *et al.* [31]. The simpler isotropic model assumes a uniform intensity of the diffuse irradiance over the sky hemisphere. Hence, I_{diff} on a surface depends on the fraction of the sky hemisphere it can see [32], which is only suitable for rough estimations or for very overcast skies [29, p. 62]. The more complex anisotropic models describe the sky diffuse radiance most accurately, with the Perez model [34] yielding the most accurate results. Adapting the model to the diffuse irradiance on the rear side of a tilted surface $I_{diff,r}$ only requires changing the module installation parameters α_M and γ_M accordingly

$$\begin{aligned} \alpha_{M,r} &= 180^\circ + \alpha_{M,f} \\ \gamma_{M,r} &= 180^\circ - \gamma_{M,f} \end{aligned} \tag{4.2}$$

Reflected irradiance

To determine the ground-reflected irradiance on the front side of the module $I_{refl,f}$, an assumption of isotropy is sufficient because the few existing anisotropic effects would introduce great complications to the calculation that are not justified, since they do not significantly improve the accuracy of the model. $I_{refl,f}$ is then given by

$$I_{refl,f} = GHI\frac{\alpha}{2}(1 - \cos\gamma_M). \tag{4.3}$$

However, this estimation delivers inaccurate results for $I_{refl,r}$. The two approaches for more accurately estimating $I_{refl,f}$, either using the view factor or via ray tracing, are introduced in the following subchapters.

4.3.1.4 View factors and the necessity of meshing

The view factor $F_{A_1 \to A_2}$ is a purely geometric quantity and the concept is known from heat transfer theory [35], describing the fraction of the radiation leaving a random surface A_1 that strikes the surface A_2 directly [36]. It can be computed as the integral of the portions of radiation leaving the differential areas dA_1 that reach the differential areas dA_2,

$$F_{A_1 \to A_2} = \frac{1}{A_1} \iint\limits_{A_1, A_2} \frac{\cos\theta_1 \cos\theta_2}{\pi r^2}\, dA_1\, dA_2 \tag{4.4}$$

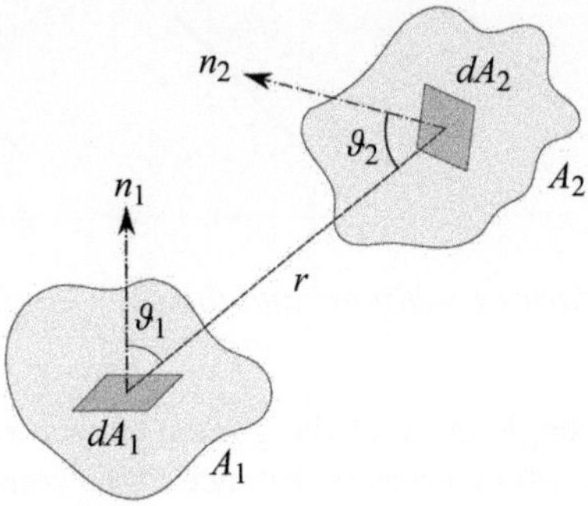

Figure 4.4 Definition of the view factor between two surfaces

where r is the distance between the differential areas dA_1 and dA_2. The angles between the normals of the surfaces and the line that connect dA_1s and dA_2 are θ_1 and θ_2, respectively, and are depicted in Figure 4.4.

Assuming a mean ground albedo α, an ideal Lambertian character[1] of the ground, and given horizontal irradiances *GHI* and *DHI*, the view factor approach can be used to calculate the incident irradiances on both faces.

Two main properties about view factors are needed. The first one is the summation rule which indicates that the sum of all view factors from a given surface is one. It physically means that the irradiation emitted by a surface necessarily hit another surface. It is written as follows:

$$\sum_j VF_{ij} = 1 \tag{4.5}$$

The other main property of the view factors is the reciprocity theorem. It gives the relation between the view factors between two surfaces, according to which surface is considered as emitting and which one is considered as receiving. The link between these two quantities is the areas A_i and A_j of both surfaces:

$$A_i VF_{ij} = A_j VF_{ji} \tag{4.6}$$

To correctly evaluate the incident irradiance on both faces, and mainly on rear face, the view factors have to be accurately calculated. For this, meshing both the ground and the modules is necessary. Indeed, to correctly take into account the shadow during the day and potential variations of albedo (like in the case of a white sheet). The modules have to be meshed due to the inhomogeneity of incident irradiance on the rear side of the bifacial module.

The calculation of the view factor is based on a geometrical assumption: the plant is made of a unique block of modules repeated in columns and rows. One only block is considered during the evaluation of the view factors. These view factors

[1]Lambertian reflection describes ideal diffuse reflecting surfaces, which will reflect light evenly in all directions, making the surface appear equally bright regardless of the viewing angle.

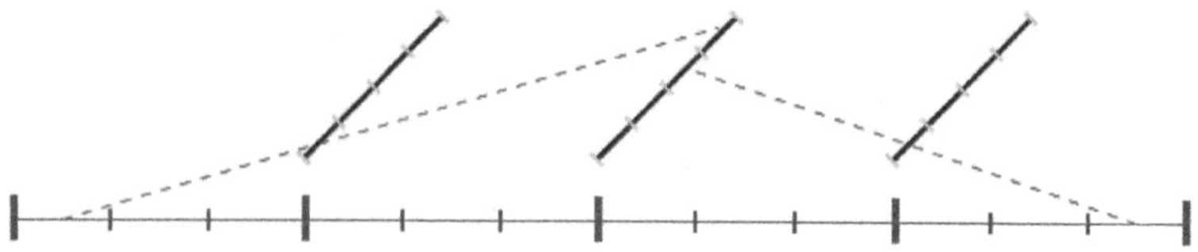

Figure 4.5 Boundaries conditions and the calculation of view factors

are evaluated from each single mesh of the ground to each mesh of the modules of the one considered block within a certain distance of the considered ground's mesh. Because these view factors depend only of the geometrical configuration of the plant, these calculations need to be done only once for each plant. This way the calculation time is reduced. The results can then be widespread to the whole plant, with a particular caution given to the boundaries conditions of the field. For instance, as seen in Figure 4.5, rows can block the ground-reflected irradiance from reaching the rear and front faces of the modules. Thanks to simple geometric relations as affine projection, the blocking effect of the other rows can be easily taken into account.

When the sun moves across the sky during the day, at each instant, the shadow on the ground and on the modules is easily and quickly obtained with the affine projection of the modules. It only remains to intercept the meshes with theses shadows and determine which single mesh belongs to a shadow and which does not. While shadowed meshes are multiplied by only the DHI because the direct portion of the solar irradiance is blocked by the shadow casted by the module on the ground, meshes outside the shadow have to be multiplied by GHI.

If the plant presents some periodicity, simulation time could be saved by taking advantage of this periodicity. Indeed, if the plant is the repetition of a same group of modules (called block) as defined in Figure 4.5 then calculation time can be saved.

4.3.1.5 Calculation of the view factor

Different view factors are needed:

- View factors from the sky to the ground
- View factors from the sky to the front and rear sides of the modules
- View factors from the ground to the front and rear sides of the modules

Some catalogues collecting computed view factors for various configurations exist but there is not a general and algebraic formulation for view factors between two differently sized plane rectangular surfaces with parallel boundaries and arbitrary position, meaning they are not necessarily parallel or perpendicular to each other and are randomly inclined. Therefore, a numerical solution has to be implemented using, for instance, the contour double integral formula. Only the vertices' coordinates of the two considered areas are needed.

Depending of the diffuse irradiance's model, this approach has to be realized in different ways. If the sky is described as anisotropic, as for example in the Perez's model, then since each part of the sky does not emit the same amount of irradiance,

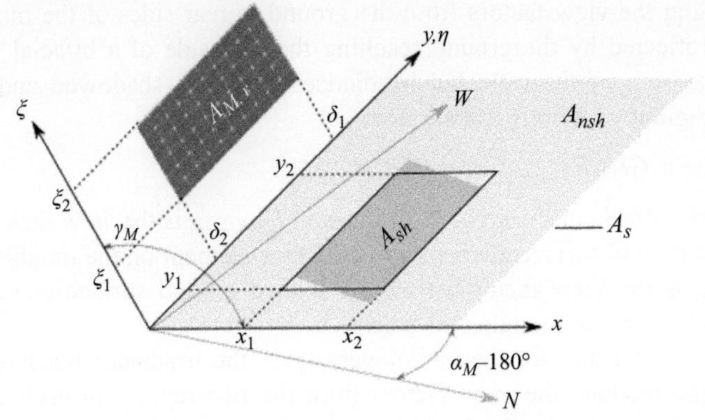

Figure 4.6 View factors from the ground to the front and rear sides

it has to be meshed too. If an isotropic sky is chosen, the intensity is uniform and the sky can be considered as a whole. This last approach in adopted for the presentation of the view factors' calculation for the sake of clarity.

First, the view factors from the ground to the front and rear sides are calculated. The meshes of the ground and the panels are well defined. Then, it just consists in applying the numerical solutions between each mesh of the ground with each mesh of the panels, as indicated in Figure 4.6. The calculation of the view factors from the sky to the ground or to the modules is more complicated since the sky is not a well-defined surface and does not have really a quantified surface. The view factors classical reciprocity theorem states that a view factor from A to B is equal to the view factor from B to A multiplied by a factor based on the areas of the two surface. Here, the sky has not a well-defined area. It is then supposed that the view factor from the sky to the ground is strictly equal to the view factor from the ground to the sky. Applying this assumption, it is still necessary to calculate the view factors from the ground to the sky. We know that all the incident beams on the ground can come either from the panels or from the sky. Applying the summation property to any mesh of the ground i, the sum of all the view factors from this mesh of the ground to all the panels plus the view factor from this mesh to the sky equals 1. We have then:

$$VF_{sky\to i} = VF_{i\to sky} = 1 - \sum_{j\in modules} VF_{i\to j} \tag{4.7}$$

The same method can be applied to the view factors from the sky to the modules except that this time the view factors from the modules to the ground are not known. Thanks to the reciprocity theorem, these view factors can be calculated and the view factor from the sky to the mesh j of the modules is given by:

$$VF_{sky\to j} = VF_{j\to sky} = 1 - \sum_{i\in ground} \frac{A_i}{A_j} VF_{i\to j} \tag{4.8}$$

Regarding the view factors from the ground to rear sides of the modules, the irradiance reflected by the ground reaching the rear side of a bifacial module is given as the sum of the reflected irradiances from the shadowed and the non-shadowed regions of the ground surface:

$$I_{refl,r} = \alpha \; GHI \; F_{A_{nsh} \rightarrow A_M} + \alpha \; DHI \; F_{A_{sh} \rightarrow A_M}. \tag{4.9}$$

where α is the albedo of the ground surface and $F_{nsh \rightarrow Am}$ is the view factor between a non-shade ground surface element and a surface element on the module rear side and $F_{sh \rightarrow Am}$ is the view factor between a shaded ground surface element and a surface element on the module rear side.

In order to account for the inhomogeneity of the irradiance reaching the rear surface of the module, the view factors from the two regions to each cell of the module are calculated individually. This process is repeated for every time step of the entire simulated time period, allowing for a calculation of the spatial and temporal distribution of the irradiance reflected onto the rear side of the bifacial module.

4.3.1.6 Ray tracing

Ray tracing is another alternative for the calculation of incident irradiances on both faces. An algorithm determines the path of many sunlight's beams, using Monte Carlo approach. Indeed, they are propagated from the sky into the panels (forward ray tracing) or from the panels to the sky (reverse ray tracing) using optical physics. Taking into account anisotropic diffuse irradiance with ray tracing approach is easier than is view factors one. Since the sky is the source of this radiation, random beams' paths have to be traced from the sky, the intensity of the beam is well-described by anisotropic models according the position of emission. By summing the contribution of each traced ray and by supposing a certain uniformity, the incident irradiance on each face of the panels can be determined. This method seems to be more appropriate for uneven shadows but the computation time can be higher too. As cited before, Radiance software is already used in the literature for implementing this method. Another software is COMSOL that includes also a package, Ray Optics Module, allowing ray tracing approach.

4.3.2 Electrical model

Using the total irradiance reaching the module's front and rear sides estimated by the optical model (either based on ray tracing or on view factor), an electrical model estimates the electrical output parameters of the bifacial module. The performance of standard and bifacial modules can be quantitatively compared using the bifacial gain which is calculated using the annual energy yields of the standard and bifacial modules being compared.

4.3.2.1 Module power

The *I–V* curve of bifacial modules taking into account irradiances on both faces and temperature has to be drawn. For this purpose, an electrical model has to be proposed. The monofacial one diode model already gives good results simulating monofacial cells' behaviour. Thus, the photo-currents of both faces are summed: the electrical

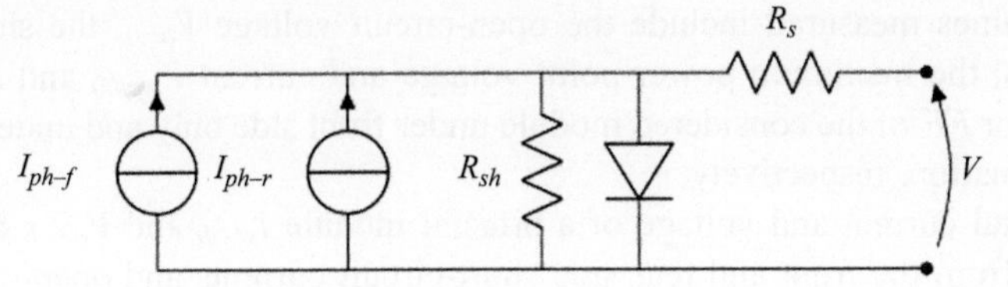

Figure 4.7 Electrical diagram of the bifacial one diode model

model of a bifacial cell used in [8] is the same as the one of a monofacial cell with two current sources, one for each face. The proposed electrical model is defined as shown in Figure 4.7.

As the well-known one diode monofacial model, five parameters have to be calculated to determine the bifacial *I–V* curve:

- The equivalent series resistance, R_s
- The parallel resistance, R_{sh}
- The ideality factor, n
- The reverse saturation current, I_0
- The photocurrent, I_{ph}

The approach is based on the conversion of monofacial indoor measurements of both faces independently to bifacial real conditions. Two flash tests are made, one for each face, usually under standard testing conditions (STCs): the irradiance at $G_0 = 1{,}000\ \mathrm{W/m^2}$, the temperature has to be at 25°C, and an AM 1.5 spectrum is used as the illumination source.

The three specific points of the bifacial *I–V* curve calculated are $(0; I_{sc-b})$, $(V_{mpp-b}; I_{mpp-b})$ and $(V_{oc-b}; 0)$. Then, the total *I–V* curve is evaluated thanks to an optimization of three of the five *I–V* parameters.

First, V_{oc-b}, I_{sc-b}, V_{mpp-b} and I_{mpp-b} are obtained thanks to a model strongly inspired from Singh *et al.* [21]. Slightly complements have been made to take into account the homogenous irradiance on the front face and the impact of the temperature's cell. The index 0 is for flash test data, f for front, r for rear, b is for bifacial, and x can refer to f or r. The first step is to consider the impact of the temperature with the temperature coefficients given by the module manufacturer:

$$\begin{aligned}
I_{sc-x} &= I_{sc-x0}\left(1 + \alpha_{I_{sc}}(\vartheta_M - 25°C)\right)\\
V_{oc-x} &= I_{sc-x0}\left(1 + \alpha_{V_{oc}}(\vartheta_M - 25°C)\right)\\
FF_x &= FF_{x0}\frac{1 + \alpha_{P_{mpp}}(\vartheta_M - 25°C)}{\left(1 + \alpha_{I_{sc}}(\vartheta_M - 25°C)\right)\left(1 + \alpha_{V_{oc}}(\vartheta_M - 25°C)\right)}
\end{aligned} \tag{4.10}$$

where $\alpha_{P_{mpp}}$, $\alpha_{I_{sc}}$, and $\alpha_{P_{mpp}}$ are the temperature coefficients of the module, respectively, for open-circuit voltage, short-circuit current and maximum power. ϑ_M, the module temperature, can be calculated using the nominal cell temperature approach described in [37].

The values measured include the open-circuit voltage $V_{oc,0}$, the short-circuit current $I_{sc,0}$, the maximum power point voltage and current $V_{mpp,0}$ and $I_{mpp,0}$, and the fill factor FF of the considered module under front side only and under rear side only illumination, respectively.

The total current and voltage of a bifacial module I_{sc-b} and V_{oc-b} have to be calculated from the front and rear side short-circuit currents and open-circuit voltages $I_{sc-f/r}$ and $V_{oc-f/r}$. With the assumption of a linear current response under different light intensities, the resultant module current can be calculated as a simple sum of the currents generated at the front and rear sides using:

$$I_{sc-b} = \frac{G_f}{G_{f0}} I_{sc-f} + \frac{G_r}{G_{r0}} I_{sc-r} \tag{4.11}$$

The gain in short-circuit current relative to monofacial front side only illumination is defined as follows:

$$R_{I_{sc}} = \frac{I_{sc-b}}{I_{sc-f}} = \frac{G_f}{G_{f0}} + \frac{G_r}{G_{r0}} \frac{I_{sc-r}}{I_{sc-f}} \tag{4.12}$$

Reasoning on open-circuit conditions, the bifacial open-circuit voltage V_{oc-b} can be deduced:

$$V_{oc-b} = V_{oc-f} + \frac{\left(V_{oc-r} - V_{oc-f}\right) \cdot \ln(R_{I_{sc}})}{\ln\left(\frac{I_{sc-r}}{I_{sc-f}}\right)} \tag{4.13}$$

Then, the bifacial fill factor under these specific irradiances can be evaluated thanks to this formula:

$$FF_b = pFF - R_{I_{sc}} \left(\frac{V_{oc-f0}}{V_{oc-b}}\right) (pFF - FF_{f0}) \tag{4.14}$$

where pFF is the pseudo fill factor and is obtained by considering no series resistance effect and given by:

$$pFF = \left(\frac{I_{sc-r0}}{I_{sc-f0}} FF_{f0} - FF_{r0} \frac{V_{oc-r0}}{V_{oc-f0}}\right) \bigg/ \left(\frac{I_{sc-r0}}{I_{sc-f0}} - \frac{V_{oc-r0}}{V_{oc-f0}}\right) \tag{4.15}$$

Finally, the output power of a bifacial module can be determined using:

$$P_{mpp} = FF_b \cdot V_{oc-b} \cdot I_{sc-b} \tag{4.16}$$

Current and voltage still need to be determined at maximum power point. Again, the assumption of a linear current response under different irradiance is made and the maximum power point current is deduced from the front and rear measures at 25°C:

$$I_{mpp-b} = \frac{G_f}{G_{f0}} I_{mpp-f0} + \frac{G_r}{G_{r0}} I_{mpp-r0} \tag{4.17}$$

Then, the maximum power point voltage is given by:

$$V_{mpp-b} = \frac{P_{mmp-b}}{I_{mpp-b}} \tag{4.18}$$

The influence in temperature is already taken into account in P_{mpp-b}. In this approach, the temperature only influences the voltage, which is acceptable in a first approach (since voltage is much more dependant of the temperature than the current).

With these parameters, one still needs to draw the *I–V* curve. Photo current is assumed to be equal to short-circuit current.

$$I_{ph-b} = I_{sc-b} \tag{4.19}$$

Furthermore, when each face was flashed separately, it was noticed that the reverse saturation current of both faces were very close. This result could have been predicted since the bifacial cell is made of one only p–n junction. Then, the bifacial reverse saturation current is supposed to be equal to the front face reverse saturation current and can be evaluated by taking $I = 0$ in the *I–V* equation of the front face with the one diode model. This gives (R_{sh} is estimated by the slope of the front *I–V* measurement):

$$I_{0-b} = I_{0-f} = \frac{I_{sc-f} - \left(V_{oc-f}/R_{sh}\right)}{\exp\left(V_{oc-f}/n\, k_b T\right) - 1} \tag{4.20}$$

The simulation model then has to make an optimization of three remaining parameters: the series resistance R_s, the shunt resistance R_{sh} and the ideality factor n. These three parameters have to be optimized to obtain an *I–V* curve which respects the previously calculated bifacial open-circuit voltage V_{oc-b} and fill factor FF_b. For instance, the optimization can be based on a monofacial *I–V* curve drawing program using Newton–Raphson method. This program can converge to different triplets according to the initial values that still give similar *I–V* curves. This approach guaranties a realistic *I–V* curve but does not imply that the five parameters will have each one a realistic value.

Since flash tests are never perfect, it has been observed that the obtained *I–V* curves can be slightly different according to the two measures used initially. As the curve is obtained from an optimization, if the calculation time is not critical, it can be a good idea to draw optimized *I–V* curves from different measurements and select the best one.

Lastly, this model gives satisfying *I–V* curves for the bifacial modules. The electrical power can be easily deduced with the *I–V* curve and the properties of the MPPT. Thus, this model permits to evaluate the output power of a module from the incident irradiances. Applied to each module of the studied field, it allows to determine the output power of the complete bifacial PV array. More accurate results can be achieved using for instance the two-diode model, but the model described by Singh *et al.* [21] 'gives already a good approximation to the expected efficiency under bifacial operations' [38].

4.3.2.2 Annual energy yield

Comparing the electrical performance of standard and bifacial modules using the front side peak power – measured indoor under STC – is not representative, since bifacial modules have a higher energy output than standard modules with the same front side peak power. Meanwhile, a new measurement standard for indoor measurements of bifacial modules has been prepared (see Chapter 3) that takes into account also for the power output generated by light incident on the rear side of the bifacial modules. A part from this, a shift from the conventional use of the peak power in W_p for the comparison of different module technologies, to the more relevant and better suited use of the annual energy yield Y in $\mathrm{kWh/kW_p}$, is imperative for a proper assessment of the performance of bifacial modules. Giving the produced energy relative to the installed peak module power not only allows for the comparison of bifacial and standard monofacial modules but also of different standard modules with varying peak powers. The annual energy yield of standard modules γ_M and of bifacial modules Y_b is given by

$$Y_{m/b} = \sum_{i=0}^{n} \frac{P_{mpp,m/b,i}}{P_{mpp,f,0}} \Delta t, \tag{4.21}$$

where the produced energy is given in reference to the front side module power $P_{mpp,f,0}$ measured at STC.

4.3.2.3 Bifacial gain

The comparison between the performances of standard and bifacial module technologies is reduced to one value, which quantifies the annual energy yield increase (or decrease) in percent based on Y_m. The so-called bifacial gain (compare also Chapter 5) given in % is thereby defined as

$$BG = 100\,\frac{Y_b - Y_m}{Y_m}. \tag{4.22}$$

4.4 Simulation results

To better understand the behaviour of bifacial modules in different configurations and conditions, simulations are carried out for varying installation parameters, including stand-alone and in-field installations. In addition to the resulting energy yield of a bifacial module in different conditions, the bifacial gain and the amount of solar irradiance reaching each cell on the rear side of the bifacial module are also considered. All the following simulations are carried out using the meteorological data from the complete year 2005 acquired at installation sites in El Gouna, Egypt, and Constance, Germany. The data are retrieved from the SoDa database [39]. This allows comparison of the performance of bifacial modules at different latitudes and weather conditions.

4.4.1 South-facing stand-alone bifacial module

Whereas the annual energy yield of a stand-alone monofacial module depends mainly on a few installation parameters, the annual energy yield of a bifacial module Y_b and the resulting bifacial gain *BG* are influenced by various other factors, including the module height h_M and the diffuse irradiance factor f_D as well as on the ground albedo. The influence of the various installation and site parameters and weather conditions on Y_b, *BG* and the rear side irradiance of the bifacial module $I_{tot,r}$ are examined separately in the following sections.

4.4.1.1 Module tilt angle and installation height

The power output of bifacial modules is more sensitive to suboptimal tilt angles and installation heights than that of standard modules. Therefore, Y_b was calculated for varying tilt angles and for varying module heights, searching for the configuration that maximizes Y_b. The resulting optimum installation parameters are given in Table 4.1 for both El Gouna and Constance. Comparable to standard modules, the optimum tilt angle of a bifacial module is larger in Constance, due to the higher latitude than El Gouna. The lower solar elevation angle at higher latitudes require PV modules to be installed at a greater γ_M, in order to receive more incident radiation.

Determining the optimal module elevation h_M is also a process of finding the compromise, at which the module is far enough from the shadow it casts, but not too far from the ground reflecting the irradiance. Due to the higher wind loads, higher module mounting heights are also costlier and mechanically more challenging, sometimes forcing project developers to opt for lower sub-optimal heights with smaller installation costs. The yield increase from the additional rear side irradiance rises significantly for higher α, due to the higher reflectance of the ground, regardless of the location of the installation. The performance of the bifacial module is maximized at a height of $h_M = 1.5$ m for all configurations of stand-alone bifacial modules.

According to the calculations (see results in Table 4.1), bifacial modules mounted in El Gouna at $h_M = 1.5$ m and the corresponding tilt angle 25°, would have a 13.46% and 33.85% higher energy yield than a comparable standard module, for $\alpha = 0.2$ and $\alpha = 0.5$, respectively. In Constance (Germany), a bifacial module would be optimally mounted at $h_M = 1.5$ m and $\gamma_M = 37°$, and would have a bifacial gain of 15.98% and 35.73% for $\alpha = 0.2$ and $\alpha = 0.5$, respectively, which is a slightly higher bifacial gain than in El Gouna. This is due to the higher portion of diffuse irradiation in Constance caused by more frequently cloudy sky conditions.

Table 4.1 Simulated bifacial gain of modules installed at various locations for albedo coefficients of 0.2 and 0.5. BF increases for larger albedo and is higher in Constance, due to higher amount of diffuse irradiance [8]

	El Gouna		**Constance**	
	$\alpha = 0.2$	$\alpha = 0.5$	$\alpha = 0.2$	$\alpha = 0.5$
Optimum module height $h_{M,opt}$ [m]	1.5	1.5	1.5	1.5
Optimum tilt angle $\gamma_{M,opt}$ [–]	25°	25°	37°	37°
Bifacial gain *BF*	13.46%	33.85%	15.98%	35.73%

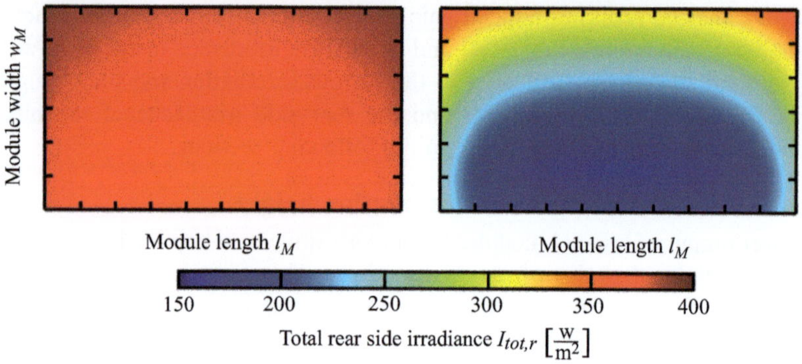

(a) El Gouna, 21.06.2005, noon, $\alpha = 0.5$, $h_M = 1$ m

(b) El Gouna, 21.06.2005, noon, $\alpha = 0.5$, $h_M = 10$ cm

Figure 4.8 Total irradiance on module rear side $I_{tot,r}$ for elevations (a) 1 m and (b) 10 cm. Lower overall $I_{tot,r}$ and stronger deviation at 10 cm, due to close proximity to shadow [8]

Figure 4.8(a) and (b) depicts the influence of the module installation height on the amount of irradiance reaching the rear side of a bifacial module for $h_M = 1$ m and $h_M = 10$ cm, respectively. Not only does the module mounted at a height of $h_M = 10$ cm receive overall less irradiance on its rear side, the rear side irradiance is also more inhomogeneous, due to the proximity of the module to the shadow on the ground.

4.4.1.2 Diffuse irradiance factor

The portion of diffuse irradiance DHI to the total GHI, given by the diffuse irradiance factor f_D, also affects the energy yield of standard monofacial modules. The influence, however, is more pronounced for bifacial modules. A PV module casts a shadow underneath itself by blocking off the direct portion of the irradiance BHI. Therefore, only DHI is available for reflection in the shadow region, and $I_{tot,r}$ is consequently decreased. Increasing f_D would consequently increase the amount of irradiation available for reflection in the shadow region, since there would be less available BHI to be blocked. The shadow intensity if hence decreased for higher f_D. This increase in f_D would also cause an increase in the bifacial gain of the module, since it would receive more rear side irradiation. This is shown in Figure 4.9(a) and (b) for El Gouna and Constance, respectively.

4.4.1.3 Date and time

Solar generated energy varies diurnally and seasonally, fluctuating strongly with the season and time dependent solar irradiation. The behaviour of equator oriented bifacial modules with optimum tilt is not any different in that aspect from the behaviour of monofacial modules with the same mounting configuration. Figure 4.10(a) and (b) shows again that the rear side irradiance strongly increases

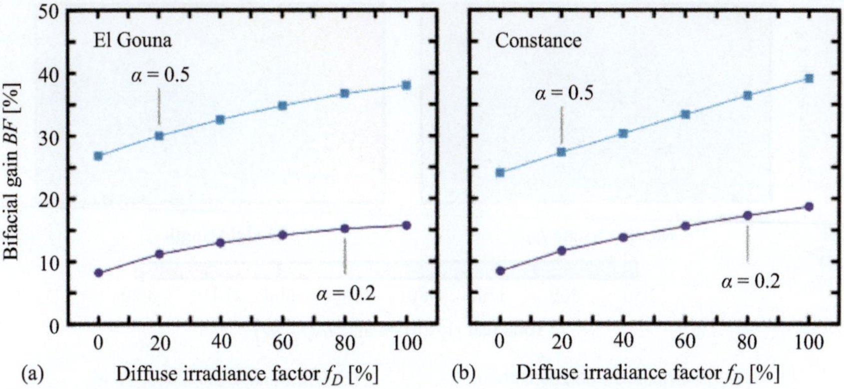

Figure 4.9 *The incident BHI is blocked by the module, reducing the solar irradiation available for reflection in the shadow region for (a) El Gouna and (b) Constance. Increasing f_D consequently causes reduction of shadow's influence and increases $I_{tot,r}$ and BF [8]*

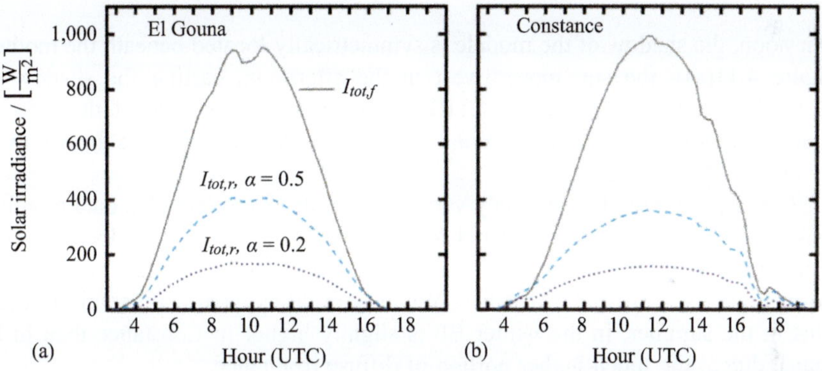

Figure 4.10 *Notice the time dependent irradiance $I_{tot,r}$ follows the same trend as $I_{tot,f}$. Whereas the peak of the irradiance is at solar noon, the amount of $I_{tot,r}$ relative to $I_{tot,f}$ is higher in the morning and evening hours, in (a) El Gouna and (b) Constance [8]*

for higher albedos. $I_{tot,r}$ additionally depends on the time of day, its curve following the same trend as $I_{tot,f}$, with its peak at solar noon of a clear day. However, where the total irradiance on the front and rear is highest at noon, the difference between $I_{tot,f}$ and $I_{tot,r}$ is also maximized at noon. The contribution of the rear side is therefore stronger in the morning and in the evening, than at noon.

The bifacial gain is further influenced by the position of the module shadow on the ground, whereby the further away the shadow is, the less it affects the module rear side irradiance. The movement of the shadow on the ground over the time of day is visible in the rear irradiances as shown in Figure 4.11(a) and (b). While at

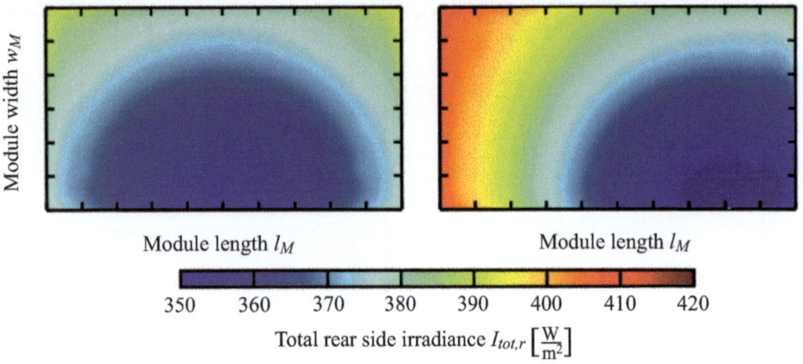

(a) El Gouna, 21.06.2005, noon, $\alpha = 0.5$, $h_M = 1$ m

(b) El Gouna, 21.06.2005, 14:00, $\alpha = 0.5$, $h_M = 1$ m

Figure 4.11 Reduction of $I_{tot,r}$ in the (b) afternoon no longer symmetrical like at (a) noon, but more to the east, closer to the module shadow. $I_{tot,r}$ increases on the module west side, due to larger distance to shadow [8]

solar noon, the shadow of the module is symmetrically located beneath the module (Figure 4.11(a)), the sun moves west in the afternoon, casting the shadow of the module further east (Figure 4.11(b)). The western rear side of the bifacial module therefore receives more ground-reflected irradiance, since it is further away from the module shadow.

The bifacial gain does not only vary on an hourly basis, but also on a monthly basis. The bifacial gain is simulated for each month separately at albedo coefficients of 0.2 and 0.5, and the results are visualized in Figure 4.12(a) for El Gouna and in Figure 4.12(b) for Constance. While both cities have comparable bifacial gains in the summer, in the winter BF is slightly higher in Constance than in El Gouna, due to the much higher portion of diffuse irradiance.

4.4.2 East-west-facing stand-alone vertical bifacial module

One of the drawbacks of solar energy is the production of most of the energy at noon, causing an hourly mismatch of demand and production, at least in residential applications. One approach to spreading the electricity production more evenly over the day without the use of yet expensive batteries, suggests the use of vertically mounted bifacial modules, as shown in Figure 4.13.

Vertically mounted PV modules are particularly interesting in combination with bifaciality. With one side of the vertically mounted bifacial modules facing East, and the other West, a vertically mounted bifacial modules produces more energy in the morning and evening, than a south-facing module, with a drop in production at noon, when the sun is shining on the edge of the module. The reflected irradiance of both the east- and west-facing sides of the module are calculated using the view factor, since the isotropic model would deliver inaccurate results.

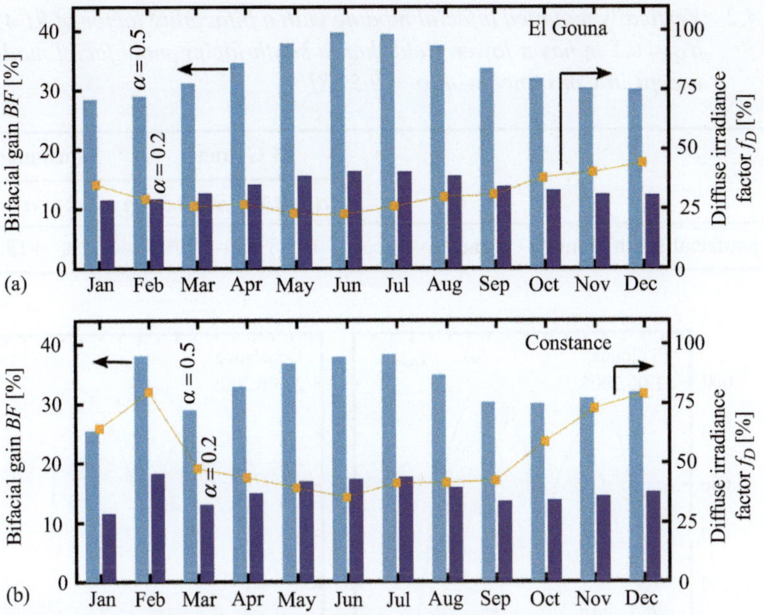

Figure 4.12 Monthly bifacial gain in (a) El Gouna and (b) Constance reaches maximum of circa 40% in the summer. BF in the winter is higher in Constance, due to higher f_D [8]

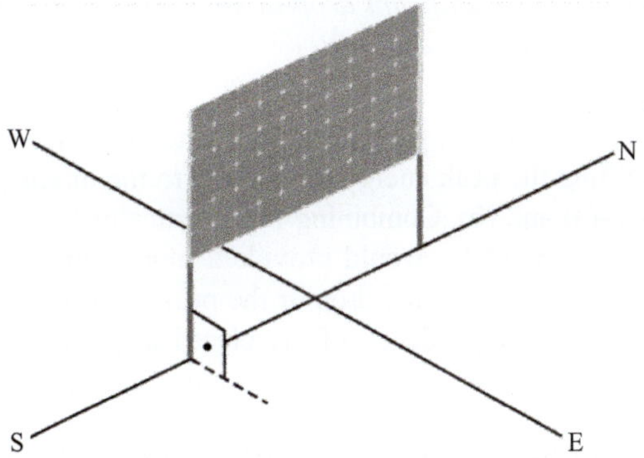

Figure 4.13 Schematic depiction of a vertically mounted, east-west-facing bifacial module [8]

Table 4.2 presents that a vertically mounted bifacial modules in El Gouna have a lower Y than a standard south-facing module, regardless of the albedo, with a loss of −14.88% and −5.99% for $\alpha = 0.2$ and $\alpha = 0.5$, respectively. Whereas the loss in the annual energy yield in Constance for an albedo of 0.2 is −4.52%, a vertically mounted bifacial modules located there would have a 15.77% higher Y for $\alpha = 0.5$.

Table 4.2 Vertically mounted bifacial module with a bifaciality factor of 91.4% at $h_M = 0.5$ m has a lower yield than a south-facing monofacial module, except in Constance with $\alpha = 0.5$ [8]

	El Gouna		Constance	
	$\alpha = 0.2$	**$\alpha = 0.5$**	**$\alpha = 0.2$**	**$\alpha = 0.5$**
BF Monofacial south-facing → Bifacial vertical	−14.88%	−5.99%	−4.52%	+15.77%

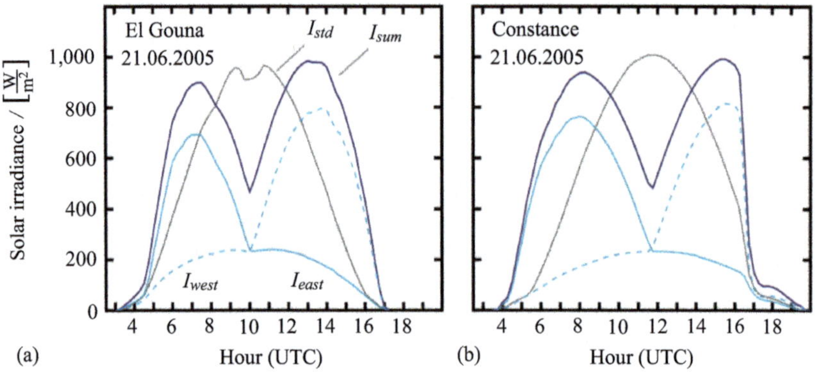

Figure 4.14 Notice the two peaks of the total irradiance I_{sum} reaching a vertical bifacial module, due to the east- and west-facing sides. I_{sum} experiences a significant drop at noon, caused by the Sun shining on the module side edge [8]

Even in case of a negative bifacial gain, a vertically mounted bifacial modules offers the advantage of shifting the peak energy production to the morning and evening as visible in Figure 4.14(a) and (b). Combining both vertically mounted bifacial modules and optimally tilted modules would provide a more homogeneous production curve over the whole day, without the drop or the peak at noon. See also Chapter 8 for a detailed study of the dependency of the electrical performance of vertically mounted bifacial modules on the geographic location of the related PV system.

4.4.3 Stand-alone bifacial module with horizontal single-axis tracking

One-axis tracking systems have the potential to contribute to a further reduction of the LCOE of solar energy. In the following, the influence of a simple, cost effective, horizontal axis tracking system (a so-called 'sun-belt tracking system') on the annual energy yield of a bifacial PV module is examined. In the sunbelt-tracking system, a module is rotated around a horizontally fixed north-south–oriented axis and tilted towards the east in the morning, horizontally upwards at noon, and towards the west in the evening, which is schematically shown in Figure 4.15.

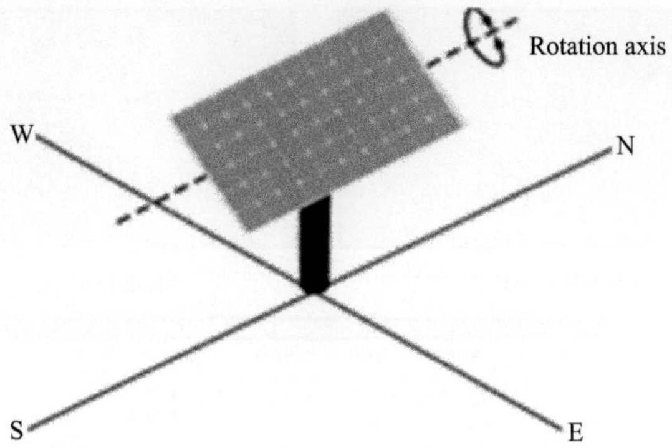

Figure 4.15 PV module mounted on a horizontal rotation axis, enabling using an east-west tracking of the Sun. Module is horizontal and facing upwards at noon [8]

Table 4.3 Adding tracking to a monofacial module increases yield by up to 18% (Nr. 1), while bifaciality increases it by up to 44% (Nr. 3). Combining bifaciality and tracking results in a gain of up to 62% (Nr. 5) [8]

	Nr.	**A**	→	**B**	**Kasese, Uganda**	
					$a=0.2$	$a=0.5$
	1	Monofacial fixed	→	Monofacial tracked	14.71%	17.93%
	2	Bifacial fixed	→	Bifacial tracked	12.82%	20.30%
$BF_{A\rightarrow B}$	3	Monofacial fixed	→	Bifacial fixed	16.47%	43.77%
	4	Monofacial tracked	→	Bifacial tracked	22.12%	37.53%
	5	Monofacial fixed	→	Bifacial tracked	40.10%	62.20%

Since this kind of tracking is optimal for regions near the Equator, here the simulation is carried out for Kasese, Uganda, quantitatively comparing the benefits of bifaciality and tracking in Table 4.3. Nevertheless the use of such systems can be beneficial also within a certain range of latitudes not closely located to the equator (see also Chapter 5).

When considering the simulation results summarized in Table 4.3, it has to be kept in mind that these apply to stand-alone single-axis tracking systems consisting of one-single module. According to the results presented in Table 4.3, while adding tracking to a monofacial module increases Y_m by up to 18% (Nr. 1), using a fixed bifacial module increases the energy yield by a significantly larger amount than tracking of up to 44% (Nr. 3). Additionally, while adding tracking to a bifacial module only increases Y_b by 20% (Nr. 2), using a bifacial instead of a monofacial module in a tracked installation increases the energy production by 38% (Nr. 4). As expected, the highest gain of up to 62%, compared to a fixed monofacial module (Nr. 5), is achieved by combining tracking and bifaciality. Using a cost-effective tracking solution like the horizontal single-axis tracker combined with a high gain

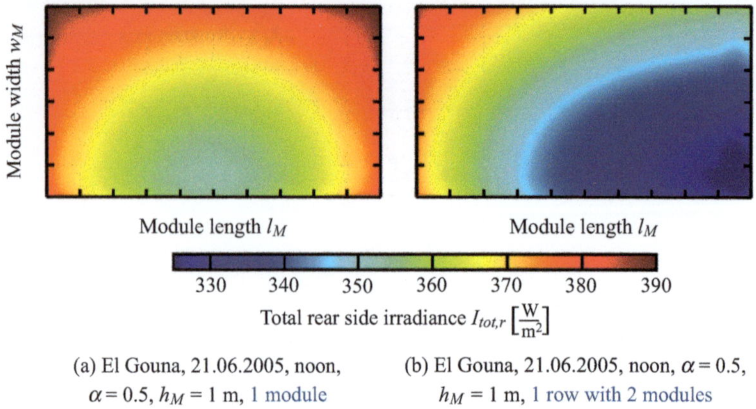

Figure 4.16 *$I_{tot,r}$ is lower for a row with two modules (b), than for a single module (a), due to the larger shadow region. The reduction is also not symmetrical, but is stronger in the area of the module adjacent to the neighbouring module [8]*

in energy yield will enable a very low cost of the electricity generated by this type of PV system in low latitude regions.

4.4.4 *Bifacial module field*

PV modules are normally either installed on a flat roof or in a field with neighbouring modules and module rows, and are rarely installed singularly as a standalone system. Some previous work on this topic can be found e.g. in [40] and [41]. The following sections will therefore be dedicated to examining the effect of adjacent modules in the same row, of additional module rows, and the performance of an entire field consisting of bifacial modules.

4.4.4.1 Adjacent modules

Whereas monofacial modules are seldom affected by further modules mounted in the same row, bifacial modules are indeed influenced by additional neighbouring modules casting further shadow on the ground. The increased shadow area on the ground reduces the reflected irradiance reaching the rear side of the module. The effect of the shadow caused by a further module can clearly be seen in Figure 4.16. The module considered in Figure 4.16 is mounted left of a second module. The reduction of the rear side irradiance is therefore not symmetrically beneath the module (Figure 4.16(a)), but further to the right (Figure 4.16(b)). This is caused by the shadow of the second module.

The greater the number of adjacent modules in a row, the stronger the reduction of the bifacial gain. However, under the given conditions, a saturation point is reached at a number of five adjacent modules, where the bifacial gain of the centre module is no longer negatively affected by additional modules. At five modules, the considered module has two modules on each side, meaning that the influence of the third module and further on the performance of a given module is negligible.

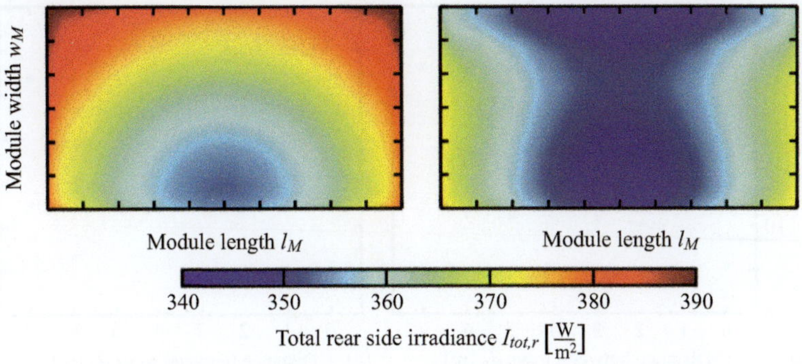

(a) El Gouna, 21.06.2005, noon, $\alpha = 0.5$, $h_M = 1$ m, 1 module

(b) El Gouna, 21.06.2005, noon, $\alpha = 0.5$, $h_M = 1$ m, 2 rows with 1 module each

Figure 4.17 *$I_{tot,r}$ drops from a) 369 W/m² to b) 356 W/m², due to additional module row. Reduction of $I_{tot,r}$ for top cell row of configuration (b) is caused by blocking effect, while for bottom cell rows it is caused by the module shadow on the ground [8]*

4.4.4.2 Additional module rows

Since PV installations rarely consist of one row of modules, but rather of several, simulations of stand-alone modules or of single module rows are insufficient. To enable an accurate prediction of the energy yield of a bifacial module field, the mutual influence of the modules rows on each other has to be taken into account. In addition to the shadowing of the module front side, occurring in the same way as for monofacial modules, additional module rows have an influence on the rear side irradiance. The ground-reflected irradiance is blocked by the rear module row, reducing the irradiance reaching the module rear side.

The blocking effect can clearly be seen in Figure 4.17(a), where $I_{tot,r}$ is shown for one bifacial module with another module behind it, compared to a stand-alone bifacial module in Figure 4.17(b). In addition to the reduction of $I_{tot,r}$ at the lower end of the module by its shadow, $I_{tot,r}$ is further reduced at the top end of the module. This additional reduction is caused by the blocking effect. Viewed geometrically, the irradiance reaching the top cell row of a module is blocked the strongest by the additional module row, which can be seen in Figure 4.17(b).

The intensity of the reduction of $I_{tot,r}$ due to blocking is effectively dependent on the distance between the module rows d_R, an important parameter in the development of any solar park, whatever the module type. To examine the influence of d_R on the energy production of bifacial modules installed in a field, the bifacial gain is simulated for varying distances between the individual rows. The resulting bifacial gain on the centre module, which is surrounded the most by other modules, is shown for $\alpha = 0.2$ and $\alpha = 0.5$ in Figure 4.18(a) and (b) for El Gouna and Constance, respectively.

As expected, mounting module rows too close to each other reduces the bifacial gain of the considered module. According to the results shown in Figure 4.18(a) and (b), a d_R of 3 m would be a reasonable distance to keep between the modules.

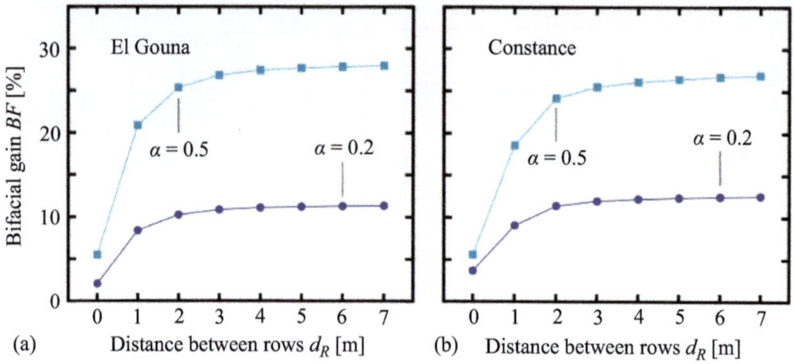

Figure 4.18 BF increases with increasing distance between rows d_R in both (a) El Gouna and (b) Constance. Saturation point reached for $d_R \approx 3$ m, further than which, additional module rows have a negligible influence on BF [8]

More than 3 m would reduce the land coverage of the solar park, without tangibly increasing the energy yield of the rear side of each module. In Chapter 6, the impact of varying row-to-row distances (translated to ground coverage) on land-related cost and consequently on the cost of the PV-generated electricity (LCOE) of monofacial and bifacial PV systems is studied quantitatively.

4.4.4.3 Bifacial module field

Compared to a stand-alone configuration, a bifacial module's rear side contributes less to the energy production in a field installation. Surrounded by other modules from every side, $I_{tot,r}$ is strongly reduced by the shadows casted by additional module and by blocking of the reflected irradiance by these modules. The discrepancies between the performances of the edge modules and the bifacial modules at the centre of the field are shown in Figure 4.19. This figure depicts the bifacial gains of all modules of a field with five rows, each with eleven modules, and a row distance of 2.5 m.

Since there are less modules in their surrounding casting shadows and blocking the reflected irradiance, the bifacial gain is higher for modules mounted at the edge of the field. However, only the first two modules from the side edge of the field have an increased *BF*, and starting from the third inner module, the bifacial gain is constant, as also shown by the simulation in Chapter 4.4.2. Compared to the inner rows, the first and last module rows exhibit an increased performance. The best and worst performing modules in a field have bifacial gains of 31.41% and 27.72% compared to 33.85% of a stand-alone bifacial module. When considering large utility scale PV systems with a size of hundreds of MW_p (corresponding to hundreds of thousands of modules) which are expected to be a typical application of bifacial modules in the future, 99% or more of the modules (i.e. all modules that have at least one module row between themselves and the edge of the PV array or that have at least two modules in the same row between themselves and the edge of

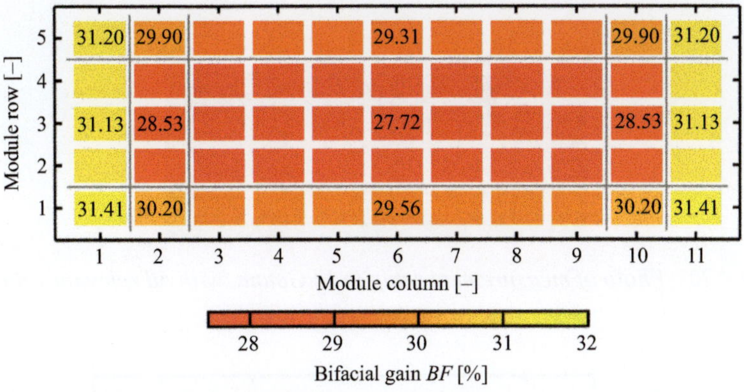

Figure 4.19 Bifacial gain of all modules in a field in El Gouna with $\alpha = 0.5$. The outer modules have a higher BF. Notice how in El Gouna the first row, has a higher BF than the last row, where the shadow of the front rows decrease $I_{refl,r}$ strongly reducing BF [8]

the array) will be optically in a location within the PV array that is corresponding to the above mentioned 'worst performing' module (in this case showing a 27% bifacial gain).

4.4.5 Result validation

To verify the accuracy of the calculation the bifacial gain according to the methods described in the previous sections (based on [8]), simulations have been carried out using the exact configuration of a module test site installed on the campus of the Technische Universität of Berlin in El Gouna (Egypt). The simulation results have been subsequently compared to the field measurement data of a monofacial and bifacial module installed on this test site obtained during 2014.

4.4.5.1 Location and setup

The measurement campaign was carried out on the campus of the Technische Universität of Berlin in El Gouna, where several bifacial and monofacial modules from various manufacturers are installed and continuously monitored. A photo of the test site is shown in Figure 4.20. The south-facing modules are mounted in pairs, each consisting of a monofacial and bifacial module. The modules mounted at a tilt angle of 20° and at a height of the lower edge from the ground of 1.2 m. In addition to the monitoring of the module output power, several other parameters, such as the global horizontal irradiance GHI, the ambient temperature, the wind speed and wind direction, are continuously measured. Two front- and back-facing ISET sensors are mounted between the modules in the same plane, measuring the front and rear side irradiance reaching the module plane, and can be seen in Figure 4.20. Using two upward- and downward-facing pyranometers, the average albedo of the ground beneath the modules, which consisted of a cement foundation surrounded by sand, is measured, delivering a result of 0.3.

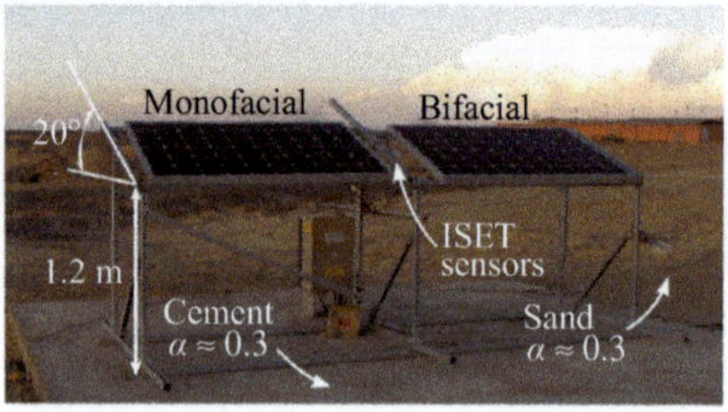

Figure 4.20 Photo of measurement setup in El Gouna, with all relevant values [8]

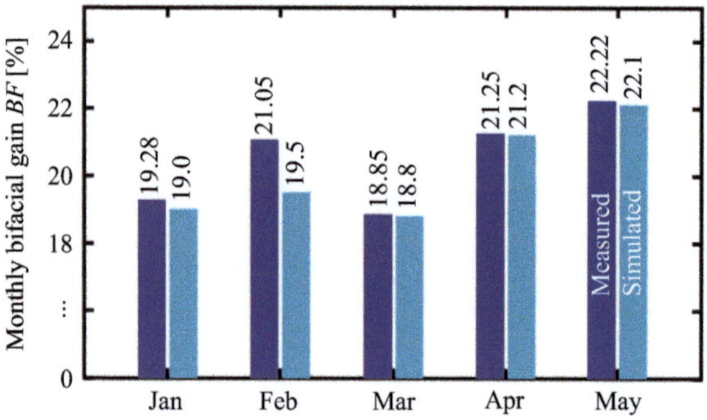

Figure 4.21 Small deviation between measured and simulated monthly bifacial gain of modules installed in El Gouna

4.4.5.2 Results

Figure 4.21 shows the measured and simulated monthly bifacial gain of the modules installed in El Gouna. The results of the simulation correlate very well with the measured monthly bifacial gain except for the month of February, where the deviation of circa 1.5% absolute is rather large and is suspected to be due to some deviation between actual diffuse irradiance fraction and the values used as input parameters for the simulation for the respective time period. The otherwise good agreement between the measured and the bifacial gain shows the reliability of the developed model under the given climatic conditions and for this system configuration, and the correctness of the assumptions made in the simulation. However, this simulation model is under continuous development at ISC Konstanz and the current version is known under the name *MoBiDiG* (modelling of distributed bifacial gain – see e.g. [42]).

In general, recent studies at the various institutions involved, regarding the energy yield prediction of bifacial modules and systems are focusing on the validation with experimental data and the improvement of the accuracy of the overall simulation results by refining the optical, thermal and electrical models and by extending the applicability of such simulation models.

4.5 Tracking of bifacial modules and systems

While the state of the art of experimental field results for tracked bifacial systems is described in a dedicated section of Chapter 5, up to now, very few studies about the theoretical prediction of the energy yield of tracked bifacial modules or systems have been published. One early example of such study is the work published by Egido *et al.* [43], where a theoretical model is presented and the resulting predicted energy yield of a bifacial module mounted on a single- and two-axis tracker is compared with the monofacial versions of the same installation configuration. Some early work on this topic can be found in [43], while recent studies regarding complete arrays of bifacial modules with single-axis tracking have been presented by Lindsay *et al.* [44] and Anoma *et al.* [45]. Further research groups are currently working on the development of more and more accurate and sophisticated models and on the validation of the simulation results with experimental results [9,42,46], and at least one commercial software provider (PVsyst) meanwhile released the first versions with the possibility of calculating the energy yield of tracked bifacial systems.

4.6 Summary/outlook

In this chapter, an optical, electrical and thermal model have been presented as the basic elements of a simulation model for the energy yield prediction of bifacial models and systems. Keeping in mind that for each of these models, several different concepts are viable and are under investigation by various research groups around the world, the model as published in [8] has been presented as an example. Thereby, an optical model for the rear side irradiance of bifacial PV modules, both stand-alone as well as in-field installed, has been established and after its implementation as a software tool, simulations of the energy yield for different scenarios have been conducted. Given appropriate weather data, simulations can be carried out for various locations. In combination with an electrical model, such tools allow for the estimation of the bifacial gain, i.e. the additional yield compared to a standard PV module, for various installation parameters, such as the tilt angle, installation height, distance between module rows and constant ground albedo coefficient.

With the model presented earlier, simulations of a 1.5 m high bifacial module in El Gouna Egypt resulted in bifacial gains of 13% and 34% for ground albedos of 0.2 and 0.5, respectively. Increasing the ground albedo to 0.85 using white reflective sheets increases the gain to up to 55%, depending on the size and setup of the white reflective sheet. In addition, a simple case of horizontal single-axis tracking of a bifacial module in Kasese, Uganda, has been examined. Compared to a fixed monofacial module, tracking of a bifacial module results in 40% gain in energy yield, assuming a ground albedo of 0.2. Modelling of a bifacial module within a field shows a reduction of the bifacial gain to 31% and 27% for the best and worst performing modules in the field, respectively, compared to a 34% gain of a stand-alone module. The parameters of an existing installation in El Gouna (Figure 4.20) were used to carry out a simulation, the results of which were compared to the actual outdoor measurements over a 5-month period. The first validation shows a good correlation between the simulation results and the outdoor measurements, providing an indication on the reliability of the developed model.

Further research in this field by the various institutions involved is targeting on an extension of the optical model, electrical and thermal model to lager application ranges such as different (and extreme) climatic zones, PV arrays of different sizes and geometries, fixed tilt mounted bifacial systems with various tilt angles (including vertically mounted modules with East/West and North/South orientation as well as horizontally mounted bifacial modules, installed e.g. in carports) and in particular also to single (horizontal and tilted) axis as well as two-axis tracking of bifacial modules and to the respective system configurations. These improvements include also a refined electrical model on module level (taking into account for the behaviour of bypass diodes) and the extension of the model to system level, taking into consideration effects caused by mismatching between the modules, inverter parameters and other factors that impact the energy yield on AC level. Furthermore, other potential improvements are related to the optical model, where it can be considered whether including a variable albedo (as done e.g. by Chiodetti *et al.* [13]), taking into account the diurnal and seasonal variation of the ground albedo coefficient, might be practical – in terms of available input data from databases – and bring some further improvement of the accuracy of bifacial energy yield predictions.

Independently from the mathematical and theoretical concepts that are implemented, it will be of utmost importance to gather more and more experimental data from relevant field installations in order to validate the accuracy of the developed simulation models and in order to further improve the models itself. On the one hand, the installation configurations of such field installations (PV systems) must be relevant regarding their geometry, i.e. in terms of module mounting height, ground cover ratio and number of modules per row and number of module rows. On the other hand, the monitored meteorological and electrical data must be complete and accurate, by using calibrated instruments. Complete means in particular, that for bifacial PV, as explained in this chapter, the monitoring of parameters such as ground albedo and diffuse irradiance is mandatory in order to obtain a full understanding of the interactions between ambient conditions and the electrical performance parameters of bifacial modules and systems. In the case of standard PV systems, ground albedo and diffuse irradiance are usually not monitored, as their impact on the energy yield of monofacial modules is less significant and their acquisition requires additional (costly) instrumentation. A noteworthy example of a flexible experimental set-up is the so-called BIFOROT set-up by ZHAW [47]; it allows the acquisition of the complete electrical and meteorological data with a high temporal resolution for an outdoor installation of a bifacial module located within a PV array in a short-time period for a large amount of different geometrical configurations (tilt angles and mounting heights). This set-up is described in Chapter 5.

The growing number of experimental and commercial set-ups is expected to give an important contribution to the improvement and validation of the accuracy of the various simulation models for the energy yield prediction of bifacial models and systems.

Another challenge is the implementation of the models as software tools allowing on the one hand for a good accuracy for a large range of installation geometries (and climate zones) and on the other hand avoiding an excessively long computation time of such tools when using standard personal computers. This will be in particular important for the integration of such models into commercial software packages.

References

[1] A. Cuevas, A. Luque, J. Eguren, and J. del Alamo. '50 per cent more output power from an albedo-collecting flat panel using bifacial solar cells'. Sol. Energy, 1982, 29. 10.1016/0038-092X(82)90078-0.

[2] A. Krenzinger and E. Lorenzo. 'Estimation of radiation incident on bifacial albedo-collecting panels'. Int. J. Sol. Energy, 1986, 4, 297–319.

[3] Y.K. Chieng and M.A. Green. 'Computer simulation of enhanced output from bifacial photovoltaic modules'. Prog. Photovolt. Res. Appl., 1993, 4, 293–99.

[4] J. Johnson, D. Yoon, and Y. Baghzouz. Modeling and analysis of a bifacial grid-connected photovoltaic system. IEEE Power and Energy Society General Meeting, 2012.

[5] U.A. Yusufoglu, T.H. Lee, T.M. Pletzer, *et al.* 'Simulation of energy production by bifacial modules with revision of ground reflection'. Energy Procedia, 2014, 55, 389–95.

[6] C.K. Lo, Y.S. Lim, and F.A. Rahman. 'New integrated simulation tool for the optimum design of bifacial solar panel with reflectors on a specific site'. Renew. Energy, 2015, 81, 293–307.

[7] S. Wang, O. Wilkieb, J. Lama, *et al.* 'Bifacial photovoltaic systems energy yield modelling'. Energy Procedia, 2015, 77, 428–33.

[8] I. Shoukry. *Bifacial Modules: Simulation and Experiment.* Konstanz, 2015.

[9] A. Pandiyan. *Automated Data Evaluation and Performance Modelling of Bifacial Solar Modules.* Freiburg: Universität Freiburg, 2017.

[10] [Online] PVsyst SA. [Cited: 23 Apr 2017]. Available from http://www.pvsyst.com/en/software/software-development.

[11] B. Wittmer. Bifacial shed simulations with PVsyst. Bifacial PV Workshop, Konstanz, 2017.

[12] Polysun Simulation Software. [Online] Vela Solaris. [Cited: 23 Apr 2017]. Available from http://www.velasolaris.com/english/product/product-overview.html.

[13] M. Chiodetti, A. Lindsay, P. Dupeyrat, *et al.* Bifacial yield simulation with a variable albedo model. European Photovoltaic Solar Energy Conference and Exhibition (EU PVSEC), Munich, 2016.

[14] B. Marion, S. Macalpine, and C.D. Nrel. A practical irradiance model for bifacial PV modules. 44th IEEE Photovolt. Spec. Conf., Washington, D.C., 2017.

[15] A. Lindsay, M. Chiodetti, D. Binesti, *et al.* Enhanced bifacial PV modelling with ray-tracing. 6th PV PMC Workshop, 2016.

[16] B. Soria, E. Gerritsen, P. Lefillastre, and J.-E. Broquin. 'A study of the annual performance of bifacial photovoltaic modules in the case of vertical facade integration'. Energy Sci. Eng., 2016, https://doi.org/10.1002/ese3.103.

[17] C. Reise and A. Schmid. Realistic yield expectations for bifacial PV systems – an assessment of announced, predicted and observed benefits. European Photovoltaic Solar Energy Conference and Exhibition (EU PVSEC), Hamburg, 2015.

[18] C. Deline, S. Macalpine, B. Marion, and J.S. Stein. Evaluation and field assessment of bifacial photovoltaic module power rating methodologies. 43rd IEEE Photovolt. Spec. Conf.

[19] J.E. Castillo-Aguilella and P.S. Hauser. 'Multi-variable bifacial photovoltaic module test results and best-fit annual bifacial energy yield model'. IEEE Access, 2016, 4, 498–506.

[20] J.E. Castillo-Aguilella and P.S. Hauser. Bifacial photovoltaic module best-fit annual energy yield model with azimuthal correction. IEEE 43rd Photovolt. Spec. Conf, 2016.

[21] National Renewable Energy Laboratory, Sandia National Laboratory, University of Iowa. Bifacial PV Project. [Online] [Cited: 23 Apr 2017]. Available from https://pvpmc.sandia.gov/pv-research/bifacial-pv-project/.

[22] J.P. Singh, A.G. Aberle, and T.M. Walsh. 'Electrical characterization method for bifacial photovoltaic modules'. Sol. Energy Mater. Sol. Cells, 2014, 127, 136–142.

[23] A. Hubner, A. Aberle, and R. Hezel. Temperature behavior of monofacial and bifacial silicon solar cells. 26th IEEE Photovoltaic Spe. Conf., 1997.

[24] U. Yusufoglu, T.M. Pletzer, L.J. Koduvelikulathu, *et al.* 'Analysis of the annual performance of bifacial modules and optimization methods'. IEEE J. Photovolt., 2015, 5, 320–8.

[25] R.S.R. Gali. *Energy Yield Model for Bifacial PV Systems: A Study and Analysis of Temperature and Rear Irradiance Models.* TU Delft, 2017.

[26] E. Özkalay. *Total Heat Input and Operational Temperature of Different Cell and Module Technologies.* TU Delft, 2017.

[27] B. Van Aken. Bifacial modules: hot or cool? Bifacial PV Workshop, Konstanz, 2017.

[28] I. Shoukry, J. Libal, R. Kopecek, E. Wefringhaus, and J. Werner. 'Modelling of bifacial gain for stand-alone and in-field installed bifacial PV modules'. Energy Procedia, 2016, 92, 600–8.

[29] V. Quaschning. *Understaning Renewable Energy Systems*. Vol. 1. Oxford: Earthscan, 2005.

[30] V. Badescu. '3D isotropic approximation for solar diffuse irradiance on tilted surfaces'. Renew. Energy, 2002, 26, 221–33.

[31] A.M. Noorian, I. Moradi, and G.A. Kamali. 'Evaluation of 12 models to estimate hourly di use irradiation on inclined surfaces'. Renew. Energy, 2008, 33, 1406–12.

[32] K.N. Shukla, S. Rangnekar, and K. Sudhakar. 'Comparative study of isotropic and anisotropic sky models to estimate solar radiation incident on tilted surface: a case study for Bhopal, India'. Energy Rep., 2015, 1, 96–103.

[33] M. Gulin, M. Vašak, and M. Baoti. Estimation of the global solar irradiance on tilted surfaces. 17th International Conference on Electrical Drives and Power Electronics (EDPE 2013), 2013, pp. 334–9.

[34] R. Perez, P. Ineichen, R. Seals, J. Michalsky, and R. Stewart. 'Modeling daylight availability and irradiance components from direct and global irradiance'. Sol. Energy, 1990, 44, 271–89.

[35] U. Gross, K. Spindler, and E. Hahne. 'Shapefactor-equations for radiation heat transfer between plane rectangular surfaces of arbitrary position and size with parallel boundaries'. Lett. Heat Mass Transfer, 1981, 8, 219–27.

[36] Y.A. Cengel and A.J. Ghajar. *Heat and Mass Transfer: Fundamentals and Applications + EES DVD for Heat and Mass Transfer*. Columbus: McGraw-Hill Education, 2010, 9780077366643.

[37] R.G. Ross Jr. Flat-plate photovoltaic array design optimization. 14th Photovoltaic Specialists Conference, 1980, Vol. 1, pp. 1126–32.

[38] G.J.M. Janssen, B.B. Van Aken, A.J. Carr, and A.A. Mewe. 'Outdoor performance of bifacial modules by measurements and modelling'. Energy Procedia, 2015, 77, 364–73.

[39] Solar Irradiation Data. [Online] [Cited: 1 May 2015.]. Available from http://www.soda-pro.com/.

[40] L. Kreinin, A. Karsenty, D. Grobgeld, and N. Eisenberg. PV systems based on bifacial modules: performance simulation vs. design factors. *Proceedings of the 43th IEEE PVSC*, 2016.

[41] J. Johnson, D. Yoon, and Y. Baghzouz. 'Modeling and analysis of a bifacial grid-connected photovoltaic system'. IEEE Power Energy Soc., 2012.

[42] D. Berrian. MoBiDiG: simulations and LCOE. Bifacial PV Workshop, Konstanz, 2017.

[43] M.A. Egido and E. Lorenzo. 'Bifacial photovoltaic panels with sun tracking'. Int. J. Sol. Energy, 1986, 4, no. 2, 97–107.

[44] A. Lindsay. Modelling of single-axis tracking gain for bifacial PV systems. 32nd EUPVSEC, Munich, 2016.

[45] A. Anoma. View factor model and validation for bifacial PV and diffuse shade on single-axis trackers. 44th IEEE PVSC, Washington, 2017.

[46] C. Deline. Bifacial PV performance models: comparison and field results. Bifacial PV Workshop, Konstanz, 2017.

[47] H. Nussbaumer. Miniaturized test array as a means to determine the energy harvest of bifacial installations. 43rd IEEE PVSC, Portland, 2016.

Chapter 5
Bifacial PV systems and yield data (bifacial gain)

Markus Klenk[1], *Yannick Veschetti*[2], *Radovan Kopecek*[3], *Hartmut Nussbaumer*[1], *Heiko Hildebrandt*[4], *and Rob Kreiter*[5]

5.1 Introduction

A major motivation for bifacial photovoltaics (PV) is an expected additional energy yield, compared to monofacial panels, due to the two-sided light sensitivity. The potential for an improved module power output and energy yield was repeatedly demonstrated by simulations [1–8], measurements on stand-alone modules [9–14] or installations [15–19] in various orientations. However, uncertainties concerning the actual output of projected systems still deter possible investors. Even in the PV community, the real quantitative benefit due to bifaciality and the best-suited technical concepts are still under discussion [20–22], as reflected by numerous publications dealing with these topics.

For monofacial modules, the clearly defined measurement of the nominal power under standardized conditions (STCs) is the accepted classification criteria, and the prediction of the system's energy output with sufficient accuracy is comparatively straightforward with available simulation tools. This is not the case for bifacial devices and systems.

The uncertainty is caused by the considerably more complicated conditions as compared to monofacial devices. Even for single stand-alone modules, there are several additional factors which have to be considered. Obviously, the additional benefit due to bifaciality is directly linked to the light intensity at the modules rear side, which is in turn dependent on the ground reflectivity (albedo), weather and season, light homogeneity and light characteristic. Even for free-standing modules, shading effects at the rear side of the module and at the reflecting ground

[1]ZHAW, Zurich University of Applied Science, School of Engineering, Switzerland
[2]INES – Institut National de l'Energie Solaire, France
[3]ISC Konstanz, Germany
[4]Next2Sun GmbH, Germany
[5]Sunfloat, The Netherlands

are unavoidable. In real, extended systems, however, the arrangement of multiple modules will have additional effects, such as direct shading by modules in adjacent rows or indirect shading of the modules surrounding, which results in a reduced rear side irradiance. Analyzing the energy output of power plants is one way to appraise the efficiency of real installations, since the energy generation can be related to the utilized area or the amount of modules of the system. However, data of larger systems are still quite rare and, because of the specific installation and shading conditions, the assignability of such results to other systems is again not straightforward.

An obvious way to visualize the benefit due to bifaciality is to analyze the "bifacial gain" which means the difference in the energy yield if bifacial and monofacial devices with similar installation situation are compared. The comparison can either include single modules or larger units of one or both device types, because typically the kWh/kWp ratio is analyzed. The kWp data usually reflect the STC front-side measurement of the bifacial module(s). In the most direct form devices of similar type and with the same front-side efficiency are compared, for example if bifacial modules with covered rear side are used as reference. Because the mounting situation has a very significant influence on the output of bifacial systems, the most general results will be obtained if free-standing modules or small arrays are measured. On the other hand, the bifacial gain is overrated then, because detrimental shading effects in larger, extended module arrays are omitted. Due to the multiple factors that have an influence on the bifacial output, it is obvious that a comparison of measurement results from different systems will show a considerable fluctuation. Nevertheless this measured real-world data should allow a rough estimation of typical power gains if bifacial devices are implemented instead of monofacial ones.

Although the main interest of bifacial PV is the extra energy yield, there are also applications which would not be feasible with monofacial modules. One of the most considered is the vertical mounting of PV systems in an east–west orientation that reaps particular benefits in snow-rich regions [18] (no sticking of snow) or desert locations (reduced or no soiling), and also contributes to a more consistent energy production throughout the day ("peak-shaving"), thus improving the alignment between electricity production and demand.

5.1.1 Key indicators to analyze the potential advantage of a bifacial system over a monofacial one

One important step on the way toward bifacial PV bankability is the collection of real-world energy-yield data which requires the monitoring of the energy production of large bifacial PV systems in different geographical locations and with various installation configurations. Ideally, part of the plant should include standard monofacial modules allowing accurate calculations of the bifacial gain.

Thereby, the following requirements have to be met in order to make the data relevant for the above mentioned scope:

- System composed of several module rows, each of them composed by several modules
- Monitoring period > several months, ideally at least 12 consecutive months in order to cover all seasons (varying diffuse light fraction)

- Subsystem with monofacial modules to be monitored simultaneously at the same site as a reference for determination of bifacial gain
- Meteo data (at least irradiance and module temperature) monitored according for example to specification given in IEA-PVPS T13-05:2014 report

Ideally, the best way of comparing bifacial versus monofacial should be done on a €/KWh basis according to an LCOE (levelized cost of energy—see Chapter 7) calculation.

This value may vary depending on the installer and land cost but the energy gain which is currently reached using bifacial module must not be lost due to additional cost such as specific mounting structure, more expensive panels or cost for artificially increasing the ground albedo.

Regarding the geographical location, apart from the total irradiance, the diffuse irradiance fraction plays an important role in the bifacial gain that can be obtained: the more diffuse light there is, the higher the irradiance of the rear side of the bifacial modules will be. On the other hand, various installation configurations enable the advantages of bifacial modules to be gained in different ways: for example, MW-size ground-mounted systems that have fixed-tilt or one-axis tracking, with natural ground or artificially enhanced albedo (white sand, reflective plates or sheets, etc.), and kW-size to MW-size PV systems on flat rooftops.

The bifacial gain is the metric that determines—together with the total cost of the installed bifacial PV system—the LCOE (€/kWh) and therefore the economic viability of bifacial PV with:

$$g_{bifacial}\,[\%] = \left(\frac{\left(e_{bifacial} - e_{monofacial}\right)}{e_{monofacial}}\right) \times 100 \tag{5.1}$$

With

- $e_{bifacial}$: specific energy yield (kWh/kWp) of the PV system with bifacial modules
- $e_{monofacial}$: specific energy yield (kWh/kWp) of the PV system with monofacial modules on the same site, with the same configuration and during the same time period

5.2 Overview about small scale bifacial systems with information concerning the bifacial gain

For a potential investor, it is important to have an estimation concerning the actual gain which can be obtained by the use of bifacial modules. Unfortunately the established simulation tools are only of limited use for bifacial systems up to now and the statements in literature about the bifacial gain cover a wide range. This obviously reflects the influence of the differing installation situations and the specific properties of the modules. In the following section a compilation of published results from smaller systems is given. This compilation may be used for comparisons but it is also tried to reveal the most significant properties which may allow a rough estimation of the bifacial gain even for very differing systems.

First, data from publications with "typical" installation conditions, which means an orientation roughly as for monofacial modules (south orientation on the northern hemisphere) and limited tilt is considered. The arrangement of the listing is chronological, starting with the newest publication date. Because for this type of installation there is a comparatively large amount of data available, the most relevant information will be condensed for a comparison.

This section is followed by a similar compilation of publications dealing with vertically installed bifacial modules, a repeatedly applied and attractive opportunity for bifacial systems. The less pronounced orientation sensitivity of bifacial modules allows broad range of potential applications. However, neither niche applications [21,23], such as low-concentration systems [24–27], nor tracker-based systems are included in this listing, because the pronounced peculiarities result in very system-specific results.

System 1

Description: Direct comparison of typical bifacial mounting situations plus vertical installation with monofacial (30° south) array.
Remark: Analysis of AC and DC output and resulting differences
Location: Chambéry, France
5 × 3 kWc arrays in different configuration, including one monofacial reference and one bifacial vertical array
BR: 90%; CEA INES PERT

Figure 5.1 Small-scale bifacial PV system with white gravels on the ground by CEA-INES [29]

Arrangement: 4 × 3 modules respectively in one plane, varying mounting, see Figures 5.1–5.3
Tilt: 30°
Height: 0.6 m (lower module edge of lowest module)
Situation: Grass, albedo 20% (est.: 15%–25% [28])
High albedo, white rocks: ~40% estimated albedo
Test duration: 12 months 09/2016–09/2017

Bifacial gain:
DC gain: 12% for white gravel (portrait)
DC gain: 13% for white gravel (landscape)
DC gain: 10% for grass ground (portrait)
DC gain: 19% for white gravel (1.6 m height)

Figure 5.2 Small-scale bifacial PV system with synthetic grass by CEA-INES [29]

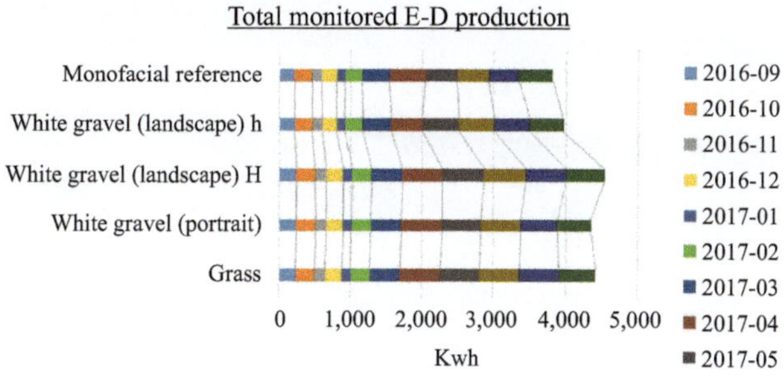

Figure 5.3 Monthly yield for the respective installation situations [30]. Graph by Eric Pilat

[29] *Y. Veschetti et al., First monitoring results of bifacial systems in different configurations, 2016*
[30] *M. Joanny et al., Bifacial systems overview, 2017*

System 2

Description: Direct comparison of a bifacial module, south, 45° tilt angle, in a small array with a monofacial module, 30° south. Also compared to vertical E/W-installation [31] (Figure 5.4).
Location: Tucson, USA
Similar setup in Germany, but limited measurement duration at this location up to now (Start June 2017).
Modules: Double glass; 22 pcs of ½ 6″ solar cells; 100 Wp
-First Solar "TetraCell"
BR: 93%
Arrangement: See Figure 5.4, vertical (E/W) and 45° tilt angle array.
Tilt: 45° and 90°, Elevation: 0.06 m (lower module edge)
Albedo, white paint: ~85% (estimation by author [31])
Test duration: November 2016–September 2017

Bifacial gain compared to monofacial 30° south:
13% (annual average; south, 45° tilt)
Seasonal variation of BG: 8%–18%
(Vertical E/W: + 15% annual average with pronounced seasonality: –15% to +40%)

[31] *L. Podlowski et al., Yield Study on Identical Bifacial Rooftop Systems Installed in the USA and in Germany, 2017*

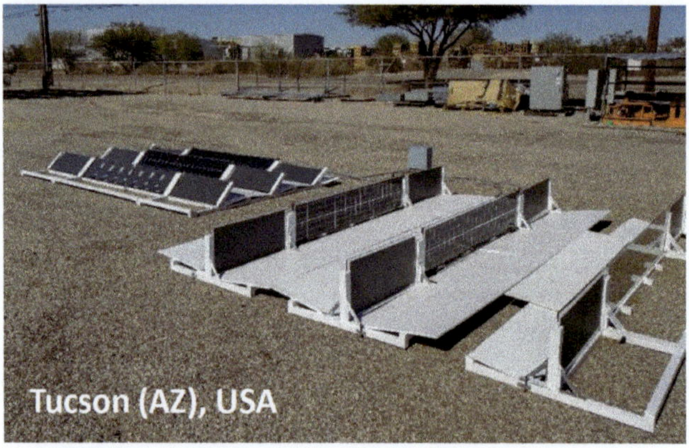

Figure 5.4 PI Berlin test array [31]. Bifacial South-oriented array with 45° tilt angle in the background, vertical installation in the foreground

System 3

Description: Nine bifacial modules mounted in landscape orientation with spacing (Figures 5.5 and 5.6)
Location: Epe, NL
Yingli Bifacial Panda, glass–glass 275 Wp (STC, front side)
1 × Sofar string inverter 3,000 W
Flat roof painted white, Albedo 63%
Tilt angle: 20°
Height: lowest point: 0.18 m

Figure 5.5 Nine bifacial modules with spacing, mounted in landscape orientation by Tempress [32]

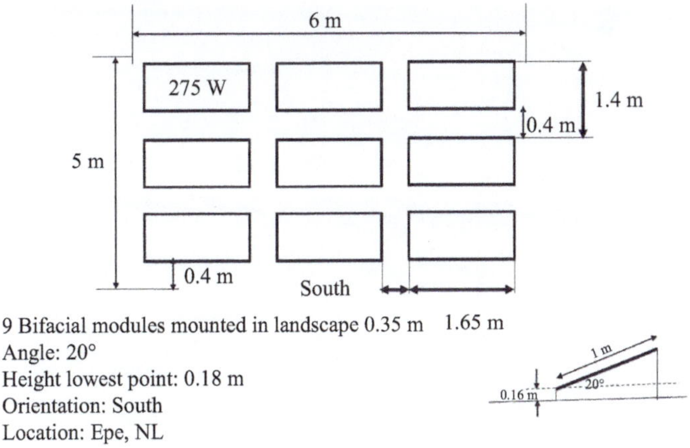

Figure 5.6 Setup of tempress test-field [32]

Orientation: South
Test duration: 3 months

Bifacial gain:
Compared to monofacial Trina 260 W black
String inverter: JSI-3000TL
Portrait on rooftop, south, 40° tilt
Measurement duration: June–September
BG: 35%–36%

Remark: Not considered in comparison (Table 5.1 and Figure 5.19), due to short test duration

[32] *W. Vermeulen, 400 kW bifacial system in NL and comparison with two other systems, 2017*

System 4

Description: Comparison of bifacial and monofacial modules. Two modules per type in one plane with lowest edge 0.7 m and 1.4 m above ground.
Location: UAE, Abu Dhabi
Modules: Monofacial, PERT bifacial and Meyer Burger HJT bifacial
Albedo: ≈24%
Average albedo measured in tilt of module with silicon sensors
Tilt angle: 22°
Height: lowest point: 0.70 m/1.4 m
Orientation: South
Test duration: 10 months

Bifacial gain:
Measurement duration: January to October 2017

Figure 5.7 Two modules per type in one plane with lowest edge 0.7 and 1.4 m above ground in Meyer Burger UAE test field

Figure 5.8 Mounting with low rear-side shading

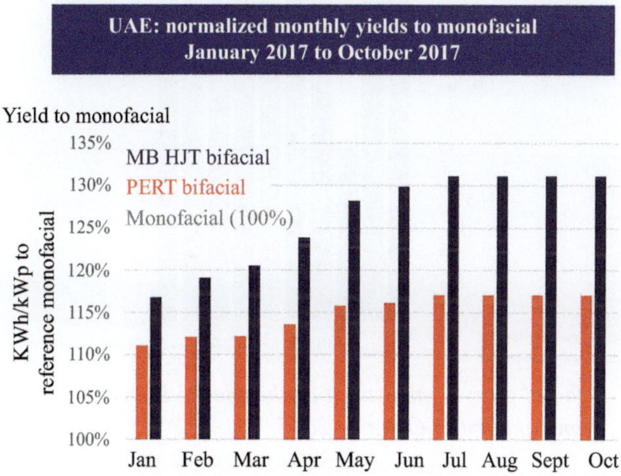

Figure 5.9 Setup of test-field [23]

BG (PERT bifacial): 11%–17%
BG (MB HJT bifacial): 17%–31%
Estimated, averaged bifacial gain for PERT and HJT bifacial modules in comparison (Table 5.1 and Figure 5.19). Conservative estimation, as higher total output in summer is underestimated.
BG PERT bifacial: 14%; BG MB HJT bifacial: 25%

Remark: Measurements show that height difference of 0.7–1.4 m (see Figure 5.7) causes energy yield difference of about 2% on bifacial modules (Figures 5.8 and 5.9).

[33] *A. Richter, Meyer Burger, 2017*

System 5

Description: Comparison Monofacial, Meyer Burger HJT (Figure 5.10)
Location: USA, Arizona
Albedo: ≈10%

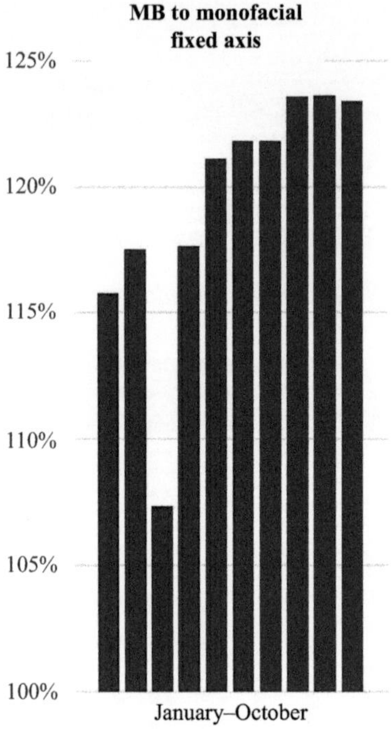

Figure 5.10 Bifacial gain as measured at the Meyer Burger test field in Arizona from January to October 2017

Average albedo measured in tilt of module
Tilt angle: n.a.; Height: lowest point: n.a.
Orientation: South
Test duration: 10 months; January to October 2017

Bifacial gain: BG: 16%–23% (March 7%–outlier)
Estimated, averaged BG for Table 5.1 and Figure 5.19: 19%. Conservative estimation, as higher total output in summer is underestimated.

[33] *A. Richter, Meyer Burger, 2017*

System 6

Description: ~0.5 kWp: 2 modules (1 bifacial, 1 reference); BR: unknown
Arrangement: bifacial test module in array with 10 modules
Tilt: 30° (est. from depictions); Elevation: unknown
Situation: Albedo 30% (est. from depiction, concrete [34])

Location: Lugano/CH
Test duration: 6 months (June to December 2014)

Bifacial gain: 13.7% (2.9% for monofacial HJT-competitor compared to c-Si Standard)

[35] *A. Richter, Bankability: Choosing right materials on module level, 2016*

System 7

Description: Respectively 8 modules with white and transparent backsheet are compared. Connected to string inverter.
Remark: transparent backsheet results in 3°C lower temperature
2 × 2.3 kWp: 2 × 8 module; BR: unknown
Arrangement: 4 × 2 in one plane on one rack,
Tilt: 45°, elevation: ~2 m (est.)
Location: China
Situation: grass, albedo 20% (est.#)
\# *grass albedo 15%–25%* [28] ⇒ *albedo 20% (est.)*
Test duration: 12 months

Bifacial gain: 3.21% per average over one year

[36] *Yu et al., A study on electrical performance of N-type bifacial PV modules, 2016*

System 8

Description: Array of ten bifacial modules compared with monofacial modules. Measurement results represent average of all 10 modules, respectively.
~ 2.5 kWp: 10 modules; BR: 95%
Arrangement: 5 × 2 in one plane on one rack
Tilt: 30°; Elevation: Minimum height (lower module edge) = 1.25 m
Location: New York/USA
Situation: 10% albedo
Test duration: >2 years

Bifacial gain: 17.7%

[37] *J.E. Castillo-Aguilella et al., Multi-Variable Bifacial Photovoltaic Module Test Results and Best-Fit Annual Bifacial Energy Yield Model, 2015*

System 9

Description: Two modules (one bifacial and one reference); BR: 95%
Arrangement: modules in one row

Tilt: 30°; Elevation: Minimum height 0.2 and 0.76 m
Location: Arizona/USA
Situation: 77% albedo
Test duration: ~5 months (May to October)

Bifacial gain (0.76 m elevation): Measured: 36.8%, extrapolated annual gain: 30.64%
Bifacial gain (0.2 m elevation): Measured: 27.3%, extrapolated annual gain: 22.75%

[37] *Castillo-Aguilella et al., Multi-Variable Bifacial Photovoltaic Module Test Results and Best-Fit Annual Bifacial Energy Yield Model, 2015*

System 10

Description: ~0.5 kWp: two modules (one bifacial and one reference); BR: 95%
Arrangement: modules in one row
Tilt: 20°; Elevation: Minimum height 0.2 m
Situation: Location: Arizona/USA
Varied albedo (70%, 68%, 22%)
Test duration: >1 year

Bifacial gain: 18.41% (70% albedo) 19.57% (68% albedo) 12.31% (22% albedo)

[37] *Castillo-Aguilella et al., Multi-Variable Bifacial Photovoltaic Module Test Results and Best-Fit Annual Bifacial Energy Yield Model, 2015*

System 11

Description: ~0.5 kWp: two modules. Bifacial module: NICE, BR: unknown
Arrangement: Single bifacial besides monofacial module. Two modules in one row. Bifacial and monofacial module landscape orientation.
Tilt: 20°; Elevation: 1 m (lower module edge) Situation: 30% albedo (est. [38]), Location: El Gouna/Egypt
Test duration: n.a

Bifacial gain: 14.3%

[39] *R. Einhaus, NICE technology for bifacial modules, 2015*

System 12

Description: kWh/kWp comparison of n-type bifacial modules (265 Wp) with multicrystalline monofacial (245 Wp).
Location: Singapore. Installation on commercial rooftop (Figure 5.11)

Figure 5.11 Test site on the rooftop of a commercial building in Singapore [40]

Bifacial: ~10 kWp, four strings; nine modules/string & inverter
Multi: ~14 kWp
Albedo: 20% (concrete)
Tilt angle: ±10°; Height: ~2 m (bifacial)
Orientation: Almost horizontal (location at equator)
Test duration: 12 months, July 2014 to June 2015

Bifacial gain: BG: 13.7%
1,540 kWh/kWp (bifi) 1,358 kWh/kWp (monofacial)
Monthly BG range: 11.6%–15.2%

[40] *H. C. Oon, A study on the performance of bifacial photovoltaic panels in the urban tropics, 2017*

System 13

Description: ~0.5 kWp: 2 × 72 cell modules (one bifacial and one bifacial with covered rear)
BR: unknown % (n-Pasha cell)
Arrangement: no direct shading
Tilt: n.a; Elevation: n.a
Situation: low albedo n.a% (dark concrete), Location: Petten/Netherland
Test duration: 1 year

Bifacial gain: 2%–6% with largest gain in summer

Remark: Not considered in comparison (Table 5.1 and Figure 5.19), due to data situation

[41] *B.B. Van Aken et al., Relation between indoor flash testing and outdoor performance of bifacial modules, 2014*

System 14

Description: ~ 0.5 kWp: two modules (one bifacial one reference); BR: unknown
Arrangement: two modules in one row. Bifacial and monofacial module in portrait orientation.
Tilt: unknown; Elevation: unknown
Situation: (albedo 30% est. from depiction, concrete [34]) Location: Taiwan (est.)
Test duration: 14 days

Bifacial gain >20%

Remark: Not considered in comparison (Table 5.1 and Figure 5.19), due to short test duration and data situation

[42] *Yu et al., 20.63% nPERT Cells and 20% PR Gain Bifacial Module, 2014*

System 15

Description: ~ 0.5 kWp: two modules (one bifacial and one reference)
BR: 91%
Arrangement: two modules in one row. Bifacial and monofacial module landscape orientation (Figure 5.12).
Tilt: 20°; Elevation: 1 m (lower module edge)
Situation: 30% albedo (est.# [38]),
Location: El Gouna/Egypt
Test duration: 8 months (January to August 2014)

Figure 5.12 Monitoring of bifacial module by ISC [12]

Bifacial gain 22.3%

[12] *C. Comparotto et al., Bifacial n-type solar modules: indoor and outdoor evaluation, 2014*

System 16

Description:
BF: n.a; Arrangement: n.a
Tilt: variable 30°, 45°, 60°, Elevation: 1 m
Situation: grass albedo 40%,
Location: Japan (est.)
Test duration: 1 month (May–June)

Bifacial gain: 17%; tilt angle 30°; 20%; tilt angle 45°; 25%; tilt angle 60°

Remark: Not considered in analysis (Table 5.1 and Figure 5.19) due to data situation and short measurement period

[43] *E. Maruyama, Recent Technological Progress of High-efficiency HIT Solar Cells, 2013*

System 17

Description: Beneath one of the arrays, crushed scallops' shells are used to enhance Situation reflection. Arrangement of the modules at comparatively elevated mounting, better situation than for typical large-scale PV power plant arrays (Figure 5.13).
2 × 3 kWp; BF: 95%
Arrangement: four rows, three modules per row,
Tilt: 35°, Elevation: see paper for arrangement.
Distance to Situation varying, dependent on position (average ~2.5 m–est.)
Situation: grass albedo ~20% (est.[#]); shells ~50% (est.[+]),

Figure 5.13 Small capacity bifacial system by PVGS [15]

* *Varying in snow rich region: 23.9% (snow), 8.6% (grass)*
\# grass albedo 18%–23% [28] ⇒ albedo 20% (est.)
\+ *Shell covered field almost constant ⇒ albedo similar to snow*
Snow albedo 40%–95% (old–new snow) ⇒ albedo ~50% (est.)
Location: Japan
Test duration: 6 months

Bifacial gain:
15.8% (grass) annual average
23.6% (shells) annual average

[15] *Sugibuchi et al., Bifacial PV power output gain in the field test using "Earth-ON" High bifaciality solar cells, 2013*

System 18

Description: Arrangement: several rows, five modules per row (Figure 5.14)
Distance: 2.5 m
Tilt: 15°, AZ: 145°, Elevation: minimum height 30 cm
Situation: 78% albedo
Location: Geilenkirchen, Germany
Test duration: 1 year

Bifacial gain: 23% annual average

[44] *N. Eisenberg et al., Outdoor bifacial module characterization: Energy generation and gain, 2012*

Figure 5.14 Flat roof bifacial system by B-Solar [44]

System 19

Description: Monofacial compared to bifacial module in array and also to stand-alone bifacial module.
Arrangement: single module in 3 × 4 array
BF: 71%
Tilt: 30°, Elevation: 70 cm (lower module edge)
Distance between rows (in S-N direction) 150 cm; between separate modules (in E-W direction) 20 cm
Situation: 50% albedo
Location: Jerusalem, Israel
Test duration: 1 year

Bifacial gain: above 16% (average yearly) for bifacial module in array
Bifacial gain: again additional ~3% to ~13% for stand-alone bifacial

[17] *Kreinin et al., Experimental analysis of the increases in energy generation of bifacial over mono-facial PV modules, 2011*

System 20

Description: Gain dependent on elevation and inhomogeneous illumination of rear side. Gain dependent on season (sun position) and on diffuse to global radiation ratio (Figure 5.15).

Figure 5.15 Flat roof bifacial system by B-Solar [45]

Arrangement: single module in array
BF: 71%
Tilt: 30°, Elevation: 70 cm (lower module edge)
Distance between rows (in S-N direction) and between separate modules (in E-W direction) 150 and 20 cm, respectively
Situation: 50% albedo
Location: Jerusalem, Israel
Test duration: ~2 months (winter); ~1.5 month (summer)

Bifacial gain:
15%–20% (summer, diffuse/ global radiation ratio 10%–70%)
7%–15% (winter, diffuse to global radiation ratio 10%–70%)
30% (both seasons, for diffuse to global radiation ratio 90%)

Remark: Not included in analysis (Table 5.1 and Figure 5.19) due to short measurement duration. Annual values for this test can be found in other reference [17]

[45] *Kreinin et al., PV Module power gain due to bifacial design. Preliminary experimental and simulation data, 2011*

System 21

Description: Arrangement: array 5 × 4 (est. from depiction)
BF: 70%
Tilt: 20°, Elevation: 30 cm (lower module edge)
Situation: 64% albedo
Location: Geilenkirchen, Germany
Test duration: 1 year (2009)

Bifacial gain: 24.3%
Monthly gain: 21.4% (September)–34.8% (December)

[46] *Sanyo module datasheet "HIT double 205", 05/2011*

System 22

Description: 4 × 9 module array (Figure 5.16)
BF: 71% (est. according to similar publications)
Tilt: 30° (est. from depiction), Elevation: 40 cm (lower module edge)
Distance between rows: n.a.
Situation: 35% albedo
Location: Berlin, Germany

Bifacial gain: 11% annual average

[47] *bsolar – results (online)*

Figure 5.16 Ground bifacial system by B-Solar [47]

System 23

Description: Arrangement: 4 × 2 modules in one plane (Figure 5.17)
BF: 71% (est. according to similar publications)
Tilt: unclear, depiction with two tilt angles
Elevation: 70 cm (lower module edge, but two line array)
Distance between rows: n.a.

Figure 5.17 Ground bifacial system by B-Solar [47]

Situation: 50% albedo
Location: Eilat-Eilot, Israel
Test duration: n.a.

Bifacial gain: 17.2% (annual average n.a.)

[47] *bsolar- results (online)*

System 24

Description: Arrangement: 4 × 2 modules in one plane (Figure 5.18)
BF: 71% (est. according to similar publications)
Tilt: unclear, depiction with two tilt angles
Elevation: unknown, also two line array)
Distance between rows: n.a.
Situation: 50% albedo
Location: Saxony, Germany
Test duration: n.a

Bifacial gain: 17% (annual average n.a.)

Remark: Not considered in analysis (Table 5.1 and Figure 5.19), due to data situation

[47] *bsolar-results (online)*

Figure 5.18 Ground bifacial system by B-Solar, setup with 50% ground albedo [47]

Table 5.1 Most relevant properties for bifacial installations; only data from publications with "typical" installation conditions (south orientation, limited tilt) and measurement duration of at least several months are considered

Ref.	Bifacial gain [%]	Bifaciality [%]	"Normalized bifacial gain" [%]	Albedo [%]	Elevation [m] Lower module edge	Tilt [°]
[36]	3.2	95	3.4	20	2.0	45
[30]	10.0	90	11.1	22	0.6	30
[47]	11.0	71	15.5	35	0.4	30
[37]	12.3	95	13.0	22	0.2	20
[30]	13.0	90	14.4	40	0.6	30
[31]	13.0	93	14.0	80	0.1	45
[35]	13.7	–	–	30	–	30
[40]	13.7	–	–	20	2	10
[33]	14.0	–	–	24	1.1	25
[39]	14.3	–	–	30	–	–
[15]	15.8	95	16.6	20	2.5	35
[17]	16.0	71	22.5	50	0.7	30
[47]	17.2	71	24.2	50	0.7	30
[37]	17.7	95	18.6	10	1.3	30
[37]	18.4	95	19.4	70	0.2	20
[30]	19.0	90	21.1	40	1.6	30
[33]	19.0	–	–	10	–	–
[37]	19.6	95	20.6	68	0.2	20
[12]	22.3	90	24.8	30	1.0	20
[37]	22.8	95	23.9	77	0.2	30
[44]	23.0	71	32.4	78	0.3	15
[15]	23.6	95	24.8	50	2.5	35
[46]	24.3	70	34.7	64	0.3	20
[33]	25.0	–	–	24	1.1	25
[37]	30.6	95	32.2	77	0.8	30

While the above compilation is an attempt to register as much published data as possible concerning the bifacial gain, a further summarization is needed for an overview. To obtain comparable data, the most relevant properties for bifacial installations are excerpted from the publications and the data is condensed in Table 5.1. Only data from publications with "typical" installation conditions (south orientation, limited tilt) and measurement duration of at least several months are considered in Table 5.1.

Obviously the very different setup of the systems hinders a simple direct comparison; large deviations concerning the bifacial gain have to be expected. Even for identical setups the different module types alone will obviously cause significant fluctuations. Some other factors are only coarsely known or estimated. So, the ground albedo is often estimated and may also show varying spectral effects [28,34]. Also the diffuse light fraction for the various system locations is not known and varies significantly from site to site. The installation height is in some cases not

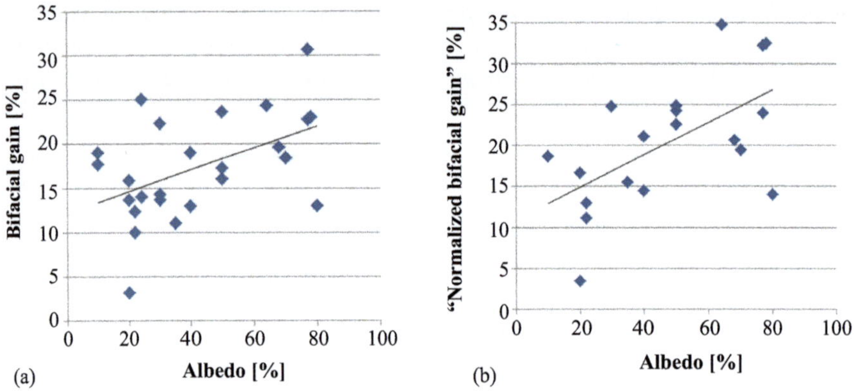

Figure 5.19 *(a) Bifacial gain plotted versus the albedo for "typical" south-oriented arrays. The trend is visible, but the fluctuation range is significant. The smallest observed bifacial gain is above of 10%, except of one outliner. (b) "Normalized bifacial gain" as an attempt to take the different bifaciality factors into account. No obvious improvement and reduced amount of data, but the concept may be useful when comparing more similar PV installations*

clearly defined, particularly if several modules are arranged in one plane with differing height above the ground. The optimum tilt is dependent on the latitude of the installations [1,2], but also on the installation height or the albedo. A correlation of the reported bifacial gain to the available parameters in Table 5.1 was only possible for the albedo.

Based on the data of Table 5.1, the bifacial gain is plotted versus the albedo in Figure 5.19(a). This type of depiction was already presented earlier [38] and is shown here with additional data. While a trend is visible in Figure 5.19(a), also the broad fluctuation range, due to the very differing setup of the published systems, is significant. In spite of the fluctuation it can nevertheless be stated that, neglecting an outliner, for all systems a bifacial gain above of 10% is observed, with increasing values for higher albedo. Published bifacial gains are in a range between 10% and 30%.

Obviously the bifacial gain should be increased for modules with higher bifaciality factor. Therefore, it seems plausible that considering the bifaciality factor could narrow the fluctuation range. As an approach to include the different bifaciality a "normalized bifacial gain" is defined as a hypothetical gain, assuming that all modules have a bifaciality of 100%.

$$\text{"normalized bifacial gain"} = (\text{bifacial gain}/\text{bifaciality}) \times 100\% \qquad (5.2)$$

The corresponding results for the systems with given bifaciality factor (not given for all modules, see Table 5.1) are depicted in Figure 5.19(b). However, no narrowing of the fluctuation range due to this approach can be observed, also the amount of

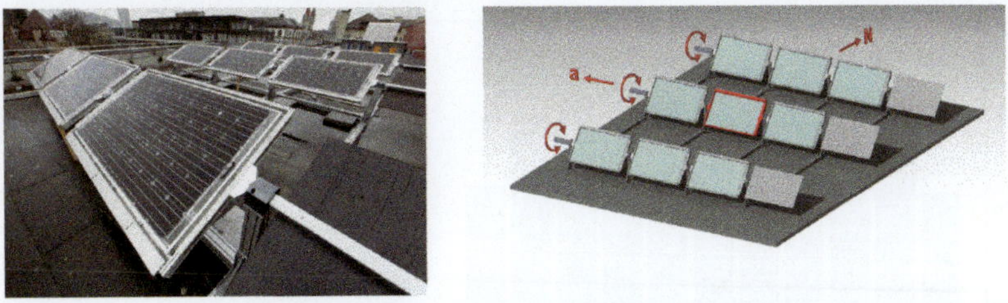

Figure 5.20 *"Bifacial Outdoor Rotor Tester (BIFROT)" is an array for the systematic measurements of bifacial systems with differing mounting conditions (by ZHAW). There is continuous, automated variation of the tilt angle. Other parameters like the distance between adjacent rows, the installation height or the ground reflectivity can be manually adjusted*

publications with reported bifaciality factors was limited. Nevertheless, this concept might be useful when comparing more similar PV installations with different module types. Similar attempts to consider the elevation or the tilt did, as in the case of the "normalized bifacial gain," also not result in a narrowed distribution.

An interesting approach is the application of empirically determined factors for the respective relevant parameters (albedo, installation height and tilt angle), as published by Castillo-Aguilella *et al.* [37] or Solarworld [48]. The effectiveness of these approaches and their generality for varying mounting situations will surely be reported in future publications with increased data sets.

Another way to obtain a better understanding is the systematic analysis of measurement results and the correlation to theoretical predictions. This is also necessary in order to develop and proof algorithms [1,2,7,8] in simulation tools for bifacial applications.

Several test-fields, partly with more extended setups than stand-alone modules in order to reflect the properties in larger PV power plants, are implemented today, particularly at renown institutes [9,29,41,49,50,51]. The results from these test-fields can be of importance for the development of those algorithms which include direct and indirect shading effects in extended arrays [5,8]. Some of these test-fields do also include grounds with differing albedo [29,30,37,49,52] at elsewise identical mounting conditions.

A special installation which belongs to this group of test-fields is the "Bifacial Outdoor Rotor Tester (BIFROT)" [53–55], shown in Figure 5.20, which is an array for the systematic measurements of bifacial systems with differing mounting conditions. The array is based on large, commercially available, 60-cell modules with a continuous, automated variation of the tilt angle of all rows. Other parameters like the distance between adjacent rows, the installation height or the ground reflectivity can be manually adjusted. Due to the setup in form of an array the typical conditions for a module in an extended power plant can be analyzed at the central module(s). Data from the other modules can be used to investigate positions at the

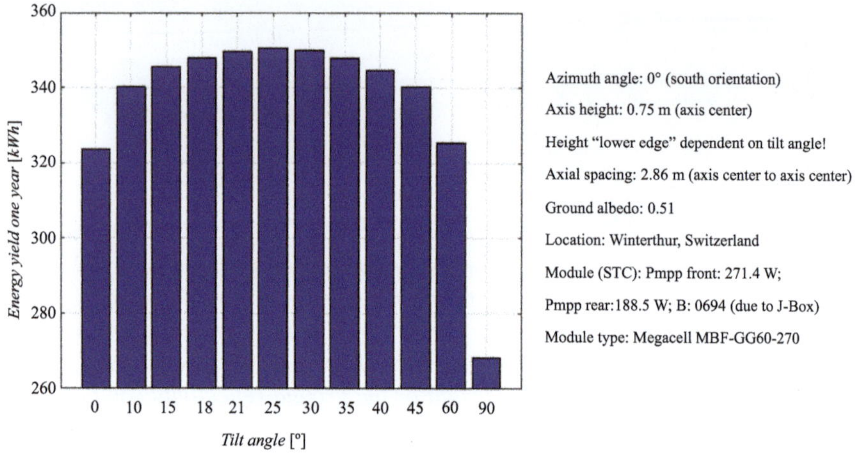

Figure 5.21 Annual energy yield in kWh for the center module in the BIFOROT array. The tilt angle was continuously varied in 12 steps per minute during the one year measurement period, starting in October 2016. For the given specific installation situation, the optimum tilt angle and the sensitivity of the power output are measured

rim of a PV system. As an example in Figure 5.21 the annual energy yield in kWh for the center module in the BIFOROT array is shown [55]. The tilt angle was continuously varied in twelve steps per minute during the one year measurement period, starting in October 2016. For the given specific installation situation, the optimum tilt angle and the sensitivity of the power output are measured. The recorded data set is currently used to test and validate the results of simulation software and algorithms for differing angles and insolation conditions.

5.2.1 Vertically installed bifacial systems

While the above compilation focuses on bifacial systems with the most often applied typical south orientation, bifacial systems are also of interest for installation schemes which would not be feasible with conventional, monofacial ones. Particularly the vertical installation, often with East/West-orientation, is an interesting option and may be a promising approach for several applications [18,19,23,52,56–61]. This type of installation avoids the maximum power generation peak at noon and instead results in a broader generation profile ("peak-shaving") [62]. Besides the options to broaden the generation profile and to minimize dust deposition (e.g. in desert applications) or snow load, simulation data and measurements on single, vertically installed modules are very promising [56,61]. Especially for vertically installed systems however, also shading is obviously very pronounced and the energy yield will heavily depend on the specific lay-out of the PV installation [3,5]. In the following sections some published systems are presented.

System A

Description:
28 kW installed in May 2015 (Figure 5.22)
Location: Germany
Albedo: ~20%
96 modules in 3 rows, 12 strings
10 m row spacing
Customized 66-cell module with n-type bifacial cells,
BF = 87%
Test duration: 3 years

Bifacial gain: +10% annual gain
Remarks in presentation:
BF < 80%: same or lower yield
BF 85%–90%: ~10% more yield
BF 95%–100%: ~15% more yield
For typical conditions in Germany

[57] *Hildebrandt, 3 MWp vertical E-W oriented system in Germany, 2017*

Figure 5.22 Vertical array with wide spacing for PV/agricultural use of land by Next2Sun

System B

Description: Vertically installed bifacial module; E/W-oriented, in a small array compared to monofacial module (30° south). Also compared to bifacial, south, 45° tilt angle [31] (Figure 5.23).

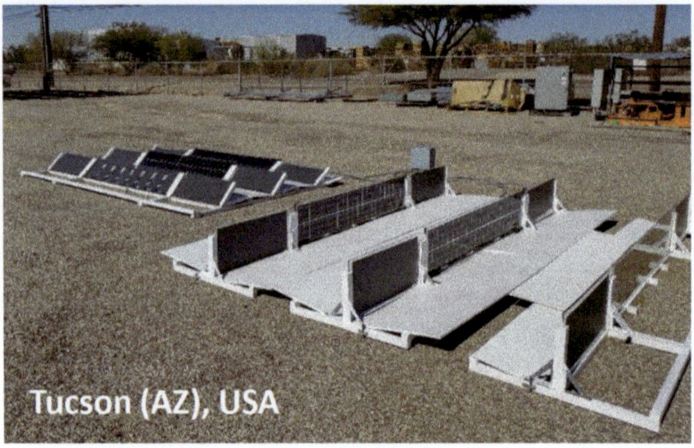

Figure 5.23 PI Berlin test array [31]. Bifacial South-oriented array with 45° tilt angle in the back-ground, vertical installation in the foreground

(see system 1 for more detailed description)

<u>Bifacial gain compared to monofacial 30° south:</u>
Vertical E/W: + 15% annual average with pronounced seasonality: –15% to +40%

[31] *Podlowski et al., Yield Study on Identical Bifacial Rooftop Systems Installed in the USA and in Germany, 2017*

<u>System C</u>

<u>Description</u>: Vertically installed 20 cell bifacial modules (Figure 5.24)
Location: Winterthur, Switzerland

Figure 5.24 Bifacial modules mounted in landscape orientation with spacing by Solarspar [52]

120 modules, approximately E/W oriented
Cell types: Megacell BiSoN, Meyer Burger HJT-SWCT
System installed on green roof. Patterns with differing planting.

Data monitoring just started

[52] *Dreisiebner, Smart Solutions for Extreme Conditions, 2017*

System D

Description: 30 kW; 315 modules
Arrangement: Vertically installed, fence-type, 30 kWp (front STC)
Vertically installed bifacial and monofacial modules with southwest/northwest orientation are compared to monofacial modules with 30° south orientation
Tilt: 90°, Elevation: ~1 m (estimation, based on depiction in publication)
Situation: fence like installation, southwest/northwest orientation, two parallel rows of about 180 m length, distance between the rows n.a.
Location: Aichi, Japan
Test duration: 1 year

Bifacial gain: 10% loss compared to monofacial 30° south oriented

[58] *Araki et al., Bifacial PV system in Aichi Airport-site Demonstrative Research Plant for New Energy Power Generation, 2009*

System E

Description: Several arrangements to test pole-mounted and fence-like applications of bifacial modules. Modules mounted with various azimuth angles.

Bifacial gain:
~90% yield compared to monofacial south with optimum tilt angle (30°)
130%–150% yield compared to monofacial south vertical

Approximately independent on orientation

[60] *T. Joge et al., Applications and field tests of bifacial solar modules, 2002*

5.3 Bifacial systems with non-standard mounting situation

Bifacial PV modules show a wider range of possible arrangement options than monofacial standard devices. For fixed monofacial modules, there is an optimum tilt angle which can be easily be determined. The two-sided sensitivity of bifacial modules allows a wider range of orientations which results in new options for innovative installation schemes, and improved efficiencies for others, such as floating applications. In the following sections, there is a closer look on vertically installed bifacial modules and floating applications.

5.3.1 Vertically installed bifacial systems

Installing bifacial modules vertically has been described many years ago. There have been mainly two perspectives: The use of bifacial modules as integrated, "dual use" devices within functional structures like noise-barriers or as objects of scientific research on the behavior of vertical bifacial systems in general. Both, increased research activities as well as fast-growing interest and associated expanding manufacturing capacities made bifacial modules more inexpensive over the past few years. Therefore the vertical installation concept also becomes more attractive and may possibly even be used in utility-scale ground-mounted systems.

5.3.1.1 Early approaches

Most experiences in applications of bifacial modules, especially in noise-barriers, were gathered in the late 1990s by TNC [18,19]. Although the yield and operational efficiency of these systems were not satisfying yet, they figured out that vertical bifacial installations have the potential to achieve at least the same or even a better yield compared to a conventional monofacial system. Unfortunately, this very early pioneer work has not been carried on. For this reason there was almost no progress in development of bifacial module technology. The only commercial bifacial module, Panasonic's "HIT double," has been expensive even though the bifaciality gain was mediocre. This situation endured for almost two decades.

Scientific approaches have been conducted by several groups, showing the basic effects and possible gain of this installation concept. A compilation of these approaches was presented in the preceding section (Systems A to E); in which the obtained bifacial gain values are compared to the also listed more common installations.

All of this scientific research has the disadvantage that important effects are not represented in the chosen test setup. In most cases, only one or two modules were installed in a non-natural albedo situation. Therefore the following effects are (more or less) neglected:

- mutual row shadowing
- real albedo situation, which is also affected by shading from adjacent rows
- shadowing from mounting system

Nevertheless, important insights have been gathered by this scientific work, as the potential of vertical installation regarding the main issues like yield potential and peak shifting was shown clearly. In practice, vertical bifacial systems have pulled some attraction for special applications like noise-barriers or fences, but there were (at least as far as known by the author) no approaches for the use of such installations within large-scale ground-mounted systems.

5.3.1.2 Next2Sun's test site

In 2015, German Startup "Next2Sun" has installed a 28 kW test site to find out how vertical bifacial installations perform in a realistic test setup for large-scale ground-mounted power plants (Figures 5.25 and 5.26).

Module type	Customized double-glass module with n-type cells
Bifaciality	87%
Pitch	11 m
Modules per row	2 × 16 modules in landscape orientation
Total height	3.0 m
Aperture height	2.0 m
Mounting system	Proprietary, steel based post-and-beam construction

5.3.1.3 General aspects

Vertical bifacial systems show some very interesting behavior regarding their yield and load profile, resulting in a broad variation of possible applications.

One characteristic of such a system is that the annual yield has only a low dependency on the chosen azimuth. Having the aperture area facing East-West or facing North-South leads more or less to the same annual yield. In fact there are differences, but these are smaller than those coming from bifaciality-factor and other effects. The same assumption can be made for any other azimuth orientation in between.

The interesting point is that the daily and seasonal production profile is completely different when comparing varying azimuth orientations.

This means that the vertical installation (of bifacial modules) provides an enormous potential for customized PV plant concepts. These concepts could be

Figure 5.25 Specific monthly yield from the vertical E/W-oriented bifacial system compared to a monofacial reference

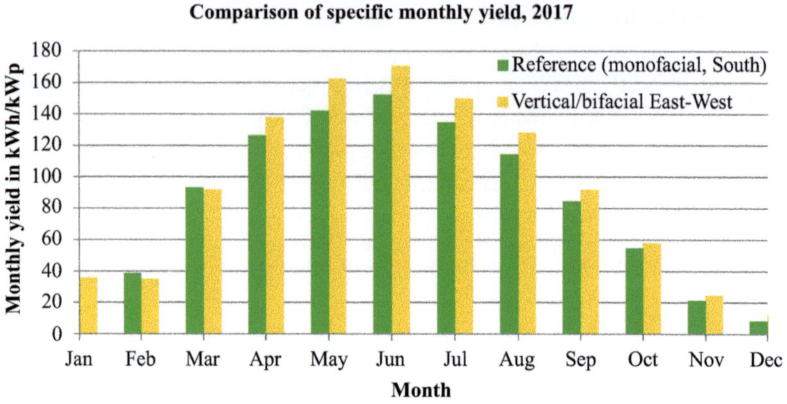

Figure 5.26 Specific monthly yield from the vertical E/W-oriented bifacial system compared to a monofacial reference

able to meet the requirements from the power grid or some special requirements of any given application. For example, daily energy consumption profiles of dairy farms can be matched almost perfectly with vertical bifacial PV-plants in East-West-alignment.

Another advantage and at the same time a disadvantage (due to higher area-related cost) is their low-impact land use for large ground-mounted vertical systems. This brings new opportunities of integrating additional functionalities into PV plants or, seen from the other side, integration of PV into structures which are primarily meant for other purposes ("dual use").

5.3.1.4 Energy yield

Compared to a South-orientated solar power plant the expected energy yield for a vertical East/West-orientated installation is around 5%–15% higher. It depends on the modules bifaciality gain, the distance between the rows, as well as the height of the installation. There are two characteristics that lead to the additional yield:

1. The harvest of diffuse irradiance at the sun-averted side of the modules.
2. The harvest of irradiation reflected by the ground at the front side as well as at the back side of the modules.

There is also a latitude dependency for vertical East/West–orientated solar applications using bifacial modules comparing to conventional plants (Figure 5.27).

In the northern hemisphere an optimal South-orientated module has an azimuth angle of 0° while compass direction West and East are defined by an azimuth angle of 90° respectively −90°. During the year the timing of sunrise and sunset changes, which means that the azimuth angle where the sun crosses the horizon is changing. An optimal south-orientated monofacial module cannot harvest direct irradiation which comes from an azimuth angle of more than 90° respectively less than −90° because the direct irradiation hits the optical inactive modules back-side.

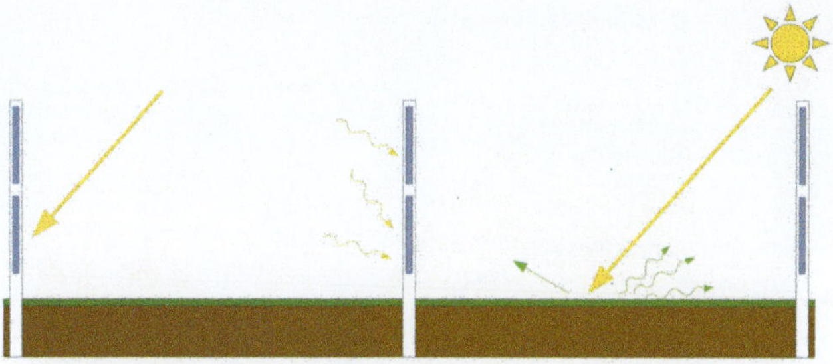

Figure 5.27 Harvest of direct and diffuse irradiance on the sun-averted side of the module and harvest of reflected light from the surrounding ground by both sides

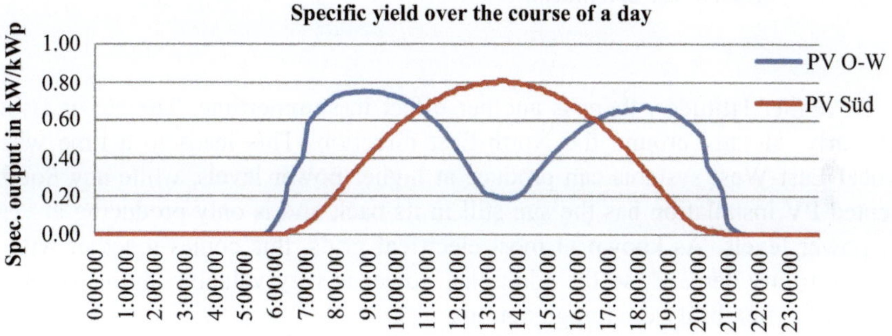

Figure 5.28 Typical symmetrical production profile for an East-West–oriented bifacial PV system with two peaks in the morning and in the evening. South-oriented monofacial systems have a single peak at noon

In comparison vertical East/West–orientated modules can use direct irradiation that comes from an azimuth angle of more than 90° respectively less than −90°. Depending on which latitude the vertical installation is erected, this effect enables a significant additional yield. Furthermore, the disadvantage of vertical installations, which need more space for the same amount of installed power than conventional plants, is decreasing for higher latitudes. The reason is conventional plants installed at higher latitudes have higher space requirements. At higher latitudes South-orientated modules have to be installed steeper and consequently in bigger distances to avoid mutual module shadowing. East-West–orientated bifacial systems are not affected by these effects.

5.3.1.5 Daily production profile

If a vertical system is orientated exactly facing East-West, it shows a symmetrical production profile with two peaks. One is in the morning and the other one in the evening. This profile is complementary to the generation of a conventional plant (Figure 5.28).

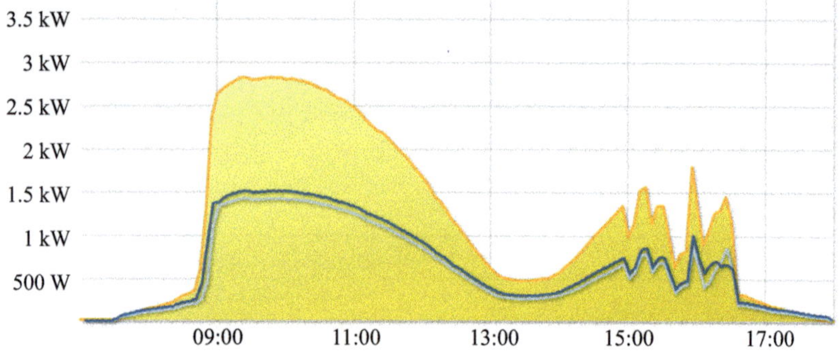

Figure 5.29 Daily production profile of a vertical bifacial test installation in Kaiserslautern (Germany), which is turned 15° from the exact East-West orientation

At higher latitudes, there is another effect in summertime. The sun is rising very early, already around the North-East direction. This leads to a time when vertical East-West systems can produce at higher power levels, while any South-oriented PV installation has the sun still in its back and is only producing at very low power levels. As known in most electrical grids, this comes together with a high electricity demand, so the additional power is very valuable and is lowering the need for flexibility or storage options.

Turning the alignment of such an installation slightly, for example from 10° to 20°, means that one of the two production peaks will slightly broaden into the early/late hours, whilst the other one will become smaller and slightly narrower. An example is shown in Figure 5.29.

This can be a very valuable option for the optimization of the electricity generation profile in small networks or self-consumption applications. The production of a PV generator can therefore be adjusted with the knowledge about the electricity consumption of one or a group of end users. This kind of optimization has only a low impact on the annual yield. The LCOE of such a system is barely affected, but the value of the energy produced can be increased.

5.3.1.6 Seasonal production profile

Going away from the short timescale of a day to a seasonal view, there are also strong differences of vertical installation in a given orientation, compared to such one in another orientation or to conventional plants. Note that this part is valid for medium and high latitudes only, as the author has no validated knowledge on this issue for near-equatorial sites.

Again looking firstly on the East-West variant, the seasonal production shows differences:

- Higher yield in summer as a result of "additional" operating hours during periods with a sun azimuth angle of more than 90° respectively less than −90°.

- The same or a lower yield in spring and autumn. Because the suns elevation angle during sun rise respectively sun set changes slower in spring and autumn the effect of mutual row shadowing becomes more significant.
- Nearby the same yield in winter. Since there are no negative impacts by snow on the modules or its supply structure there is even an additional yield caused by reflected irradiation at the ground. So for snow-prone regions or snowy periods, there is a large advantage of the vertical system.

From the grid operators point of view in most cases this seasonal behavior of vertical solar plants is not an advantage. Still, solar energy production is volatile, often the grid demand is changing seasonal or during the course of the day and, of course, the energy production by the solar system is much higher in summer than during the winter.

However, having in mind that orientation of a vertical system does not strongly affect the annual yield, this can be mitigated by changing from east-west orientation to South-North orientation. While the advantage regarding the daily production profile will vanish, a higher portion of the annual yield comes in the winter half of one year. Naturally, the characteristic of solar irradiation cannot be changed and the yield in summer will still exceed the yield in winter by far, but the balance can be shifted significantly toward wintertime.

5.3.1.7 Dual use application: "Agro-PV"

So-called "agrophotovoltaics-concepts," which use of the same area for farming and PV, have aroused interest in the last few years. The most popular approach for Agro-PV is the elevated installation of PV modules with a height of at least 4 m from the ground. This enables agricultural machines to drive through and use the land underneath. The module array causes shadowing on the ground, what can be advantageous or disadvantageous, depending on the climate conditions and the crops to be cultivated.

Vertical bifacial PV plants open up a new path for Agro-PV-concepts. There is almost no coverage on the ground area and nearly no influence on the distribution of irradiation and rainfall. Therefore, the impact on growth and maturation could be small. Usually these kind of installations have a row pitch of at least 10–20 m, which enables the use of regular agricultural machinery, especially when using modern GPS-controlled machines.

At the moment, there is a lack of experience on agrophotovoltaic-concepts in general and especially for such based on vertical bifacial systems. Scientific and practical knowledge has to be gathered and evaluated, so a real "dual use" of farmland with PV and agriculture can become reality.

5.3.1.8 Dual use application: "Natural PV"

Natural habitat protection is becoming more and more important as a restriction for the development of large-scale PV plants. Especially highly developed countries with a high population density have a lack of areas without intensive human usage.

Conventional PV plants are usually in conflict with habitat protection. A ground coverage of more than 50% leads to a strong interference of nature and environment, because the previous vegetation cannot be obtained. In most cases substantial habitat functionalities are lost, which further leads to an occupation of farmland, so these ecological functionalities can be equivalent restored.

Most of these issues can be solved or reduced by installing the modules vertically. With their low impact on the environmental conditions of the vegetation, interference of those plants on the flora is low in general. In most cases, a PV site was used as farmland before. Changing such areas from intense agricultural use to a more extensive use, for example as grassland, combined with a vertical PV plant will typically increase the ecological value of the area. Fertilization and erosion will be reduced, coming together with a higher biological diversity and increased habitat functionality.

In countries with a progressive legislation on environmental protection, there is a kind of a "market" for the valorization of ecological functionalities and habitats. These instruments could be used to generate additional returns from the operation of a vertical PV system by developing and maintaining valuable ecological habitat structures on the same area.

5.3.1.9 Dual use application: PV integration into functional structures

Basically there are many vertical structures which can be used for PV integration. Following examples of use are of particular relevance:

- Noise barriers
- Fences
- Railings

Even though first applications of bifacial modules in vertical alignment were their integration into existing structures like noise barriers, these examples are still niche applications. Obviously, economical advantages coming along with the double use of a structure are counterweighted by other economical disadvantages or nontechnical barriers. Those barriers have to be overcome to make use of this high potential for electricity generation without any additional land use.

Often a general obstacle for such "dual-use" applications is that the functional requirements of the primary structure and additional PV generation differ. For example, noise barriers need a rough surface for good absorption and high weight for good barrier functionality. In contrast solar modules are weight-oriented designed and have a low-structured glass surface.

Other issues making real double-use applications challenging are:

- Administrative hurdles: Long authorization processes (e.g. for noise barriers along highways) as well as a general lack of interest for PV applications by approval authorities and local politicians.
- Safety aspects: Adding a PV system into a public building turns a simple building structure into an electrical installation which has much higher safety requirements.
- Functional constraints: Since vertical installations rather harvest sunlight with low incident angles they are more sensitive for shadowing from surrounding objects.

Figure 5.30 *Vertical green roof installation of special-designed bifacial modules in Winterthur, Switzerland*

Nevertheless, it is important to face these challenges and follow up the dual-use approach because in industrial countries land use becomes more and more restricted. Furthermore, in most industrial countries land prices rise constantly.

5.3.1.10 Vertical bifacial on rooftops

Although their typical application may be ground-mounted plants or integration into existing vertical structures, there are also approaches of using vertical bifacial PV on rooftops. One advantage compared to ground-mounted plants is the possibility of reaching far higher albedo values. But, on the other hand, mechanical engineering is more challenging when using standard modules, as their exposed location leads to a strong momentum resulting from the wind load.

A pilot project is being conducted in Winterthur, Switzerland, by ZHAW (Zurich University of Applied Science) and Swiss society "Solarspar" [52], where some special-designed bifacial modules are installed on a green roof (see Figure 5.30). For vertical installations the distance between the rows is a decisive factor with regard to the mutual shading losses. Due to their narrower module geometry the 20-cell modules can be placed with a row distance of 1 m, which results in the same losses as for broader 60-cell modules and 3 m row distance.

The special module design avoids the high wind loads coming along with the use of standard modules with 1 m in height. Also the visibility from the ground is reduced and the general appearance is improved. Most important, the vertical installation of the modules suppresses the conflict between green roofs and PV, due to the often almost complete covering of the roof area with modules. Because of the better accessibility also the maintenance of the green roof is improved compared to typical dense PV installations where the plants are located below extended module areas.

5.3.2 Floating bifacial PV

5.3.2.1 Floating PV applications in general

Using water as basis for PV is relatively new, but strongly upcoming application area. The main reason for turning to floating PV is the land use of ground-mounted PV systems. In many areas of the world land is scarce and PV might compete with e.g. agricultural use. In other geographical areas, there simply is not enough usable land to supply renewable energy locally.

Taking into account that 71% of the earth's surface area consists of water, it seems a logical step to consider floating PV. Large patches of freshwater are potentially available, as long as the original function of the water surface is not compromised. In such dual use applications, energy generation can go hand in hand with other use of the water. Clear examples are industrial or agricultural water basins, sand or gravel mining pits, lakes and canals. Next to these freshwater applications, the seas and oceans provide a vast potential of surface area. Developing such solar PV at sea projects is on the roadmap of several system and project developers.

A clear advantage of floating PV is the potentially large scale of projects. Many water areas are far larger than available land areas, leading to larger project scales and lower cost/kW. A floating system in general leads to somewhat higher cost compared to ground-mounted systems. However, it is expected that larger project scales and additional benefits of water will result in a lower overall LCOE for water-based systems.

5.3.2.2 Beneficial effects of water on PV systems

Apart from the available surface area, a number of benefits can be identified for water as basis for solar PV installations: additional cooling, higher insolation of water areas, and reflection of light on water and thus higher irradiation of the modules.

1. The cooling effect is ascribed to the lower temperature of the water body, due to temperature inertia of the water mass, compared to a roof-top or ground. This leads to lower average temperatures for PV modules and thus to higher performance when air temperatures rise. In general, this effect will be largest for systems that position the PV module close to the water surface.
2. Higher insolation of water areas is caused by the difference in heating of land and water. Above the warmer land area's moist air is driven upwards, where it cools down leading to cloud formation. This effect is much slower above the cooler water surface. Finally, on the edge of water and land the moist air is drawn away from the water toward land. Therefore, clouds form much more readily above land and the water area sees more open skies.
3. The beneficial effect of reflection on water is discussed in detail below for bifacial PV modules. Monofacial modules probably do benefit from reflected light from the water surface onto the front of the module. However, this effect is much lower than the bifacial gain measured on water.

As an added benefit of installing a PV system on water some system manufacturers mention the shadowing effect of the PV system. This leads to lower water

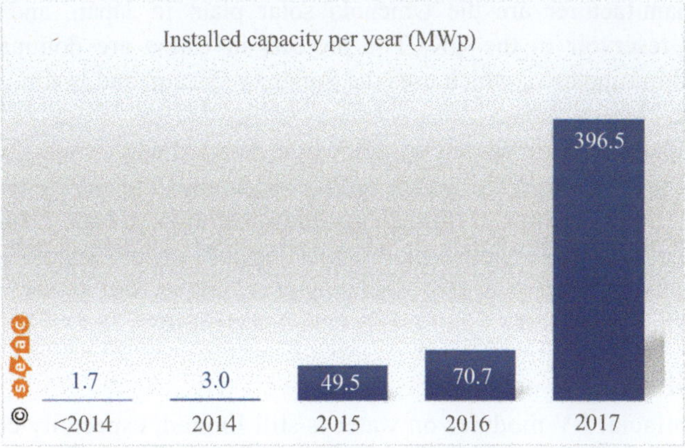

Figure 5.31 Installed floating PV capacity per year (MWp); taken with permission from seac [63]

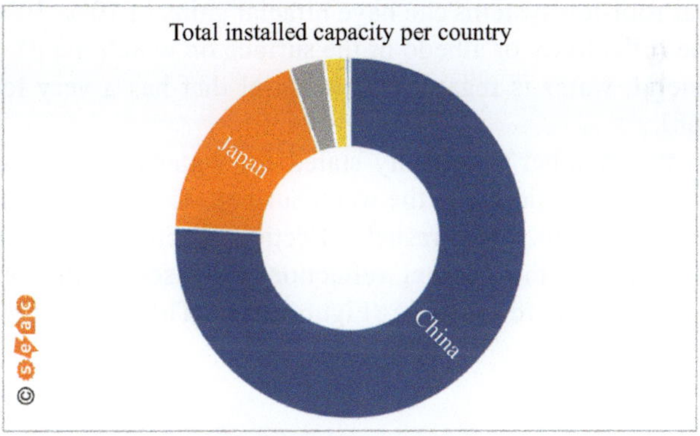

Figure 5.32 Division per country of currently installed floating PV capacity worldwide; taken with permission from seac [63]

temperatures and thus less evaporation. This is possibly beneficial for drinking water or agricultural water basins. However, in natural areas shadowing could have a negative effect on ecology.

5.3.2.3 Status of floating PV systems installed

Currently installed floating PV capacity amounts to well over 500 MWp worldwide, of which most capacity was installed in the last three years (Figure 5.31) [63]. About 90% of the capacity was installed in Japan and China, as a small number of very large projects (Figure 5.32).

A floater system used in these large-scale projects is the Ciel & Terre Hydrelio® system that uses hollow plastic floaters onto which the PV modules are mounted close to the water surface area at a low angle. The type of floater leads to enclosed fields of floaters, which cover most of the water surface. Typical projects from this manufacturer are the Umenoki solar plant in Japan, and the Queen Elizabeth II reservoir in the UK. The projects in China are dominated by the Huainan coal mining area, which uses the Sungrow system, and is similar in design to the Ciel & Terre system.

Next to these systems, based on fields of connected and closed floaters, other systems are more open to the water surface. Examples are the system by NRG Energeia [64] and the Koine Multimedia systems that involve open floaters systems and mirrors or trackers for added light harvesting [65] or the Solaris Synergy system that is marketed as having self-regulating panel angles [66]. Other systems also include tracking such as the SunProjects island system [67].

5.3.2.4 Bifacial PV on water

The use of bifacial PV modules on water is still limited, especially compared to the large-scale floating PV projects described above. Sunfloat developed a dedicated floating system for bifacial modules. Next to the advantages of floating PV already mentioned, bifacial modules add an extra and significant advantage, the bifacial gain. As shown earlier in this chapter, bifacial modules in ground mounted or flat roof-top systems can have bifacial gains of 10%–30%, depending strongly on the reflectivity or albedo of the surface on which the PV modules are placed. In general, water is regarded as material that has a very low albedo of below 10% [68].

However, this number commonly stated for water albedo is valid at a 0° incident angle, or perpendicular to the water surface. This would correspond with an overhead sun position. The Fresnel reflection function shows that at larger incident angles (>65°) the Fresnel reflection increases in the range of 0.05 steeply up to 1 at 90° incident angle (Figure 5.33). Thus, reflection of light on

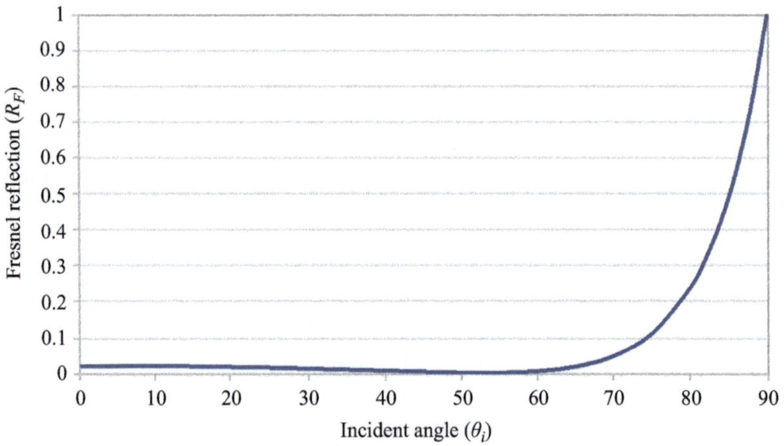

Figure 5.33 Fresnel reflection curve for air-water interface at different incident angles

water should be especially pronounced at low incident angle, e.g. at the edges of the day (dusk and dawn). In addition, in the Northern hemisphere the reflection is higher during autumn, winter, and spring, when incident angles are larger during the full day.

These numbers are calculated for ideal conditions, ignoring waves and fouling of the water. Waves lead to more diffuse reflection patterns, and temporarily distorted incident angles. This could have an enhancing effect on the light reflected from the water surface. In addition, floating particles in water cause light scattering in different directions. This also might have a positive effect on the reflection of sunlight.

Therefore, practical reflection of sunlight from a water surface is much higher than the textbook albedo number would suggest. This is especially true when averaged over the year, because of variations in incident angle along with the changing seasons.

5.3.2.5 Example for floating PV system—Sunfloat

Sunfloat designed a system with bifacial PV modules in mind. Key in this design is a floating structure that is open to the water surface. Module tilt angles of 25°–45° were chosen, to allow for maximum use of the light reflection on water. In addition, the open structure has little or no effect on the water ecology. No permanent shading occurs and oxygen entry into the water is not limited by the floating structure (Figure 5.34).

Solar modules on moving floats will endure more mechanical stress over their lifetime. Accordingly, glass–glass bifacial modules are used as PV modules for their increased durability. Glass–glass modules are more resistant to micro crack formation and generally come with a longer performance warranty. Another advantage of double-glass modules is their recyclability after their useful life [69].

Figure 5.34 Pilot test of 5.5 kWp Bifacial PV system Sunfloat

Figure 5.35 Artists impression of the Solar Lagoon

The Sunfloat system is scalable in modular units of around 1 kWp. Currently, several projects of 0.5–1 MWp are under development. In addition, Sunfloat has launched the concept of the Solar Lagoon in the Dutch lake IJsselmeer (Figure 5.35). This Solar Lagoon is regarded as an ideal interplay between nature, recreation, and renewable energy production. The solar PV capacity of a stretch along the Afsluitdijk (sea barrier) in this lake would allow for installation of up to 6 GWp of floating PV capacity.

5.3.2.6 Performance

The performance of the representative floating bifacial pilot test system of 5.5 kWp (front-side flashed power) described above was measured (Figure 5.36). Daily energy yield was compared to the theoretical performance of an equivalent monofacial ground-mounted system based on local weather conditions. Daily full load sun hours during the monitoring period were obtained from a national database for the nearest measurement location, and were used to predict this performance.

The data show a comparable to slightly lower energy yield at low full load sun hours, and significantly higher energy yield when the number of full load sun hours is average to high. The observed bifacial gain in energy yield ranges up to 90% during some of the days with better sun conditions. On average, during the short monitoring period, a bifacial gain in energy yield of over 40% was observed. As this data has been acquired during winter time, the high incident angles of the sunlight could cause a more pronounced bifacial gain. Due to the factors mentioned above, it is difficult to predict the yearly bifacial gain by theoretical modeling. Further insights are expected from field results from the expected increasing number of floating bifacial PV installations.

5.3.2.7 Conclusion

Floating PV offers clear advantages over ground-mounted PV systems. On top of the general advantages, bifacial modules add the bifacial gain as extra advantage. Based on optics theory, bifacial gain on water is expected most at higher incident angles or lower sun positions. Experiments in the field show that—for flat light angles of the incident light—average bifacial energy gains of well over 30% can be reached with bifacial PV on water, making it comparable to white rock or white painted roof-tops. Therefore, water is a highly suitable surface for large-scale bifacial PV plants.

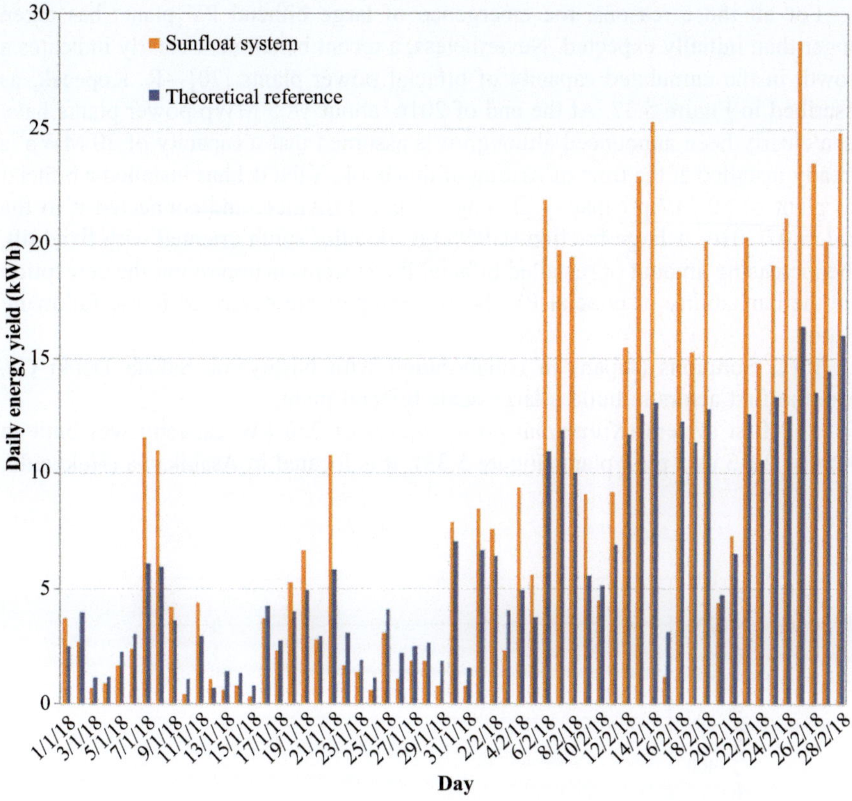

Figure 5.36 Daily energy yield (kWh) for a 5.5 kWp Sunfloat system in the period 1/01/2018–28/02/2018, compared to a theoretical monofacial reference

5.4 Overview of large-scale bifacial systems and growth perspectives

As mentioned previously, a bifacial gain between 10% and 20% is generally reached on small systems depending on the parameters such as albedo and module height. Nevertheless, there is a clear requirement to demonstrate the economic interest of bifacial concept through the construction of large PV plants (> MW). First of all, it is necessary to quantify accurately the bifacial gain on a large power plants but it is also mandatory to prove on an economic basis that the bifacial gain is not annihilated by additional costs at the system level such as:

- Bifacial module integrating bifacial cells
- Structure frame specificities to increase module height and limit rear shadowing
- Ground preparation for increase albedo
- Specific inverters with adapted power dimensioning.

For all those reasons, the emergence of large bifacial PV plants has taken longer than initially expected. Nevertheless, a recent benchmark clearly indicates a growth in the cumulated capacity of bifacial power plants [70]—R. Kopecek, as described in Figure 5.37. At the end of 2016, about 18.5 MWp power plants have been clearly been announced although it is assumed that a capacity of 30 MWp is already installed at the time of writing of this book. YINGLI has installed a bifacial PV plant of 50 MWp capacity Datong, Shanxi Province, and connected it to the grid in 07/2016. A large fraction (>95%) is modules south oriented with fixed tilt. The increasing amount of installed bifacial PV systems is improving the perception and the bankability of bifacial PV. Some examples are described in the following section.

PVG Solutions (Japan) in collaboration with Nishiyama Sakata Denki Co. were the first actors to build a large-scale bifacial plant.

The first system (*Kuranuma power plant*) of 250 kW capacity was built in October 2013 as a pilot plant (Figure 5.38). It is located in Asahikawa (Hokkaido,

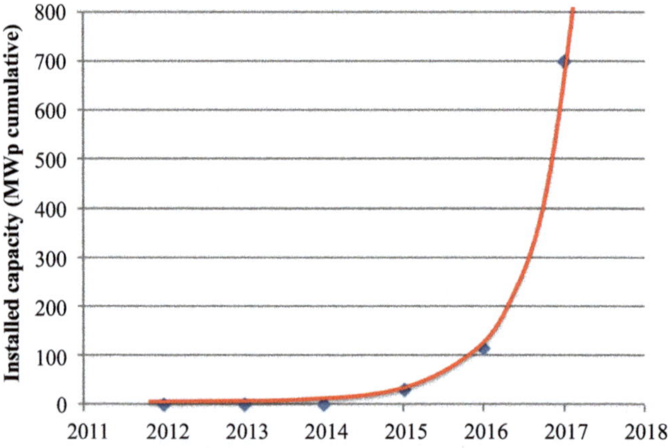

Figure 5.37 Installed cumulated capacity of bifacial PV plant since 2011

Figure 5.38 Kuranuma power plant (250 kW, PVGS & Nishiyama Sakata Denki Co.)

Japan). The plant, specially adapted for snowy regions, has the following characteristics:

- 1064 modules (PST254EarthON60, front-side power in STC: 254 W)
- Orientation: landscape, south, fixed tilt (40°)
- Height: minimum 1.5 m.
- Arrangement: 4 m long (four modules respectively along vertical axis)
- Number of modules per inverter: unknown
- Monofacial reference system: not included
- Albedo: variable (20% (Bare Soil [34]) up to 90% (fresh snow [34])

The mounting structure shown in Figure 5.38 integrates metallic frames on the rear side. Although those frames will induce rear shadowing, their presence can be justified by cost issues and be required for mechanical strength toward environmental impact (wind, snow, etc.). This feature is generally not seen for small systems which optimization is mostly based on an increase in the bifacial gain, independently of cost issues. Currently, a very limited amount of studies concerning the impact of shadowing on the bifacial gain have been released in the literature. One was proposed by PVGS in collaboration with University of Miyazaki (Japan) [71]. It shows that the impact of a metal plate (width = 75 mm) located on the rear side of the module induces a power loss rate between 0.6% and 4.8% depending on its distance from the module. In this system, the 60-cell module had three by-pass diodes. The experiment also showed the formation of a slight hot spot with a maximum increase of 10°C in temperature at the rear sided shaded area.

Another complete study was also recently released by van Aken *et al.* [72]. Diffuse rear irradiance was created by placing scattering white panels at 1 m distance behind the module. A black, white or "aluminum-colored" object of 10 cm width and 25 cm height was placed at a variable distance between the module and the scattering panels and positioned to shade two cells or four cells from the same string. The rear-side irradiance per cell was measured under the same conditions. It was shown that with decreasing distance the observed drop in current due to indirect light shading increases. For thin objects close to the module, no current drop is observed for cells not directly behind the object, but a small reduction is observed when the object is 10–20 cm from the module. The drop in current is in good quantitative agreement with the measured reductions in rear side irradiance. This study also shows that increasing the reflectance of the near-field object, strongly reduces the drop in current (Figure 5.39).

These results combined with indicated bifacial gain for large systems indicates that the optimization of the mounting structure to avoid rear shadowing is not as critical as for the front side.

Kuranuma pilot plant was used as a reference model to build a larger plant in May 2013 of 1.25 MW capacity (Hokuto Solar Power Plant) in the same location (Figure 5.40). Over a period of 32 months, an energy yield over 1,200 kW/year is obtained although latitude 43.5N and heavy snowfall in winter. Based on estimated generated power for a monofacial system at the given location, the bifacial gain is considered to be over 20% [73]. Another significant advantage relies on increased

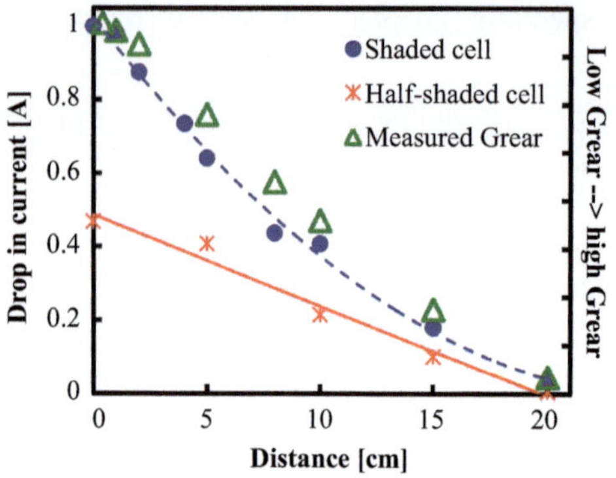

Figure 5.39 The drop in current versus the shade distance. Triangles indicate the measured irradiance, with numbers in reverse direction [72]

Figure 5.40 Hokuto Solar Power Plant (1,250 KW, PVGS & Nishiyama Sakata Denki Co.)

Figure 5.41 Power Plant for fixed tilt (2.5 MW, "La Hormiga," Chile)

performance of the bifacial system in a snowy environment. Indeed, the rear side produces more energy due to higher albedo and accelerates the snow melting on the front side due to rear irradiance (thermalization effect).

Two additional bifacial power plants were installed in Chile in 2016 by MEGACELL and ENEL groups. The power plants have the following characteristics:

- "La Hormiga" power plant (MegaCell/Imelsa) (Figure 5.41):
  - Capacity: 2.5 MWp
  - ~ 9,090 modules (BiSoN solar module, front-side power in STC: 275 W)
  - Orientation: landscape, north, fixed tilt

Figure 5.42 Power Plant with single axis tracking (1.25 MW, la Silla, ENEL)

- Height: n.a m.
- Arrangement: n.a × 3 modules respectively horizontal and vertical axis
- Shadowing: limited as the mounting structure is not under the solar cells (see picture)
- Monofacial reference system: included
- Albedo: white quartz (~40% (desert sand—[74]))

- La Silla power plant (ENEL) n°2 (Figure 5.42):
  - Capacity: 1.7 MWp
  - ~ 6,070 modules (BiSoN solar module (MBA-GG60 270/280 Wp), front-side power in STC: 280 W)
  - Orientation: portrait, north, horizontal single-axis tracking
  - Height: minimum n.a.
  - Arrangement: 4 × 2 modules respectively along horizontal and vertical axes
  - Shadowing: limited as the mounting structure is not under the solar cells (see picture)
  - Monofacial reference system: not included
  - Albedo: variable (~40% (desert sand—[74]))

The owners of these plants want to demonstrate the potential of bifaciality with these bifacial fields. Currently, these power plants have been connected to the grid, but no detailed data about the energy yield is yet available. As according to the system configurations a bifacial gain of 30% for the fixed tilt system (la Hormiga) and 40% are respectively expected by the fixed tilt plant and for the single-tracking plant [75].

Currently, a very large bifacial PV plant (12.8 MWp) was installed by SUNPREME and is located in New Jersey (USA). The 12.8 MW installation project (Figure 5.43), which began in mid-2015 was commissioned in February 2016.

Figure 5.43 Power Plant SUNPREME (12.8 MW, New Jersey, USA)

Initial energy production numbers are showing the results expected with an 8%–10% additional energy harvest. Different albedos will be evaluated to further maximize the energy harvest of the system [76].

Power plant characteristics:

- Capacity: 12.8 MWp
- Sunpreme modules (MAXIMA GxB 310 W Bifacial Module, integrated HJT cells—bifacial double-glass power)
- Orientation: portrait, south, single axis tilt
- Height: n.a.
- Arrangement: four modules along vertical axis
- Rear shadowing: unknown (see picture)
- Number of modules per inverter: n.a
- Monofacial reference system: unknown
- Albedo: variable (~40% (desert sand—[74]))

The largest bifacial PV system that has been completed and connected to the grid by 05/2017 is the 50 MW "Top Runner" project in Datong City, Shanxi Province that has been built using 186,120 bifacial modules from Yingli with 285 W front-side power per module [77].

At the time of writing of this book, several big cell and module producers such as e.g. Trina and Longi are switching a part of their capacity to bifacial PERC+. Both companies published information about the setup of 20 MW bifacial PV systems to be built with bifacial PERC+ modules.

Another important application that is also suited for large-scale ground-mounted plants that allow the simultaneous utilization of the ground for agricultural purposes (e.g. as farmland) is the vertical installation of bifacial modules. Figure 5.44 shows an example that has been implemented by the company Next2Sun and that plans to implement such systems on a large scale.

Figure 5.44 Vertical installation of bifacial modules, by Next2Sun

"Datong" power plant (Yingli):

- Capacity: 50 MWp
- ≈17,900 modules (Yingli bifacial Panda, front-side power in STC: 280 W, bifaciality ≈78%)
- Orientation: portrait, south-facing, fixed tilt of 38°
- Height: 1.1 m
- Row pitch: 6 m
- Arrangement: 11 × 2 modules (width × height) per rack
- Shadowing: two parallel beams of the support structure are behind each module at ≈ 7 cm (see Figure 5.33)
- Monofacial reference system: 50 MWp fixed tilt system with 270 W multicrystalline modules
- Albedo: not measured, photographs show the underground to be grass/yellow sand

The Datong PV power plant was realized under the Chinese "Top runner" program in Datong-city, Shanxi, China. It consists of two 50-MW systems, one with bifacial, N-PERT modules, the other with monofacial multicrystalline p-type modules. Both systems are designed identical. The bifacial system has produced an annual yield of 1,593 kWh/kWp, very close to the expected value of 1,600 kWh/kWp, supplying in total over 80 GWh per year. The PV plant was connected to the grid in June 2016. The Datong PV plant is reported to generate just over 10% higher energy yield (kWh/kWp) compared to the neighboring 50-MW monofacial system, despite the rear structure limiting the irradiance on the rear. Also no special measures, like white gravel, were taken to increase the albedo (Figure 5.45).

Figure 5.45 Overview of Yingli's 50-MW Datong PV plant (left). Rear view of the structure showing the horizontal support beams behind all modules (right)

A second PV plant in the "Top runner" program has been announced in November 2017. This project will build a 100-MW bifacial system in Wuhai-city, Inner Mongolia, China with Panda bifacial modules, to be grid connected in June 2018.

5.5 Horizontal single-axis tracked bifacial systems

In the last 5 years, horizontal single-axis tracking (HSAT) became a very important technology in regions close to the equator in order to maximize the energy yield and to minimize the electricity generations costs (LCOE). Nevertheless, bifaciality in combination with tracking was for a long time thought not to be compatible, because installers believed that they are cannibalizing one of the advantages when combined. However, recently companies have realized that the combination of tracking with bifacial modules makes very much sense and lead to very high power generations. HSAT in combination with tracking became one of the biggest booms in large PV system installations as can be observed at many conferences, workshops and in the field installations.

Enel—as a first company has presented at the EUPVSEC 2017 in Amsterdam [78] and bifiPV workshop bifiPV2017 in Konstanz [79] corresponding results from its La Silla system which is shown in figure 5.42. With a ground albedo of 23%, a gain of ca. 13% can be observed when comparing a HSAT monofacial system with a HSAT bifacial one with nPERT (BiSoN) modules bifacial factor 90%) as can be seen in Figure 5.46.

TRINA, when promoting HSAT with bifacial modules, is calling this combination even "$1 + 1 > 2$" as they have realized that tracking systems are much better suited for bifacial modules than most fix tilt systems, as in fixed tilt monofacial systems

- modules are mounted often close to the ground and
- many modules are mounted next to each other.

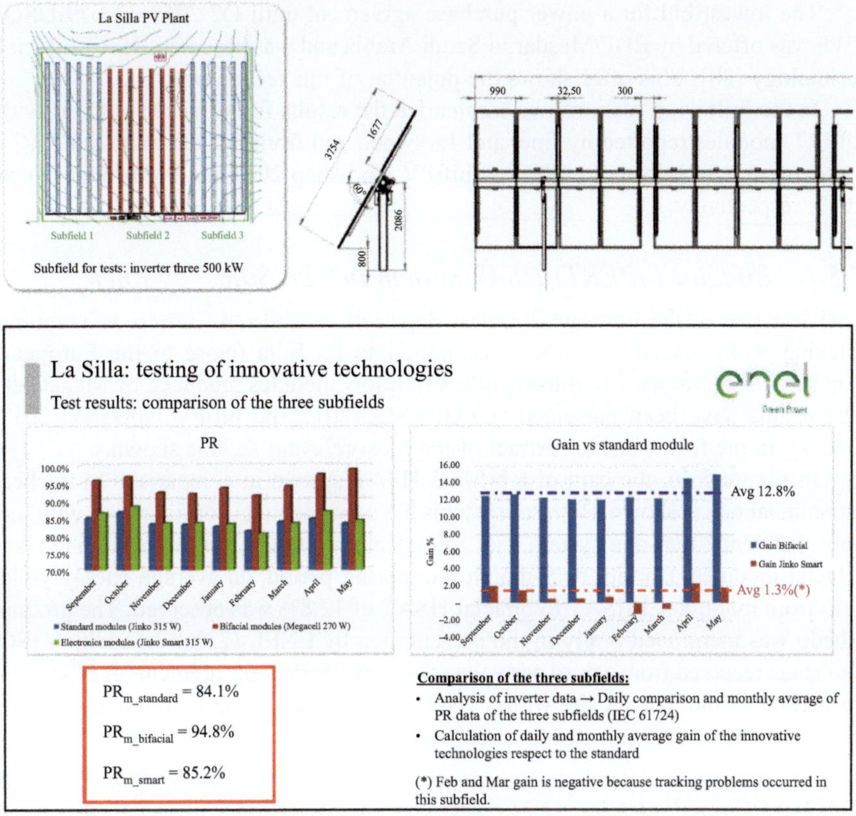

Figure 5.46 Installation in Chile in La Silla (ESO) by Enel; latitude: 29.3°S, bifacial double-glass nPERT (BiSoN) modules, 90% bifacial factor, albedo 23%

So if the invest for a tracking system is planned anyhow, it makes very much sense to use bifacial modules in many cases—e.g. if the ground albedo is high enough—about 20% and higher: which is the case for sandy desert regions. For such regions the albedo is between 20% and 40% which is already high enough to enhance the performance of bifacial tracked system compared to tracked mono-facial to of around 10%–15%. Whether it is around 10% or rather 15% (or even higher) depends mostly on

- used module's bifaciality factor
- actual albedo
- installation geometry
- adapted tracking system toward bifaciality
- geographical location

The lowest bid for a power purchase agreement until Q2 2018 of 1.79 USct/kWh was offered by EDF/Masdar in Saudi Arabia and was based on HSAT bifacial technology [80]. This offer shows the potential of this technology

In the following sections, we summarize the results from tracked systems with nPERT modules reported by Enel and Jolywood and from tracked bifacial PERC+ systems from TRINA and LONGi at bifiPV workshop 2017 and PV Module Tech 2017, respectively.

5.5.1 Bifacial (nPERT) HSAT system in "La Silla" by Enel

Enel was one of the first one, besides MegaCell in Chile in Currico, to combine tracking with bifaciality in their installation in La Silla (close to the European Southern Observatory, ESO) using nPERT BiSoN modules produced by MegaCell. The results have been presented at EUPVSEC 2017 and bifiPV workshop 2017 [78,79]. In the following, an extract of the most relevant facts is shown.

In Figure 5.46, the gain of a bifacial HSAT system in comparison to tracked monofacial one is shown. Both subsystems have the identical configuration with the only difference that one system uses monofacial modules and the other one uses bifacial modules. During the 9-month monitoring period, an average energy yield gain from monofacial HSAT to bifacial HSAT of 12.8% was observed. The ground albedo was mentioned orally in the presentation by ENEL (23%); in addition ISC Konstanz received from a third party the results of albedo measurements at a location very close to the la Silla site with a visually identical ground morphology—these measurements showed an albedo of 23% as well.

5.5.2 Bifacial nPERT HSAT PV system by Jolywood using their own nPERT modules

Jolywood is using their bifacial modules in tracking systems; however, there are not many data available yet. There is a 40 MWp large installation in China and many smaller ones which combine PV with agriculture as depicted in Figure 5.47, which were presented at the bifiPV workshop 2017 in Konstanz as well [81].

5.5.3 Fixed tilt and single-axis tracking of bifacial PERC+ modules by TRINA

By time of writing, there are only a few bifacial PERC+ systems installed. This is expected to change in the future as, due to the competitive module cost (in USD/Wp), there are many plans for such bifacial systems in China, North Africa, South and Latin America. Currently, Trina, JA Solar and Longi are the main producers of bifacial PERC+ modules. Trina has presented some of their results at PV Module Tech 2017. In summary, compared to monofacial fixed tilt, Trina obtained a combined gain of bifacial with single axis tracking of 10%–33%. A monofacial SAT reference system has not been reported [82].

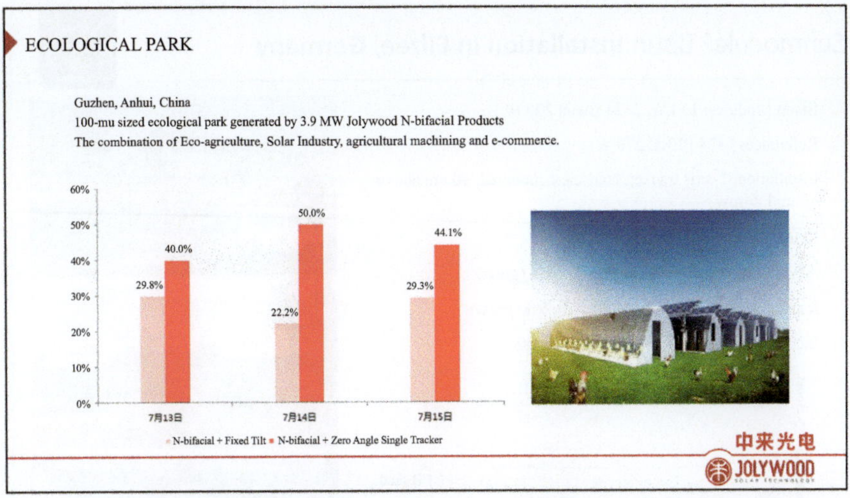

Figure 5.47 Jolywood's installation in China; technology: nPERT (Jolywood), bifacial factor > 90%, Albedo not indicated (high due to green houses), Bifacial tracked gain: 40%–50% (compared to monofacial fixed tilt)

5.5.4 Fixed tilt and tilted single-axis tracking system with bifacial PERC+ by Longi

Longi has also presented some of their results regarding the energy yield obtained on a fixed tilt system and a tilted single axis tracking PV system using Longi's bifacial PERC+ modules at PV Module Tech 2017 in Kuala Lumpur. In summary, Longi was reporting a bifacial gain (bifacial fixed vs. monofacial fixed) of 12% for a system mounted over a sand surface. For bifacial tracking, they report a combined gain of bifacial + SAT (i.e. gain from monofacial fixed to bifacial SAT) of 30%–46%, thereby it has to be noted that the tracking axis is tilted and not horizontal.

5.5.5 Tilted vertical single-axis tracking system with bifacial PERC+ by Solar World

Solar World has evaluated another tracking with vertical single-axis tracking systems and mounted slanted monofacial and bifacial PERC+ modules "Bisun." Figure 5.48 shows the results of three different systems mounted in Germany.

Reference monofacial PERC modules have been compared with 2 Bisun systems on grass and sand. The measured albedo was 14% for grass and 30% for sand. The monitoring has been done during 7 months and the average energy boost was 5% for grass and 10% for sand. The high boost for January 2017 was due to snow coverage on the front side of the tracked systems.

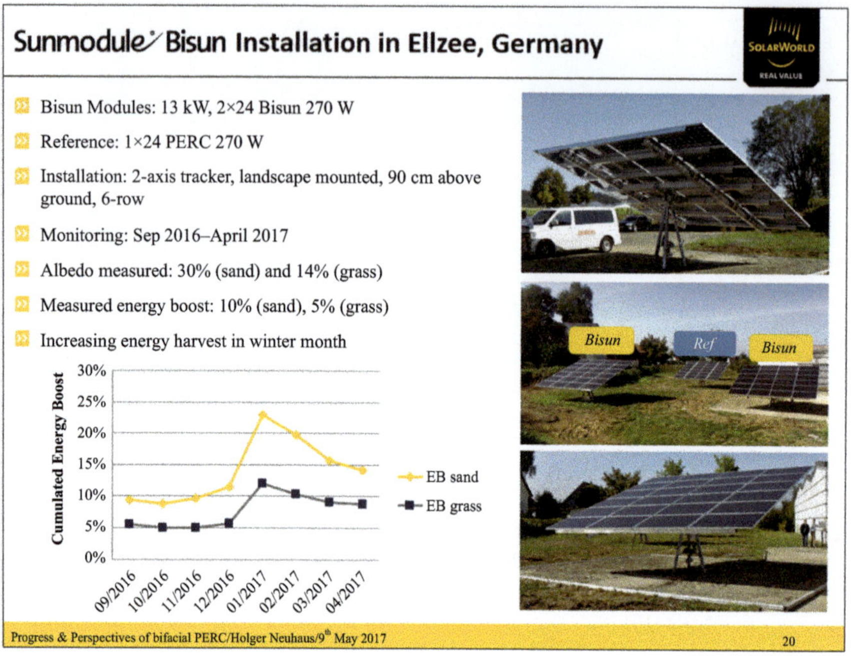

Figure 5.48 Solar World's testing site and results for tilted vertical single-axis tracking in Germany

5.5.6 Summary of tracked bifacial PV systems

Up to now, amongst the publicly available data, the most relevant and most complete field results are the one reported by Enel for their HSAT system located in la Silla (Chile). More and more systems are installed in a similar configuration showing impressive gains compared to fixed tilt monofacial systems. However, by the time of writing very few data regarding energy yield are publicly available.

The world's largest bifacial solar project up to Q2 2018 was connected as part of a 100 MW installation end of 2017 in Golmud in China's western province of Qinghai. The project, which was developed by State Power Investment Corporation's Huanghe Hydropower Development Co. Ltd., includes 20 MW bifacial modules from LONGi Solar, 20 MW of bifacial modules from Trina, 20 MW modules from Jinzhou Yangguang Energy using bifacial cells from Jolywood and 11 MW bifacial modules from JA, while the rest are monofacial modules. In total, 71 MW of bifacial module capacity was installed [83] (Figure 5.49).

The long-term data about energy yield from this system will allow for the comparison between fix tilt monofacial and bifacial HSAT systems—as well as for the comparison between different bifacial module technologies (bifacial PERC+ compared to nPERT).

In the following, a simple calculation, which explains the "$1+1>2$" advertisement from Trina, is performed. First of all we assume an energy performance in

Figure 5.49 The 100 MW PV project in Golmud, China includes the world's largest bifacial installation of 71 MW with bifacial modules supplied by leading PV companies Longi, Trina, JA Solar (photo credit: Mr. Dong from SPIC Solar)

dependence of system front-side power kWh/kWp for simple technologies like (a) fixed tilt monofacial, (b) tracked monofacial and (c) fixed tilt bifacial systems:

Single technology:

- Monofacial fixed tilt: 1,000 kWh/kWp
- Monofacial HSAT: 120 kWh/kWp (+20%)
- Bifacial fixed tilt: 1,150 kWh/kWp (+15%)

By combining bifaciality with tracking and assuming that the tracking system is adapted toward bifacial applications, the combination of both technologies results in a larger gain than for the single technologies.

> **Combined tracking with bifaciality compared with fix tilt monofacial:**
> 1,000 kWh/kWp × 1.20 × 1.15 = **1,380.5 kWp (+38%)**
> The reason is quite simple—you have to multiply and not to add the gains.
> **Pushing the limits:**
> If you would e.g. condition the ground with white stones, the bifacial gain could be increased to 25%. One more axis for tracking could also add ca. 5%. That means that
> 1,000 kWh/kWp × 1.25 × 1.20 × 1.05 = **1,575 kWp (+57.5%)**
> could be reached: 57% more power as compared to the monofacial fixed tilt system. Depending on the additional costs for two-axis tracking systems and for artificially increasing the ground albedo, such a system configuration could lead to the lowest LCOEs achievable with a currently commercially available PV module technology.

5.6 What does bifacial gain tell us? How to transfer this to lowest LCOEs?

As shown in Chapters 2 and 3 as well as in this chapter, bifaciality can be implemented by varieties of architectures for solar cells, modules and in addition there

(a) (b) (c)

Figure 5.50 (a) La Hormiga fixed tilt bifacial PV plant in St Felipe, Chile, (b) vertical bifacial PV plant by Next2sun in Germany and (c) a tracked bifacial PV plant in La Silla, Chile

are even many more applications on system level. This makes bifacial PV a complex technology. In the following, we will describe how bifacial gains are defined, what bifacial gains can be expected and what this means for real applications.

Bifacial systems offer a very promising possibility to reduce the LCOE for many PV system applications. As shown in the previous sections of this chapter, there is a huge application field—such as large ground-mounted systems, flat reflective rooftops, sound blocking systems, floating systems or even in utility-scale systems using trackers. As mentioned before, the last application is very interesting, these days achieving the lowest LCOE for many cases (see e.g. [84]).

Not only are there many potential application fields, there are also various mounting geometry possibilities: from standard slanted systems over horizontal to even vertical bifacial installations with almost zero ground coverage ratios. Three prominent examples are depicted in Figure 5.50.

5.6.1 Definition of bifacial gain

An obvious way to visualize the benefits of bifaciality is to analyze the "bifacial gain," which means the difference in the energy yield if bifacial and monofacial devices with identical installation configurations are compared. The comparison can either include single modules or larger units of one or both device types, because typically the energy yield in kWh/kWp ratio is analyzed. The kWp data usually reflect the STC front-side measurement of the bifacial module(s). In the most direct form, devices of similar type and with the same front-side efficiency are compared, for example if bifacial modules with covered rear sides are used as reference.

The bifacial gain is usually defined as (compare introduction):

$$g_{bifacial}\,[\%] = \left(\frac{\left(e_{bifacial} - e_{monofacial}\right)}{e_{monofacial}}\right) \times 100 \tag{5.3}$$

With

- $e_{bifacial}$: specific energy yield (kWh/kWp) of the PV system with bifacial modules
- $e_{monofacial}$: specific energy yield (kWh/kWp) of the PV system with monofacial modules on the same site, with the same configuration and during the same time period

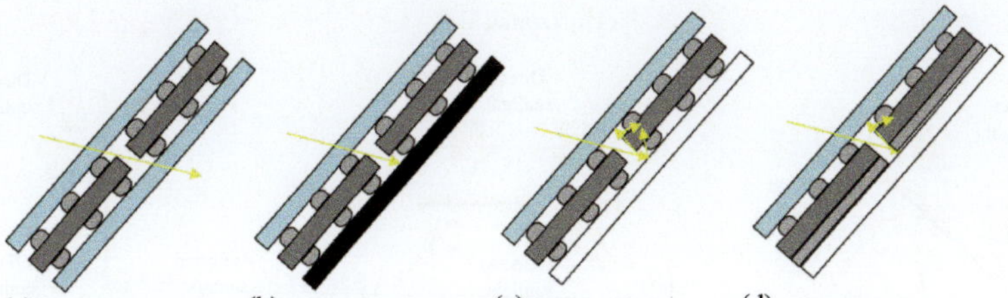

Figure 5.51 Schematic cross section of a (a) bifacial module and three possible monofacial reference modules with (b) bifacial cells and black backsheet, (c) bifacial cells and white backsheet and (d) monofacial cells and white backsheet

As the bifacial gain is another way to indicate the energy yield, it is the metric that determines—together with the total cost of installing and operating the bifacial PV system—the LCOE (€/kWh) and therefore the economic viability of bifacial PV.

The above mathematical definition of bifacial gain is quite simple—however, there are different possibilities in terms of what module type can be chosen for the monofacial reference. Therefore sometimes the reported bifacial gains already differ there—even if at a first glance identical conditions are applied. Figure 5.51 depicts in (a) the bifacial module and three different monofacial references (b)–(d) which are very often used.

Many groups use standard white backsheet modules with monofacial cells for reference (Figure 5.51(d)), some use monofacial white backsheet modules with the same bifacial cells (Figure 5.51(c)) and some monofacial black backsheet modules with the same bifacial cells (Figure 5.51(b)). All three references will lead to different results, as the white backsheet is causing additional reflection of the front-incoming light into the solar cells. Even if the monofacial solar cell has similar properties as the bifacial (e.g. front-side power, voltage and temperature coefficient) the front-side power of the module is increased by ca. 2% at STC (standard test conditions: 25°C, 1,000 W/m^2, AM 1.5 spectra) because of the additional reflection of light to the front side and during field measurements the energy harvest is increased more. An increased level of power can also be seen in the case of the bifacial cell and white backsheet: the total additional energy yield (kWh/kWp), also due to the scattering of the light into the solar cell rear side, can be up to 5%, as observed, for example, in LG NeON modules.

Therefore, if you want to observe bifacial gain only, as a reference the same bifacial cell in a module with a black rear cover or black backsheet is required.

This comparison reveals precisely what additional energy is provided by the rear side only. If you take for example a monofacial module with a bifacial solar cell and white backsheet as a reference, you will underestimate the bifacial gain by ca. 5%, as the rear side is already contributing in field measurements. Therefore, the choice of different references leads already to different results reported in various publications.

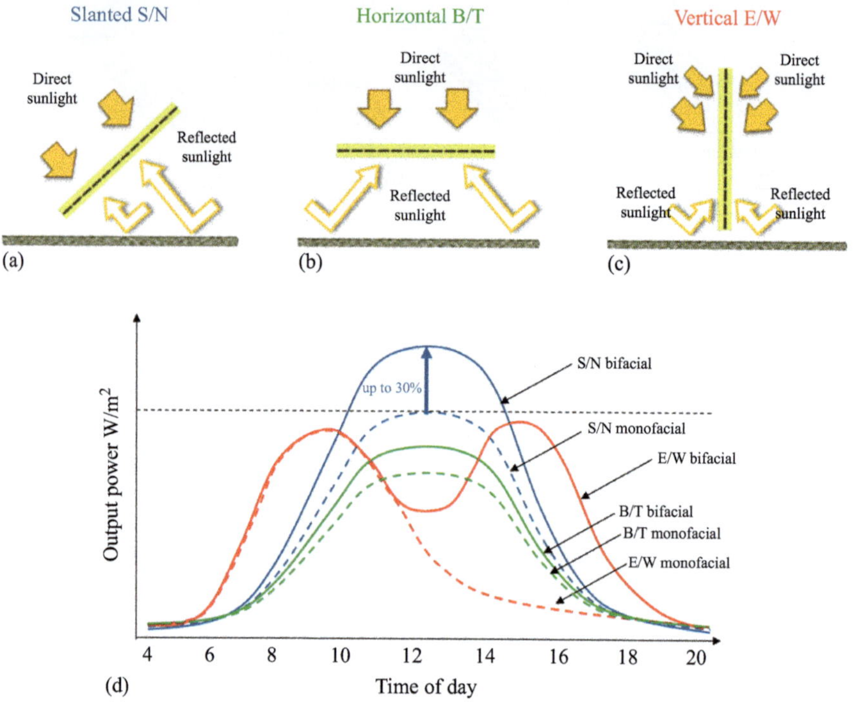

Figure 5.52 (a)–(c) Possible applications for bifacial modules and (d) resulting daily power generation curves compared to monofacial ones in the same configuration

Another important point is that the temperature coefficient of the monofacial reference module should be in the same range as for the bifacial ones. Otherwise, for example when comparing bifacial heterojunction modules (temperature coefficient for P_{mpp} around 0.30%/°C) with standard monofacial aluminum back surface field (Al-BSF) c-Si modules (temp coeff around 0.45%/°C), a significant part of the gain attributed to bifaciality will be due to the reduced temperature losses of the HJT module. Here, as a reference, the same HJT module with a black back cover would be the best choice leading to an "apple to apple" comparison.

5.6.2 Examples of bifacial gains: comparison of apples with apples

Not only the choice of different references, but obviously also different mounting geometries will lead to different bifacial gains—and as we will show, these can be even more than 100% in some cases. Figure 5.52 depicts different mounting geometries: (a) slanted S/N (south/north)-oriented mounting, (b) horizontal B/T (bottom/top) and (c) vertical E/W (east/west)-oriented mounting.

The slanted S/N-oriented mounting leads to the highest powers of the applied bifacial modules as the front side produces the highest possible power and the rear, depending on the albedo of the ground, can contribute up to 30% additional electricity. Here, a 300 Wp module can behave as a module with an effective power of close to 400 Wpe. This relationship can be seen in Figure 5.52(d) between the dotted and solid blue curve.

Horizontal B/T-oriented installations, used in car ports, for example, demonstrate very similar behavior, only that the absolute energy production is reduced, as the module is—apart for sites located nearby the equator—not oriented at an optimal angle toward the sun. The monofacial and bifacial generation curve is demonstrated by the green dotted and solid lines respectively. The shape for all installations so far discussed is very similar, having a peak intensity around noon.

A completely different form (camel and dromedary curve) is generated by a vertical E/W-oriented installation. When you install a bifacial module with a high bifacial factor (b: rear power/front power >0.9/e.g. an nPERT BiSoN (Bifacial Solar Cells on N-type) or "HJT module" from Sunpreme) you end up with the solid red line. Much more electricity is generated during morning and evening as compared with the S/N-oriented case. During midday there is a generation dip, as the direct sunlight is shining on the frame and only diffuse light is hitting the module front and rear side. However, due to the ground coverage ratio close to zero and due to the broader generation peak this installation geometry is very interesting. Now: if you install a monofacial module in such a mounting geometry the generation peak moves to a dromedary-like (green dotted line) shape with generation energy less than 50% compared to the bifacial one. Here the bifacial gain is therefore higher than 100%. However such a comparison does not make much sense. In this case the vertical bifacial modules have to be compared with a monofacial S/N-oriented module. Depending on the installation latitude the bifacial gain can be even negative—in this case, if modules are installed vertically in sunbelt regions. However this might make also sense in some cases, if the soiling can be reduced by the vertical installation.

Table 5.2 summarizes several examples of various installation geometries and resulting "bifacial gains" for BiSoN nPERT modules. Because in the large bifacial systems in Chile, standard monofacial modules with white backsheet are used as a reference by developers MegaCell and Enel, the real physical bifacial gains would differ from there slightly.

In the case of the fixed-tilt S/N module system, there are already many cases reported all around the world with different albedi. Depending on the ground albedi (25% for natural sand and 75% for white stones) bifacial gains from 15% to 30% can be achieved.

When it comes to vertical E/W systems things become more complex and also not so many reference systems exist. In these cases, not only are the module type and albedo of importance but so are the mounting geometry of the reference module and the installation latitude. If you compare with a vertical installed monofacial module, a bifacial gain of more than 100% can be observed. This comparison makes only little sense—here a comparison with a slanted South-oriented monofacial module is more interesting as well. If you install such systems

Table 5.2 Bifacial gains for BiSoN (nPERT) modules with various installation geometries

Bifacial module	Bifacial installation geometry and latitude	Installation geometry of monofacial reference	Albedo	"Bifacial gain" (rounded to 5% steps)
nPERT ($b > 0.9$)	Slanted fixed tilt in San Felipe, Chile (32°S)	Slanted fixed tilt	25%	15% [85]
nPERT ($b > 0.9$)	Slanted fixed tilt in San Felipe, Chile (32°S)	Slanted fixed tilt	65%–75%	30% [85]
nPERT ($b > 0.9$)	Vertical installation, USA	Vertical installation	Unknown	100+% [86]
nPERT ($b > 0.9$)	Vertical installation in Winterthur, Switzerland (47°N)	Slanted fixed tilt	25%	10% [11]
nPERT ($b > 0.9$)	Vertical installation in Saar, Germany (49°N)	Slanted fixed tilt	25%	10% [87]
nPERT ($b > 0.9$)	Vertical installation in el Gouna, Egypt (27°N)	Slanted fixed tilt	25%	–5% [88]
nPERT ($b > 0.9$)	Single-axis tracked in La Silla, Chile (29°S)	Single-axis tracked	25%	15% [78]

at high latitudes, where the amount of diffuse sunlight is higher and where the vertical mounting is less far away from the optimum slanted angle, an electrical gain of 10% is observable—however, at low latitudes even an electrical loss of −5% was observed. Still this application remains interesting because of several reasons: the ground coverage is close to zero, the generation peak is broader and vertical installations have less soiling problems. However, also some challenges have to be solved as the wind loads are high using this mounting configuration.

Within the last few months bifacial systems using single-axis tracking have gained more and more attention, as experimental results in large systems showed that the bifacial gain in those cases is also very high. This is because many tracking mounting systems are almost ideal for bifacial modules as they are mounted high from the ground with high row spacing. Therefore, the bifacial gains—in this case, the gains compared to monofacial single-axis tracking—are very similar as for the fixed-tilt systems. The first one to report this behavior was Enel in la Silla [78].

A combination of single-axis tracking with bifacial modules in systems with high albedo result in electrical gains of over 40% compared to fixed-tilt monofacial modules [81].

5.6.3 Bifacial applications in reality: comparison of apples with oranges

We have learned that bifacial gains, as they are defined, can reach values of more than 100%. However, this information is not very practical for system designers.

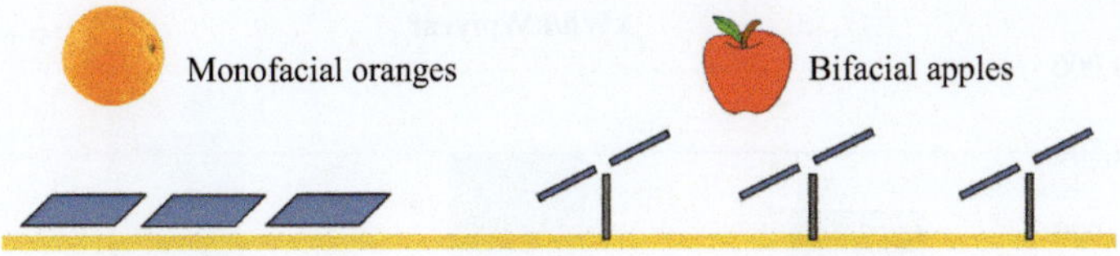

Figure 5.53 Schematic drawing of (a) a monofacial S/N-oriented system and (b) an E/W-oriented bifacial single-axis tracked system

The only interesting question for them is: how can a PV system with the lowest LCOEs be designed? Then, the best possible monofacial installation has to be compared with the best bifacial one, as depicted, for example, in Figure 5.53.

Many PV system designers are using commercial software packages such as PVsyst [89] or Polysun [90], which allow also for the simulation of bifacial fixed tilt PV systems, for this purpose. With all the necessary import parameters such as module properties, system geometry and data for specific local conditions, the energy output and—in combination with the information about the system cost—the LCOE can be calculated. Using a simulation model developed at ISC Konstanz [91] which is capable of conducting simulations for bifacial tracked systems as well, the energy yield for a monofacial fixed tilt, a monofacial HSAT and a bifacial HSAT system has been calculated for a given location in Chile and—with cost assumption for 2017—the LCOE has been calculated as well. The result is summarized in Figure 5.54.

5.6.4 Summary

Bifacial gains show how bifacial modules increase the electrical performance of a system when bifacial modules instead of reference monofacial modules are mounted. Depending on the choice of reference modules, these values can differ by more than 5% (rel.), even when choosing the same installation configuration for the bifacial and the monofacial system. In order to determine the real bifacial gain—the additional power that the rear side is generating—the same bifacial module covered by a back sheet should be used as reference. Bifacial gains are also dependent on module bifacial factor b. Bifacial PERC+ modules at the moment have $b < 80\%$, nPERT and HJT $b > 90\%$. Therefore, it has to be also stated which modules with which b were used in corresponding modeling or experiment.

In special configurations, bifacial gains of more than 100% can be measured, when e.g. bifacial vertical installations are compared with monofacial vertical installations. However in practice, for the optimal design of PV systems, it makes only sense to compare the energy output for an optimized monofacial versus an optimized bifacial system and at the end compare the resulting LCOEs. The meaning of "optimized" can be influenced by restrictions imposed by the specific application and by the available installation site.

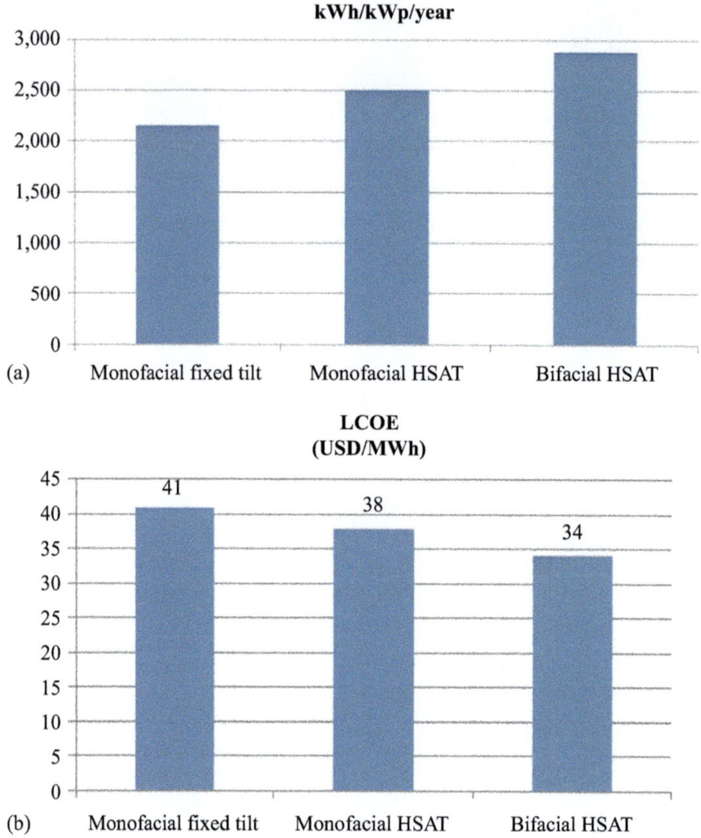

Figure 5.54 Examples of (a) energy yield and (b) resulting LCOE for different module and system technologies when installed in Chile (assumption for monofacial installed fixed-tilt system cost: US\$0.92/Wp and US\$1.00/Wp for monofacial and bifacial horizontal single-axis tracker) with a ground albedo of 25%. In this case the tracking gain (monofacial horizontal axis tracking compared with monofacial fixed tilt) is 17%. Using bifacial instead of monofacial modules on the HSAT system results in an additional 14.7% (rel.) gain, leading to a combined gain (tracking + HSAT) of 34% (= 1.17 × 1.147)

5.7 Conclusion

Over the past 10 years, some data regarding the energy yield of several bifacial PV systems have been published, for demonstration purposes, by various academics, and manufacturers of bifacial PV cells and modules, such as PVGS, bSolar and Sanyo/Panasonic. A summary and analysis of such data found in the literature is given in this chapter. This data show, on the one hand, that even under conditions

Figure 5.55 View on the three biggest bifacial power plants with different technologies until Q1 2018

that are not ideal (ground albedo less than 20%, which corresponds to, for example, grassland), the bifacial gain of a system is usually higher than 10%; on the other hand, if measures are taken to increase the ground albedo to more than 60%, bifacial gains of 20%–30% are possible compared to monofacial PV systems with the same nominal (front side) peak power, installed at the same site. The increase of the module height over the ground is also a key parameter influencing the bifacial gain (a minimum height of 1.5 m should be sufficient → details are given in next chapter). Simulations show, that already with an albedo of 20%, compared to fixed tilt monofacial, up to 35 to 40% energy yield can be gained with bifacial HSAT (see in example figure 30 of [92] as compared with a monofacial module: at least 10% of that comes from the bifacial performance gain and the other 25% from the HSAT. If a single-axis tracking system were applied to a bifacial PV system, depending on the tracking costs, an LCOE of 4 US$/kWh calculated for a large ground-mounted system with a yearly global horizontal irradiance of 2,200 kWh/m^2 (e.g. for southern Europe, North Africa and India) would already be possible today. More details on the assumptions for the LCOE calculations are given in Chapter 7. Accordingly, for many potential installation sites, sufficiently high bifacial gains can be achieved without additional investment for modification of the ground surface properties. Also, the bifacial concept could show stronger advantage in case of high land cost. Regarding large PV plants, the recent growth in the cumulated capacity confirms the interest of the industrials for the concept although some extra time would be required to estimate precisely the energy gain and associated production cost. In addition, a standardized setup for measuring the power of bifacial modules is needed. Even if—from the technical and scientific point of view—there is no doubt regarding the fact that a significant additional energy yield can be obtained with bifacial PV, the bankability still requires more data from the fields and reliable simulations models (see also Chapter 4) that have proven their accuracy compared to measured data. More insight about the bankability of PV systems is given in Chapter 6.

So far, the PV plant from Hokuto (125 MW, PVGS & Nishiyama Sakata Denki Co.) remains the best documented large-scale PV system and has shown a bifacial gain of 19% based on monitoring data over almost 3 years. More and more much larger power-plants are set up and results reported. Figure 5.55 depicts the three largest bifacial PV systems until Q1 2018 in respect to their corresponding technologies.

However, they will not stay the largest for a long time and in 2 years people will even laugh about this—as bifaciality will become standard and power-plants with new dimensions in size are constructed. There are several plans already to set up 100 MW bifacial plants in 2018 from Yingli with nPERT in Mongolia [93], and NSP with bifacial PERC+ in Taiwan [94] and many others.

References

[1] U. A. Yusufoglu, T. M. Pletzer, L. J. Koduvelikulathu, C. Comparotto, R. Kopecek, and H. Kurz, "Analysis of the annual performance of bifacial modules and optimization methods," *IEEE J. Photovolt.*, vol. 5, no. 1, pp. 320–8, Jan. 2015.

[2] U. A. Yusufoglu, Lee, T. H., Pletzer, T. M., *et al.*, "Simulation of energy production by bifacial modules with revision of ground reflection," *Energy Procedia*, vol. 55, pp. 389–95, 2014.

[3] H. Nussbaumer, Klenk, M., Schär, *et al.*, "PV installations based on vertically mounted bifacial modules evaluation of energy yield and shading effects," Hamburg, 2014.

[4] N. Kasahara, K. Yoshicka, and T. Sailoh, "Performance evaluation of bifacial photovoltaic modules for urban application," in *Proceedings to the 3rd World Conference on Photovoltaic Energy Conversion*, Osaka, Japan, 2003, vol. 3, pp. 2455–8.

[5] J. Appelbaum, "Bifacial photovoltaic panels field," *Renew. Energy*, vol. 85, pp. 338–43, Jan. 2016.

[6] J. P. Singh, T. M. Walsh, and A. G. Aberle, "Performance investigation of bifacial PV modules in the tropics," in *Proceedings of the 27th European Photovoltaic Solar Conference and Exhibition*, 2012, pp. 3263–6.

[7] G. J. M. Janssen, B. B. Van Aken, A. J. Carr, and A. A. Mewe, "Outdoor performance of bifacial modules by measurements and modelling," *Energy Procedia*, vol. 77, pp. 364–73, Aug. 2015.

[8] I. Shoukry, J. Libal, R. Kopecek, E. Wefringhaus, and J. Werner, "Modelling of bifacial gain for stand-alone and in-field installed bifacial PV modules," *Energy Procedia*, vol. 92, pp. 600–8, Aug. 2016.

[9] B.B. Van Aken and A. J. Carr, "Relating Indoor and outdoor performance of bifacial modules," in *Proceedings of the 40th IEEE Photovoltaic Specialist Conference*, Denver, USA, 2014, pp. 1381–3.

[10] J. Lossen, "Increased energy yield for Bosch's n-type solar cells in bifacial application," presented at the bifi PV 2012, Konstanz, 2012.

[11] H. Nussbaumer, "New opportunities for a better power distribution by the use of bifacial modules in future PV systems," presented at the bifi PV 2014, Chambery, 2014.

[12] C. Comparotto, M. Noebels, L. Popescu, *et al.*, "Bifacial n-type solar modules: indoor and outdoor evaluation," in *Proceedings to the 29th European Solar Energy Conference EUPVSEC*, Amsterdam, 2014, pp. 3248–50.

[13] A. Moehlecke, F. S. Febras, and I. Zanesco, "Electrical performance analysis of PV modules with bifacial silicon solar cells and white diffuse reflector," *Sol. Energy*, vol. 96, pp. 253–62, Oct. 2013.
[14] F. Baumgartner, A. Büchel, H. Nussbaumer, *et al.*, "Bifacial modules used in cable based PV carport applications," in *Proceedings to the world PV Conference*, Kyoto, Japan, 2014.
[15] K. Sugibuchi, N. Ishikawa, and S. Obara, "Bifacial-PV power output gain in the field test using 'EarthON' high bifaciality solar cells," in *présenté à 28th European Photovoltaic Solar Energy Conference and Exhibition, Paris*, 2013.
[16] V. Poulek, A. Khudysh, and M. Libra, "Innovative low concentration PV systems with bifacial solar panels," *Sol. Energy*, vol. 120, pp. 113–6, 2015.
[17] L. Kreinin, N. Bordin, A. Karsenty, A. Drori, and N. Eisenberg, "Experimental analysis of the increases in energy generation of bifacial over," in *26th European Photovoltaic Solar Energy Conference and Exhibition*, Hamburg, 2011.
[18] T. Nordmann, T. Vontobel, and R. Lingel, "Exploiting high solar irradiation in alpine regions using bifacial PV modules," *presented at the 28th European Photovoltaic Solar Energy Conference and Exhibition*, 2013, pp. 3248–50.
[19] T. Nordmann, "15 years of experience in construction and operation of two bifacial photovoltaic systems on Swiss roads and railways," *presented at the bifi PV*, Konstanz, Germany, 23 Apr. 2012.
[20] T. Nordmann, "Technical wrap up: are we doing the right things right? Critical summary and outlook for successful bifacial future," *presented at the bifi PV*, Konstanz, Germany, 26 Oct. 2017.
[21] R. Guerrero-Lemus, R. Vega, T. Kim, A. Kimm, and L. E. Shephard, "Bifacial solar photovoltaics – a technology review," *Renew. Sustain. Energy Rev.*, vol. 60, pp. 1533–49, Jul. 2016.
[22] R. Kopecek, "Bifaciality: One small step for technology, one giant leap for kWh cost reduction – The stoty continues," *presented at the Workshop on: New materials and Innovative Solar Cell Architectures*, Paris, Oct. 2015.
[23] R. Hezel, "Novel applications of bifacial solar cells," *Prog. Photovolt: Res. Appl.*, vol. 11, pp. 549–56, 2003.
[24] G. João, "Testing bifacial PV cells in symmetric and asymmetric concentrating CPC collectors," *Engineering*, vol. 5, no. 1, pp. 185–90, 2013.
[25] A. Moehlecke, F. S. Febras, and I. Zanesco, "Electrical performance analysis of PV modules with bifacial silicon solar cells and white diffuse reflector," *Sol. Energy*, vol. 96, pp. 253–62, Oct. 2013.
[26] P. Ooshaksaraei, K. Sopian, R. Zulkifli, M. A. Alghoul, and S. H. Zaidi, "Characterization of a bifacial photovoltaic panel integrated with external diffuse and semimirror type reflectors," *Int. J. Photoenergy*, vol. 2013, pp. 1–7, 2013.
[27] Y. S. Lim, C. K. Lo, S. Y. Kee, H. T. Ewe, and A. R. Faidz, "Design and evaluation of passive concentrator and reflector systems for bifacial solar

panel on a highly cloudy region – a case study in Malaysia," *Renew. Energy*, vol. 63, pp. 415–25, Mar. 2014.

[28] M. Chiodetti, M. A. Lindsay, P. Dupeyrat, *et al.*, "PV bifacial yield simulation with a variable albedo model," in *Proceedings of the EU PVSEC 2016*, Munich, pp. 1449–55.

[29] Y. Veschetti, "First monitoring results of bifacial systems in different configurations," *presented at the bifi PV 2016*, Myiazaki, Japan, 2016.

[30] M. Joanny, J. Libal, R. Kopecek, Y. Veschetti, and H. Colin, "Bifacial systems overview," *presented at the bifi PV Workshop*, Konstanz, Germany, 26 Oct. 2017.

[31] L. Podlowski, S. Wendlandt, and D. Cormode, "Yield study on identical bifacial rooftop systems installed in the USA and in Germany," *presented at the bifi PV Workshop*, Konstanz, Germany, 25 Oct. 2017.

[32] W. Vermeulen, "400kW bifacial system in NL and comparison with two other systems," *presented at the bifi PV Workshop*, Konstanz, Germany, 25 Oct. 2017.

[33] A. Richter, "Meyer burger presentation," *presented at the 9th PVPMC Modeling Workshop*, Weihei, China, 05 Dec. 2017.

[34] M. P. Brennan, A. L. Abramase, R. W. Andrews, and J. M. Pearce, "Effects of spectral albedo on solar photovoltaic devices," *Sol. Energy Mater. Sol. Cells*, vol. 124, pp. 111–6, May 2014.

[35] A. Richter, "Bankability: choosing right materials on module level," *presented at the bifi PV 2016*, Miyazaki, Japan, 2016.

[36] B. Yu, S. Dengyuan, S. Zhonggang, *et al.*, "A study on electrical performance of N-type bifacial PV modules," *Sol. Energy*, vol. 137, pp. 129–33, Nov. 2016.

[37] J. E. Castillo-Aguilella and P. S. Hauser, "Multi-variable bifacial photovoltaic module test results and best-fit annual bifacial energy yield model," *IEEE Access*, vol. 4, pp. 498–506, 2016.

[38] J. Libal and R. Kopecek, "Review of bifacial PV systems," *presented at the bifi PV 2016*, Miyazaki, Japan, 2016.

[39] R. Einhaus, "NICE technology for bifacial modules," *presented at the bifi PV Workshop 2015*, Antofagasta, 2015.

[40] H. C. Oon, K. W. NG, and H. Khoo, "A study on the performance of bifacial photovoltaic panels in the urban tropics," *presented at the World Engineers Summit on Climate Change (WES)*, Singapore, 2015.

[41] B. B. Van Aken and A. J. Carr, "Relating indoor and outdoor performance of bifacial modules," in *2014 IEEE 40th Photovoltaic Specialist Conference (PVSC)*, 2014, pp. 1381–3.

[42] S.-H. Yu, C. J. Huang, P. T. Hsieh, *et al.*, "20.63% nPERT cells and 20% PR gain bifacial module," in *2014 IEEE 40th Photovoltaic Specialist Conference (PVSC)*, 2014, pp. 2134–7.

[43] E. Maruyama, "Recent technological progress of high-efficiency HIT solar cells," *presented at the PV*, Japan, Tokio, 2013.

[44] N. Eisenberg, "Outdoor bifacial module characterization: energy generation and gain," *presented at the bifi PV 2012*, Konstanz, 2012.

[45] L. Kreinin, N. Bordin, A. Karsenty, A. Drori, D. Grobgeld, and N. Eisenberg, "PV module power gain due to bifacial design. Preliminary experimental and simulation data," in *Photovoltaic Specialists Conference (PVSC), 2010 35th IEEE*, 2010, pp. 2171–5.
[46] "Data Sheet SANYO_HIT-210_205__200_DNKHE1__DE_-1.pdf."
[47] "http://www.b-solar.com/technology.aspx?Sel=Field%20Results." [Online]. Available from: http://www.b-solar.com/technology.aspx?Sel=Field%20Results.
[48] "Calculating the additional energy yield of bifacial solar modules," *Calculating the Additional Energy Yield of Bifacial Solar Modules*, 2016. [Online]. Available from: http://solarkingmi.com/assets/Calculating-Additional-Energy-Yield-Through-Bifacial-Solar-Technology-SW9002US.pdf.
[49] M. Chiodetti, "Bifacial PV plants: performance model development and optimization of their configuration," Independent thesis Advanced level (degree of Master (Two Years)), KTH, School of Industrial Engineering and Management (ITM), Energy Technology, Heat and Power Technology, 2015.
[50] B. B. Van Aken, M. J. Jansen, A. J. Carr, G. J. Janssen, and A. A. Mewe, "Relation between indoor flash testing and outdoor performance of bifacial modules," in *29th European Photovoltaic Solar Energy Conference, Amsterdam*, 2014.
[51] C. Reise and A. Schmid, "Realistic yield expectations for bifacial PV systems – an assessment of announced, predicted and observed benefits," in *Proceedings of the EU PVSEC 2016*, Munich, 2016, pp. 1775–9.
[52] A. Dreisiebner, "Smart solutions for extreme conditions," *presented at the bifi PV Workshop*, Konstanz, Germany, 25 Oct. 2017.
[53] F. P. Baumgartner, "Bifacial outdoor rotor tester BIFROT," *presented at the EU PVSEC 2016*, Munich, 2016.
[54] M. Klenk, H. Nussbaumer, F. P. Baumgartner, N. Keller, and T. Baumann, "Bifacial outdoor rotor tester – BIFROT," *presented at the bifi PV 2016*, Myiazaki, Japan, 2016.
[55] M. Klenk, "BIFOROT – experimental data for LCOE appraisal of bifacial systems," *presented at the bifi PV Workshop*, Konstanz, 25 Oct. 2017.
[56] S. Guo, T. M. Walsh, and M. Peters, "Vertically mounted bifacial photovoltaic modules: a global analysis," *Energy*, vol. 61, pp. 447–54, Nov. 2013.
[57] H. Hildebrandt, "3 MWp vertical E-W oriented system in Germany," *presented at the bifi PV Workshop*, Konstanz, Germany, 25 Oct. 2017.
[58] I. Araki, M. Tatsunokuchi, H. Nakahara, and T. Tomita, "Bifacial PV system in Aichi Airport-site demonstrative research plant for new energy power generation," *Sol. Energy Mater. Sol. Cells*, vol. 93, no. 6–7, pp. 911–6, Jun. 2009.
[59] S. Obara, D. Konno, Y. Utsugi, and J. Morel, "Analysis of output power and capacity reduction in electrical storage facilities by peak shift control of PV system with bifacial modules," *Appl. Energy*, vol. 128, pp. 35–48, Sep. 2014.
[60] T. Joge, Y. Eguchi, Y. Imazu, I. Araki, T. Uematsu, and K. Matsukuma, "Applications and field tests of bifacial solar modules," in *Photovoltaic*

Specialists Conference, 2002. Conference Record of the 29th IEEE, 2002, pp. 1549–52.

[61] B. Soria, E. Gerritsen, and P. Lefillastre, "Vertical facade integration of bifacial PV modules outdoor testing & optical modelling.pdf," *presented at the 28th European Photovoltaic Solar Energy Conference and Exhibition*, 2013, pp. 3216–21.

[62] S. Obara, D. Konno, Y. Utsugi, and J. Morel, "Analysis of output power and capacity reduction in electrical storage facilities by peak shift control of PV system with bifacial modules," *Appl. Energy*, vol. 128, pp. 35–48, Sep. 2014.

[63] M. de Jong, "Presentation," Bussum, The Netherlands, Solar Energy Application Centre (SEAC), 07 Nov. 2018.

[64] "nrg-energia," *nrg-energia*. [Online]. Available from: http://www.nrg-energia.it/floating-pv-systems.html.

[65] "Koine Multimedia," http://www.koinemultimedia.eu/wp/. [Online]. Available from: http://www.koinemultimedia.eu/wp/.

[66] "Solaris Synergy," *Solaris Synergy*. [Online]. Available from: http://www.solaris-synergy.com/home.html.

[67] "sun-projects," *sun-projects*. [Online]. Available from: http://www.sun-projects.nl.

[68] J. A. Coakley, J. R. Holton, and J. A. Curry, "Reflectance and albedo, surface," *Encyclopedia of the Atmosphere*. Academic Press, pp. 1914–23, 2003.

[69] M. A. A. Goris, "Production of recyclable crystalline Si PV modules," *presented at the European PV Solar Energy Conference and Exhibition (EUPVSEC)*, Hamburg, Germany, 2015.

[70] R. Kopecek, "Closing Session, 3rd BIFI workshop," *presented at the 3rd BIFI Workshop*, Myazaki, Japan, 2016.

[71] S. Goda, *presented at the nPV workshop*, 's-Hertogenbosch, Netherlands, 2014.

[72] B. van Aken and K. de Grot, "Near field partial shading on rear side of bifacial modules," *presented at the 7th International Silicon PV Conference*, Freiburg, 2017.

[73] N. Ishikawa, *presented at the 3rd BIFI Workshop*, Myazaki, Japan, 2016.

[74] "Albedo Measurements," *Climate Data Information*. [Online]. Available from: http://www.climatedata.info/forcing/albedo/ [Accessed 30 May 2017].

[75] R. Kopecek, J. Libal, I. Shoukry, C. Comparotto, E. Cabrera, and E. Wefringhaus, "c-Si desert module for minimizing the costs for electricity production in PV systems," *presented at the Photovoltaic Technical Conference*, Marseille, 2016.

[76] "Case Studies – Sunpreme," *Sunpreme Case Studies*. [Online]. Available from: http://sunpreme.com/casestudies/ [Accessed 30 May 2017].

[77] D. Song, "High efficiency N-type silicon bifacial solar cells: from industrial production to application in large scale PV plants," *presented at the SNEC*, 19 Apr. 2017.

[78] A. Di Stefano, "La Silla PV plant as a utility-scale side-by-side test for innovative modules technologies," *presented at the 33rd European Photovoltaic Solar Energy Conference and Exhibition (EUPVSEC)*, Amsterdam, 2017, pp. 1978–82.

[79] F. Bizzarri, "ENEL Innovative tracked bifacial PV plant at la Silla observatory in Chile," *presented at the Bifacial PV Workshop*, Konstanz, Germany, 2017.

[80] A. Dipaola, "Saudi Arabia gets cheapest ever bids for solar power in auction," *Bloomberg.com*, 10 Mar. 2017. [Online]. Available from: https://www.bloomberg.com/news/articles/2017-10-03/saudi-arabia-gets-cheapest-ever-bids-for-solar-power-in-auction.

[81] L. Yong, "JOLYWOOD nPERT technology and its application," *presented at the Bifacial PV Workshop*, Konstanz, Germany, 2017.

[82] "Testing bifacial solar PV modules outperforming energy output expectations," *Trinasolar*, 08 Jul. 2017. [Online]. Available from: http://www.trinasolar.com/us/resources/blog/testing-bifacial-solar-pv-modules-outperforming-energy-output-expectations.

[83] "71 MW Bifacial Solar Project in China World's largest Bifacial Solar Power Project Installed in China's Qinghai Province," *taiyangnews*, 01 Oct. 2018. [Online]. Available from: http://taiyangnews.info/technology/20-mw-bifacial-solar-system-in-china/.

[84] J. Parnell, "Bifacial technology was likely reason world's lowest ever solar bid was rejected," *PV_tech.org*, 2017. [Online]. Available from: https://www.pv-tech.org/editors-blog/bifacial-technology-was-likely-reason-worlds-lowest-ever-solar-bid-was-reje.

[85] R. Kopecek and J. Libal, "Complex problems require simple solutions: how to measure bifacial devices correctly?," *Photovolt. Int.*, vol. 37, pp. 105–12, 2017.

[86] "Bifacial solar cells-the two sides of the story 050515." [Online]. Available from: https://www.solarchoice.net.au/blog/news/bifacial-solar-cells-the-two-sides-of-the-story-050515.

[87] Next2Sun, "internal communication."

[88] I. Shoukry, "Bifacial modules – simulation and experiment," Master Thesis, University of Stuttgart, Stuttgart, 2015.

[89] "PVSyst Photovoltaic Software." [Online]. Available from: http://www.pvsyst.com/en/.

[90] "Polysun Simulation Software." [Online]. Available from: http://www.velasolaris.com/.

[91] D. Berrian, "MoBiDiG: simulations and LCOE," *presented at the Bifacial Bifi PV Workshop*, Konstanz, Germany, 26 Oct. 2017.

[92] A. Lindsay, "Modelling of single-axis tracking gain for bifacial PV systems," *presented at the 32nd EUPVSEC 2016*, 2016, pp. 1610–7.

[93] J.S. Hill, "Yingli solar begins construction of 100 megawatt top runner project in inner Mongolia," *CleanTechnica*, 20 Nov. 2017. [Online].

Available from: https://cleantechnica.com/2017/11/20/yingli-solar-announces-begins-construction-of-100-mw-top-runner-project-in-inner-mongolia/.

[94] M. Osborne, "Neo Solar Power's mono PERC bifacial modules selected for 100 MW PV plant in Taiwan," *PV Tech*. [Online]. Available from: https://www.pv-tech.org/news/neo-solar-powers-mono-perc-bifacial-modules-selected-for-100mw-pv-plant-in.

Chapter 6
Impact of bifaciality on the levelized cost of PV-generated electricity

Joris Libal[1]

6.1 Levelized cost of electricity for photovoltaic systems

6.1.1 Introduction

The cost of electricity generated by photovoltaic (PV) systems is an important criteria that determines the competitiveness of PV in general compared to other – fossil and renewable – methods of electricity generation and that serves also to determine the best choice – from the economic point of view – in terms of PV module technology and system configuration for a given application and given specific geographical location. The reduction of the cost of PV-generated electricity is the driving force behind all research and development activities along the whole value chain of PV manufacturing, starting from the purification of the silicon feedstock and ending with the design and construction of PV systems and their components as well as of their efficient operation and maintenance (O&M).

The levelized cost of electricity (LCOE) is a widely used metric that aims to include, on the one hand, the complete cost (e.g. in euro or USD) related to the construction and operation of a PV system and on the other hand, all factors that have an impact on the total electricity generated (in kWh) during the lifetime of the PV system. Generally speaking it can be represented by the following relationship:

$$\text{LCOE} = \frac{\text{total life cycle cost}}{\text{total lifetime electricity generation}} \quad (6.1)$$

In the following, the concept of the LCOE and its application to PV will be shortly introduced and the impact of the use of bifacial – instead of monofacial – PV modules on LCOE will be discussed.

6.1.2 Parameters involved in the calculation of the LCOE

In this chapter, we define the LCOE of PV-generated electricity as the average generation cost, taking into account the complete cost required for generation of PV-generated electricity. Costs for transmission to and through the grid, as well as

[1]ISC Konstanz, Germany

profit margins of sales of electricity are excluded. As they vary strongly with the location of the considered PV system and as they are always exposed to the possibility of policy changes at the considered location, taxes and feed-in-tariffs (or other political support mechanisms for PV) are not taken into consideration as well, the same is valid for the residual value of the PV system at a given time. In addition, taking into account feed-in-tariffs or similar would distort the comparison with other energy sources that are supported differently or not at all.

Summarizing, the LCOE concept considered here, rather than supplying absolute numbers (that are in continuous evolution), has the scope of comparing different PV technologies – in our case, standard monofacial with bifacial module technologies – from the economic point of view and to shed light on important factors that have an impact on the respective LCOE. In addition, as an important application where bifacial PV is expected to feature the highest potential for an LCOE reduction, only utility scale ground mounted PV systems are considered.

The calculation of the actual LCOE of a specific project (e.g. with the scope to compare it to actual grid electricity prices) requires additional information such as taxes, tax credits and feed-in-tariffs. A publicly available LCOE calculation tool allowing for the choice between various financial models and for the implementation of the most important financial parameters is included in the System Advisor Model (SAM), developed by National Renewable Energy Laboratory, Sandia National Laboratory and the U.S. Department of Energy [1].

On the cost side (numerator in (6.1.)) of the PV system, the actual prices (i.e. cost + profit margin) of the following items have to be considered (see also Chapter 7) and the peculiarities of bifacial modules have to be taken into account:

- PV modules: there might be a price difference between comparable monofacial and bifacial modules, due to differences in production cost and/or due to a price premium for bifacial modules according to their higher energy yield potential;
- balance of system (BOS) hardware: mounting racks, cabling, inverters, etc.: taking into account only the nominal power at standard test condition (STC) for both module types, due to the higher average power output of the module array, a higher total inverter capacity (depending on the expected additional energy yield), will be required for bifacial PV systems;
- installation, land preparation, etc.;
- project development;
- land: optimum row-to-row distance tends to be higher for bifacial compared to monofacial PV systems, leading to a lower ground cover ratio for bifacial PV systems;
- O&M: if no measures have been taken to artificially increase the ground albedo, no difference in O&M cost is expected;
- financing: depending on the maturity (and track record) of a given bifacial PV technology (and the specific module supplier – see Chapter 7 for more details), the financing terms (discount rate – see the next section) can be less beneficial for PV systems based on bifacial modules compared to systems using standard monofacial modules.

Regarding the total amount of electricity generated during the complete system lifetime (denominator in (6.1.)), the following factors will have to be considered (see also Chapter 4):

- geographic location (yearly solar irradiance and temperatures, albedo, soiling): in contrast to the situation for monofacial systems, the ground albedo has a significant impact on the energy yield of a bifacial system. Depending on the system type (ground mounted or flat roof top) and specific site, various methods of artificially enhancing the ground albedo can be taken into consideration;
- module technology: when maintaining all the other characteristics (efficiency, temperature behaviour, yearly degradation rate) fixed, the additional energy yield is directly proportional to the bifaciality factor of the module;
- system configuration (tracking/fixed tilt, row-to-row distance, mounting height, tilt, azimuth, etc.): in contrast to monofacial systems, for bifacial systems, the mounting height is of significant importance for the energy yield;
- considered system lifetime: the useful system lifetime is determined by the yearly degradation rate which in turn depends strongly on the bill of materials of the module. A meaningful comparison between monofacial and bifacial technology will consider the same laminate structure for both; i.e. either glass/glass for both (using the same encapsulant) or glass/backsheet for both technologies. In summary, the same lifetime should be considered for bifacial and monofacial PV systems.

It has also to be mentioned that most of the above-listed input parameters, such as e.g. the solar irradiance as well as the financing conditions, are subject to different levels of uncertainties (see also Chapter 7), resulting in a range of possible (and probable) LCOE values rather than one single value (a comprehensive study on this topic has been presented in [2]). In this chapter, a quantitative comparison between monofacial and bifacial PV systems based on standard technologies in terms of LCOE will be performed. For the purpose of clarity, the above-mentioned uncertainties will not be taken into account, but a sensitivity study, involving some selected important parameters, will be presented.

A complete formulation of the LCOE concept includes also the financing conditions and takes into account of the fact that money that will be spent in the future has a lower value than money that is spent today. The latter concept is one of the fundamentals of financial mathematics and is implemented by the concept of 'net present value'. Accordingly, the LCOE is defined as the energy price (euro/kWh) for which the Net Present Value of the total project cost is zero. In other words, the LCOE is the averaged ('levelized') energy price (euro/kWh) over the complete lifecycle of the project for which the project reaches the financial break-even.

$$\sum_{t=1}^{N}\left(\frac{LCOE_t}{(1+d)^t}\times E_t\right)=\sum_{t=1}^{N}\frac{C_t}{(1+d)^t} \tag{6.2}$$

with t is the year of lifetime of the power plant (1, ..., N), N the economic lifetime of the power plant, $LCOE$ the levelized cost of energy (USD/kWh) in the year t,

d the real discount rate (without inflation), E_t the energy (kWh) produced in year t and C_t is the total expenditures (debt and equity service, O&M, etc.) in year t with

$$\frac{C_t}{(1+d)^t} \tag{6.3}$$

representing the net present value of an expenditure C_t paid in the year t.

As also shown e.g. in [3], (6.2) can be rearranged resulting in the following equation for an LCOE that has a constant value over the complete plant lifetime:

$$LCOE = \sum_{t=1}^{N} \frac{C_t/(1+d)^t}{E_t/(1+d)^t} \tag{6.4}$$

The fact that not only the future expenses C_t, but also the energy generated in the future is discounted by the discount rate d, can be explained by the fact that, from the economic point of view, the energy generation corresponds to revenues from sales of electricity and as such, future revenues have to be discounted as well and their net present value has to be used in the LCOE calculation.

In this chapter, the LCOE will be calculated based on the expenses for each year of the lifetime of the power plant, thereby, the following elements will be considered:

- repayment of debt and equity;
- operating expenses, taking into account the inflation rate.

Accordingly, (6.4) can be written as:

$$LCOE = \sum_{t=1}^{N} \frac{(I_t + O_t)/(1+d)^t}{E_t/(1+d)^t} \tag{6.5}$$

with I_t is the repayment for debt and equity in the year t and O_t is the expenses for O&M in the year t.

The energy generation will be calculated by determining the initial energy yield kWh/kWp, which depends on the plant location (meteo data, ground albedo, etc.), its configuration and on the used module technology. Multiplying the initial energy yield with the nominal plant capacity (W_p at STC) will deliver the total electricity production for the first year of operation. A yearly degradation rate of performance of the PV system is then assumed to calculate the electricity production for every year of the power plant's lifetime. Using this information, the LCOE is then calculated by applying formula (6.5).

6.1.3 Risk management in bifacial PV systems

Risk in the development of PV systems can include financial, economic, security and technological risk. All of these risks will impact the cost of capital and the cost of operations and maintenance. We assume that the financial, economic and security risk are the same for a bifacial system and the comparable monofacial system. Technical risk in a PV project is determined by the long-term performance record of similar technologies already in the field for 2–5 years (see also

Chapter 7). However, a new technology inherently has increased risk as there is less long-term field data. Bifacial PV, at the time of the writing of this chapter, lacks an extensive global track record. However, as discussed in Chapters 2 and 3 of the present book, bifacial PV technology is based on similar cell technology as monofacial PV cell technology. The larger risk in a bifacial PV project is the lack of a clear and reliable calculation of the expected bifacial gain, and therefore of the total power output of a system. Bifacial gain at the system level is dependent on many factors including system design, ground albedo and geographic location. This is being addressed by the development of reliable bifacial modelling software (see Chapter 4), but validation of these models by comparing their predictions with the field data monitored on multiple commercial bifacial PV systems is required in order to improve the accuracy and reliability of these models.

6.1.4 Importance of the weighted average cost of capital

Normally, financing of a utility scale PV system will be secured by a combination of debt and equity. While the interest rates for debt (provided by financial institutions, such as banks) will depend on the current, country-specific economic situation combined with the country-specific risk, the interest rates for equity (capital provided by private or institutional investors) will depend mostly on the perceived and evaluated project-specific risk (see Chapter 7). A project with a high technical risk (e.g. related to a technology without an extensive track record, such as e.g. bifacial PV at time of writing of this book) will have to bear a higher interest rate requested by the equity provider, at the same time, for such a project, the banks will limit their risk exposure by limiting the share of debt they contribute to the projects. Both factors will increase the weighted average cost of capital (WACC), which is defined as follows:

$$WACC = e \times i_e + d \times i_d \tag{6.6}$$

with e and d are the share of equity (e) and share of debt (d) and i_e and i_d are the interest rates for equity (i_e) and for debt (i_d).

In addition to determining the cost of capital for the power plant project under consideration, the WACC is most commonly used as the discount rate in (6.2)–(6.5) and has therefore a significant impact on LCOE as will be shown in the next section with a focus on the differences between bifacial PV as new technology compared to established (monofacial) PV technologies. As it is difficult to forecast inflation rates, in the following LCOE calculations the real discount rate (without inflation) instead of a nominal – inflation corrected – discount rate will be used. In this way, the LCOE value resulting from the calculations represents a real value denoted in the currency at its current value (at the time of performing the calculation).

6.2 Sensitivity study for LCOE of bifacial PV

The aim of this section is to use the concept of LCOE described in the previous section to calculate the LCOE for PV power plant projects based on standard, monofacial PV technology and on bifacial PV technology. Thereby, a sensitivity analysis will be performed in order to highlight the impact of the following parameters that represent the most important characteristics of bifacial PV:

- energy yield (kWh/kWp) translated to 'bifacial gain' compared to the yield of a monofacial PV system installed with a comparable (or better: identical) configuration at the same location (see also Chapter 5);
- CAPEX required for installing the PV system including the purchase of all hardware in (€/Wp), assuming different levels of 'price premium' (additional price) for the bifacial system compared to the standard (monofacial) PV system;
- WACC, assuming that bifacial PV potentially has to bear a 'risk premium' (see Chapter 7) that is expected to diminish within the near future;
- ground cover ratio, taking into account that – depending on the installation site and on the module technology – in order to fully exploit its potential, bifacial systems might require a lower ground cover ratio (i.e. a higher land consumption) than monofacial systems.

In addition, the dependencies between some of the above aspects will be examined also for different irradiance levels, corresponding to different geographic locations.

6.2.1 General assumptions and LCOE of monofacial PV

To generate a benchmark, the LCOE of a utility scale, ground-mounted PV system based on fixed tilt mounted monofacial modules is calculated. Thereby, the following assumptions (representative for a utility scale, ground-mounted system in 2018) are made:

- monofacial module price: 0.31 €/Wp
- module P_{mpp} at STC (60 cells module): 300 Wp
- CAPEX (capital expenditures) for installed PV system: 0.79 €/Wp
- system lifetime: 25 years (glass-backsheet modules)
- performance ratio of the system: 0.82
- yearly degradation rate for Pmpp: 0.4%
- OPEX (operating expenditures): 15 €/Wp

6.2.1.1 Definition of performance ratio of a PV system

According to the IEC 61724 standard, the performance ratio of a PV system is defined as follows:

$$PR = \frac{Y_{AC}(t)}{E_{POA}(t)} \tag{6.7}$$

where $Y_{AC}(t)$ is the cumulative-specific AC energy yield (kWh/kW) of the PV system in a given time period and $E_{POA}(t)$ is the cumulative plane of array irradiance (kWh/m^2) during the same time period divided by the irradiance at STC (1,000 kW/m^2). The PR of a PV system characterizes all relevant losses of a PV system such as:

- thermal losses
- shading losses (e.g. mutual shading by the module rows)

Table 6.1 Examples for GHI (global horizontal irradiance), POA (plane of array) irradiance for optimum tilt angle and specific yearly energy yield for fixed tilt monofacial PV systems at some selected geographic locations

	Yearly GHI (kWh/m^2)	**Yearly irradiance POA (kWh/m^2)**	**Yearly energy yield (kWh/kWp)**
Lübeck (Germany)	1,080	1,270	1,041
Valencia (Spain)	1,810	2,130	1,746
Atacama desert (Chile)	2,467	2,640	2,164

- DC and AC cabling losses
- inverter losses
- losses due to mismatch of series connected modules within the same string
- losses due to decreased module efficiency at low irradiance
- and other loss factors

Accordingly, the bifacial energy yield gain can also be expressed as the ratio between the performance ratio of the studied bifacial system and the performance ratio of a monofacial reference system:

$$bifacial\ gain\ [\%] = \left(\frac{PR_{bifi}}{PR_{mono}} - 1\right) \times 100 \tag{6.8}$$

Based on the above assumptions, the LCOE is now calculated for different locations of the power plant, characterized mainly by different solar irradiance levels resulting in different specific energy yields (kWh/kWp) and – due to different country-specific risk levels – different WACC. As a simplification, although these parameters are also likely to change with geographic location, performance ratio (depending on the module temperature under operating conditions) and OPEX will be kept constant for the various scenarios. Table 6.1 shows some examples of the global horizontal irradiance values for geographic locations and the corresponding achievable energy yield. For simplicity, for all locations a performance ratio of 82% for the monofacial reference system has been assumed. In practice, considering a comparable system configuration (e.g. same shading loss due to mutual row-to-row shading for all locations), due to the fact that Lübeck has a cooler climate than Valencia and the Atacama desert, Lübeck will show less thermal losses and consequently have the highest performance ratio of all the three locations under consideration.

Figure 6.1 shows the results of the LCOE calculations for a monofacial PV system, in dependence on the specific yield (kWh/m^2) for a WACC ranging from 4% (e.g. best cases in Germany) to 12% (e.g. worst cases in Greece). Accordingly, apart from the solar irradiance and the related achievable energy yield (kWh/kWp), the WACC has a significant impact on LCOE: comparing a PV system located in a well-established market such as e.g. Germany, featuring a low WACC of 4% but also a low energy yield (e.g. around 1,000 kWh/m^2, due to the low irradiance), with a PV system located in a country with higher risk (12% WACC), the irradiance in

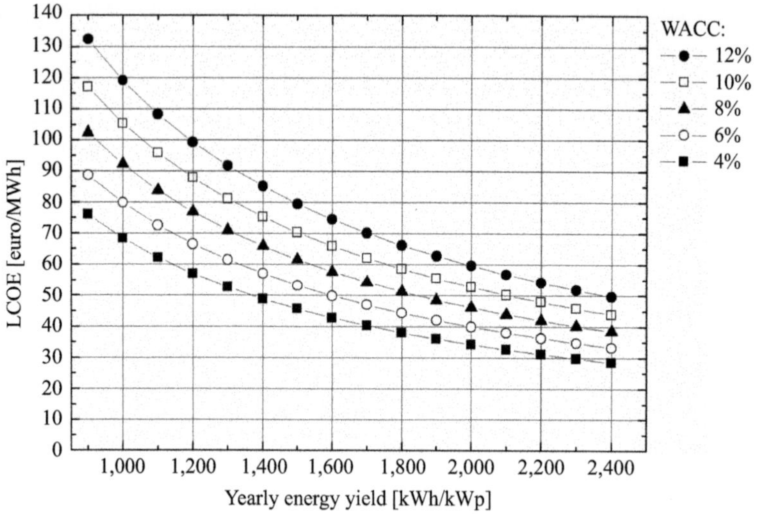

Figure 6.1 Results of LCOE calculations for fixed tilt, monofacial (utility scale, ground mounted) PV system for various WACC and different energy yields corresponding to different geographic locations. At the time of writing, a WACC of 4% is representative for best cases in Germany and 12% can represent some less favourable situations, e.g. in Greece

the high risk country must be high enough in order to allow for specific energy yields of at least 1,700 kWh/kWp.

6.2.2 LCOE of bifacial PV and monofacial PV: sensitivity study

In this section, the results of LCOE calculations for bifacial systems will be presented and its sensitivity against bifacial gain, system CAPEX and WACC will be investigated. Using the general assumptions listed in the previous sections and a WACC of 6%, Figure 6.2 shows the results of the LCOE calculations of bifacial systems depending on the additional energy yield ('bifacial gain') and on a potential price premium for bifacial systems that can be expected partly due to higher module cost and partly due to higher BOS cost (e.g. higher inverter capacity required). The monofacial case (0% bifacial gain, 0% price premium) has been calculated for a system CAPEX of 0.792 €/Wp and a monofacial yearly energy yield of 1,746 kWh/kWp.

Due to the fact that, in this calculation, for the financial conditions (discount rate, WACC) as well as for the main technical data (such as system lifetime and degradation rate), the same values have been assumed for monofacial and for bifacial systems, assuming a price premium of 10%, the installation of the bifacial system must be done in a way to guarantee at least 10% additional energy yield compared to the monofacial system. For 15% price premium, the bifacial gain must be at least 15% and so forth. When considering the price premium, it has to be

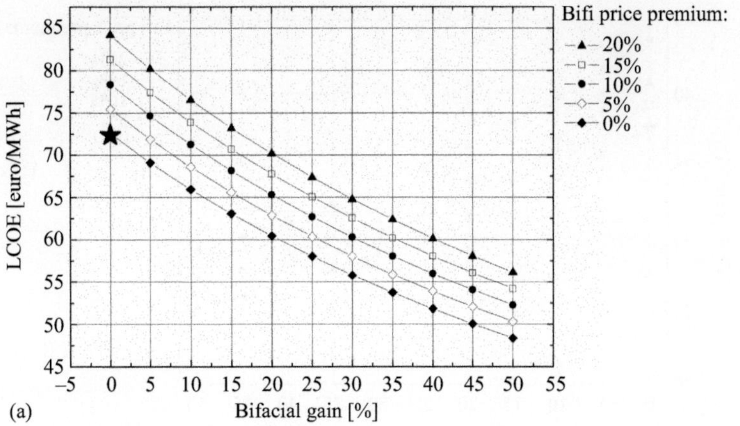

(a)

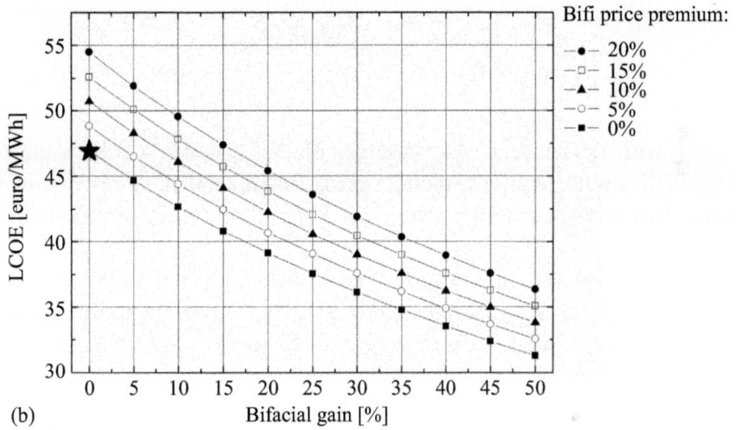

(b)

Figure 6.2 *(a) Results of LCOE calculations for bifacial fixed tilt, utility scale, ground-mounted PV systems at a location featuring 1,100 kWh/kWp (e.g. north of Germany) for monofacial systems and a CAPEX of 0.79 €/Wp for the monofacial system. (b) Results of LCOE calculations as in (a) but for a location featuring 1,750 kWh/kWp for monofacial systems (e.g. south of Spain). (c) Results of LCOE calculations as in (a) but for a location featuring 2,110 kWh/kWp for monofacial systems (e.g. Atacama desert in Chile). In (a)–(c), the LCOE of the monofacial PV system is respectively marked with a star, corresponding to 0% price premium and 0% bifacial gain. Note that, in these cases, the same WACC has been used for monofacial and bifacial systems and that the price premium can be – at least partially – due to a higher module price but here the percentage is based on the value of the bifacial system (and not of the bifacial module)*

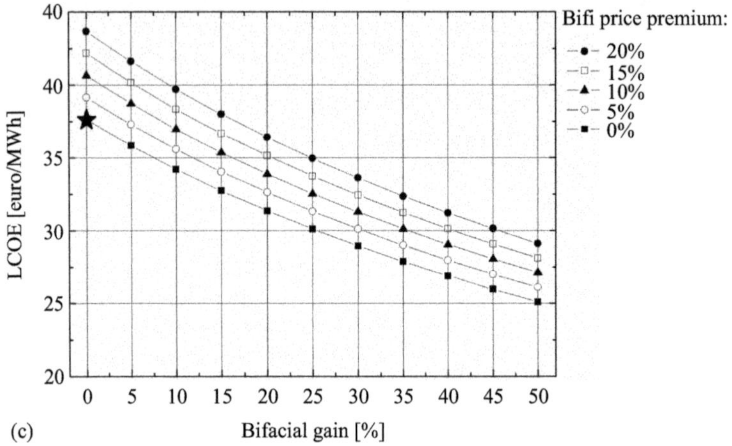

(c)

Figure 6.2 (Continued)

noted that – with the current cost structure of PV systems – a 15% increase in system CAPEX would require a module price for bifacial modules that is around 30% higher than for the monofacial modules. In the future, these ratios are expected to be subject to variations.

Considering as an example, the situation of a 10% price premium in the scenario shown in Figure 6.2(b), a bifacial gain of 20% that has been demonstrated for some large ground-mounted farms (see Chapter 5), the LCOE will be reduced from 47 €/MWh for the monofacial to 42 €/MWh for the bifacial system, corresponding to a relative LCOE reduction of around 10%.

However, as described in Chapter 7, at the time of writing, for bifacial PV systems, a higher WACC must be assumed in order to reflect the technical risk perceived by the lenders (banks and equity providers). Accordingly, a sensitivity analysis of the LCOE for bifacial PV systems depending on the WACC has been performed and the result is presented in Figure 6.3 for a location with a yearly energy yield of 1,700 kWh/kWp for a monofacial reference system. Apart from varying the bifacial gain and the WACC, compared to the monofacial case, a 15% higher module price has been assumed for the bifacial systems. Based on the current cost structure of ground-mounted PV systems, this corresponds to a 6% increase of the CAPEX of the installed system; accordingly, the CAPEX of the monofacial system has been assumed to be 0.79 €/Wp (and WACC to 4%), while the CAPEX for the bifacial system has been set to 0.84 €/Wp. The results are summarized in Figure 6.3 and show the following conclusions for the considered scenarios:

- when comparing a bifacial system with a monofacial reference system and assuming the same WACC of 4% for both, it will be sufficient to have a bifacial gain exceeding 6% in order to compensate for the 6% higher system CAPEX in order to achieve a lower LCOE for the bifacial system;

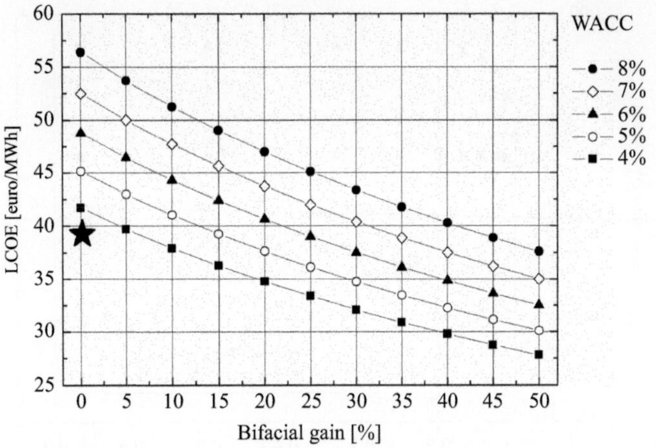

Figure 6.3 Sensitivity analysis for the LCOE of bifacial PV systems (system CAPEX: 0.84 €/Wp) depending on the bifacial gain and the WACC. The LCOE for the monofacial system (0.79 €/Wp) with a WACC of 4% is indicated by the star symbol. The LCOE has been calculated for a location where the monofacial system yield is 1,700 kWh/kWp per year

- however, assuming that the perceived project risk of the bifacial PV system would result in a WACC that is 1% (abs.) higher compared to the standard monofacial reference system, an additional 10% bifacial gain would be required in order to reduce the bifacial LCOE to match the LCOE of the monofacial reference system. For the scenario selected here (6% higher system CAPEX for bifacial system), a 15% bifacial gain is required for the bifacial system to achieve a lower LCOE than the monofacial system.

6.2.3 Sensitivity analysis: bifacial gain versus ground cover ratio and resulting LCOE

As mentioned earlier, assuming a sufficiently high albedo, the bifacial gain will depend also from the ground cover ratio (related to the number of modules on each racking tables and the row-to-row pitch). When increasing the row-to-row distances, starting from very narrow row-to-row distances (high ground cover ratio), as used for example at low latitudes due to the low optimum tilt angles for monofacial PV systems, initially, the energy yield will increase for monofacial as well as for bifacial systems. This is due to the reduction of shadowing losses from mutual row-to-row shading. At a certain point it is expected that there will be a row-to-row distance where a further increase will lead to almost no increase of energy yield for the monofacial system, while the energy yield of the bifacial system will further increase due to the increased area of non-shaded ground surface that can be 'seen' by the rear sides of the modules (see also the view factor concept explained in Chapter 4 about energy yield simulations). Energy yield simulations

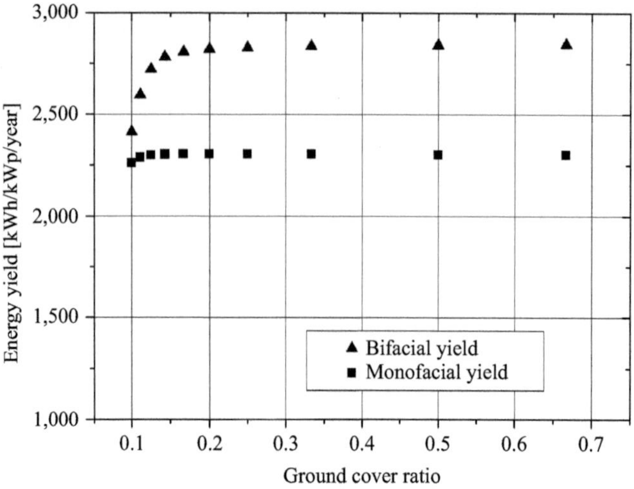

Figure 6.4 Variation of the energy yield of an equator oriented bifacial PV system in dependence of the ground cover ratio compared to a monofacial reference system with the same ground cover ratio and the same azimuth. The tilt angles are optimized for the bifacial as well as for the monofacial system as well as for the various ground cover ratios

have been conducted using the commercial software PVsyst (version 6.6.8) for a monofacial and bifacial PV system with the following characteristics:

- location in Chile (nearby the European South Observatory 'la Silla')
- ground albedo 40%
- mounting height of the modules: 1.5 m
- assumption of infinite number of neighbouring modules in a row and infinite number of module rows
- the tilt angle has been optimized according to the row-to-row pitch (ground cover ratio) in order to minimize on the one hand the losses due to mutual row-to-row shading and on the other hand to maximize the irradiance collected by the front and rear side of the modules.

The results of these simulations are summarized in Figure 6.4. Accordingly, it can be seen that for monofacial systems, decreasing the ground cover ratio to values below 0.5 does not result in a significant additional energy yield, whereas for the bifacial system – from the point of view of energy yield – a decrease down to a 0.25 might be reasonable. The saturation of the bifacial gain at a ground cover ratio of 0.25 can be also observed in Figure 6.5 where, based on the calculated values for the energy yields of the monofacial and the bifacial systems, the bifacial gain has been calculated and is represented accordingly.

Taking into account that land has a certain cost as well as its preparation and the related civil works, these cost factors are strongly country specific and are also related to the level of labour cost as well as to the typology of the land. As the cost

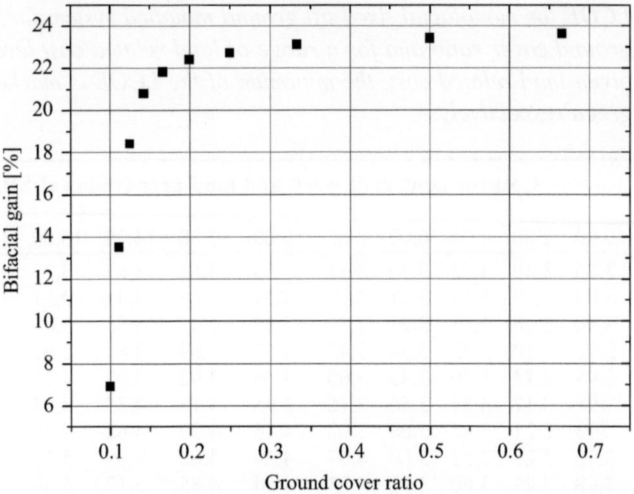

Figure 6.5 Variation of bifacial energy yield gain in dependence of the ground cover ratio of an equator oriented fixed tilt bifacial system compared to a monofacial system with the same ground cover ratio

contribution of land and land preparation cost is proportional to the total land area that is consumed for the construction of a given PV system, it is directly related to the ground cover ratio (inversely proportional) and to the area specific cost (€/m^2).

Using the considerations made earlier, regarding the relation between system cost, energy yield and bifacial gain, as well as of the dependency between ground cover ratio and the energy yield of bifacial and monofacial systems, an optimization of the ground cover ratio for both typologies in terms of minimizing the LCOE can be made. The following numbers, calculated for ground-mounted large-scale systems, should be taken only to illustrate trends and correlations, whereas the absolute values will be strongly dependent on the specific project (country, technology, size, financial boundary conditions, etc.) and, in particular, are expected to be in continuous evolution in the future. It has also to be highlighted that the above-calculated trends for the bifacial gain will change for situations where the climatic conditions (diffuse light fraction) and/or the ground albedo will be different. The impact of these factors is studied in Chapter 8.

In addition to the ground cover–dependent energy yields calculated earlier, the assumption has been made that – for the case of 10 USD/m^2 for the cost of land and land preparation – the total cost for the installation of the monofacial fixed tilt system is 0.80 USD/Wp (including the modules at 0.41 USD/Wp), when assuming a ground cover ratio of 0.5. While for the bifacial modules, a 15% higher price (0.47 USD/Wp) has been assumed, leading to a system cost (again for a ground cover ratio of 0.5) of 0.86 USD/Wp.

Under these assumptions, and in the case of a WACC of 6.5% for both system typologies, the LCOE has been calculated for monofacial and for the bifacial system for various ground cover ratios and for a range of land-related cost (from 0 to

Table 6.2 LCOE for monofacial fixed tilt ground mounted system for varying ground cover ratio and for a range of land related cost levels. For a given land related cost, the minimum of the LCOE is marked in green respectively

LCOE (USD/kWh)	**Cost for land, civil work and land preparation (USD/m^2)**										
GCR	0.00	2.00	4.00	6.00	8.00	10.00	12.00	14.00	16.00	18.00	20.00
0.90	3.36	3.40	3.45	3.49	3.54	3.58	3.62	3.67	3.71	3.76	3.80
0.67	3.03	3.09	3.15	3.21	3.27	3.33	3.39	3.45	3.51	3.57	3.63
0.50	3.00	3.08	3.15	3.23	3.31	3.39	3.47	3.55	3.63	3.71	3.78
0.33	2.98	3.10	3.22	3.34	3.45	3.57	3.69	3.81	3.92	4.04	4.16
0.25	2.98	3.13	3.29	3.45	3.60	3.76	3.92	4.07	4.23	4.39	4.54
0.20	2.98	3.17	3.37	3.56	3.76	3.95	4.15	4.35	4.54	4.74	4.93
0.17	2.98	3.21	3.45	3.68	3.92	4.15	4.38	4.62	4.85	5.09	5.32
0.14	2.98	3.25	3.52	3.80	4.07	4.34	4.62	4.89	5.17	5.44	5.71
0.13	2.98	3.29	3.60	3.92	4.23	4.54	4.85	5.17	5.48	5.79	6.11
0.11	2.98	3.33	3.68	4.04	4.39	4.74	5.09	5.45	5.80	6.15	6.50
0.10	2.98	3.37	3.76	4.15	4.54	4.94	5.33	5.72	6.11	6.50	6.89

Table 6.3 LCOE for bifacial fixed tilt ground mounted system for varying ground cover ratio and for a range of land related cost levels. For a given land related cost, the LCOE values that are lower than the optimized monofacial system (see Table 6.2) are highlighted in green respectively

LCOE (USD/kWh)	**Cost for land, civil work and land preparation (USD/m^2)**										
GCR	0.00	2.00	4.00	6.00	8.00	10.00	12.00	14.00	16.00	18.00	20.00
0.90	3.59	3.63	3.68	3.72	3.76	3.80	3.85	3.89	3.93	3.98	4.02
0.67	3.13	3.18	3.24	3.29	3.35	3.41	3.46	3.52	3.57	3.63	3.68
0.50	2.91	2.98	3.05	3.12	3.19	3.26	3.33	3.39	3.46	3.53	3.60
0.33	2.78	2.87	2.97	3.07	3.17	3.27	3.37	3.47	3.57	3.67	3.77
0.25	2.72	2.85	2.98	3.11	3.24	3.37	3.49	3.62	3.75	3.88	4.01
0.20	2.69	2.85	3.01	3.17	3.33	3.49	3.66	3.82	3.98	4.14	4.30
0.17	2.68	2.87	3.06	3.25	3.45	3.64	3.83	4.02	4.21	4.40	4.59
0.14	2.67	2.89	3.12	3.34	3.56	3.79	4.01	4.23	4.46	4.68	4.90
0.13	2.67	2.92	3.17	3.43	3.68	3.94	4.19	4.44	4.70	4.95	5.21
0.11	2.66	2.94	3.23	3.52	3.80	4.09	4.37	4.66	4.94	5.23	5.51
0.10	2.66	2.97	3.29	3.61	3.92	4.24	4.56	4.87	5.19	5.51	5.82

20 USD/m^2). The numbers are shown in Tables 6.2 and 6.3 and allow for the following conclusions:

- Amongst the ground cover ratios studied here, for monofacial systems there is an optimum value that allows for the lowest LCOE and this ground cover ratio depends on the land-related cost. For the present scenario, it ranges between 0.5 and 0.65.

- A higher packaging than this optimum value will save land-related cost; however, this savings will be counterbalanced by a strong reduction of the energy yield due to mutual row-to-row shading losses and consequently lead to an increase in LCOE compared to the minimum possible values.
- A lower than optimum packaging will increase the land-related cost and – as stated above – after a certain limit, a further increase in spacing between the rows does not result in any significant energy yield gain for monofacial systems.
- For the bifacial systems it can be observed that for very low land-related costs the lowest ground cover ratios with the highest bifacial energy yield (and highest bifacial gain) show the lowest LCOE values.
- The same as for monofacial systems, with increasing land related cost, for bifacial systems, the optimum values for the ground cover ratio shift to higher values – in this case to around 0.33 and 0.5 but – in contrast to the monofacial system, do not reach 0.67.
- The advantage of bifacial systems in terms of LCOE is reduced for high land-related cost, for scenarios where – due to high land cost or other constraints – very high ground cover ratios are required. In such cases, depending on the price difference between monofacial and bifacial modules (the present calculation is based on a 15% higher price for the bifacial module), a monofacial system will be more beneficial in terms of LCOE.

6.2.4 Summary

In this chapter, the concept of LCOE has been described and applied to compare the LCOE achievable with utility scale, ground-mounted bifacial PV systems to the LCOE of a comparable standard monofacial PV system as a reference. Thereby the following main findings can be summarized:

- For monofacial modules, considering WACC values from 4% to 12%, the LCOE is reduced by more than 50% when moving the installation site from a region with low irradiance (e.g. north of Germany with 1,100 kWh/kWp yearly energy yield) to a region with highest energy yield (e.g. the Atacama desert in Chile with 2,100 kWh/ kWp yearly energy yield) (see Figure 6.1), while maintaining constant the WACC.
- This significant impact of irradiance on LCOE is the reason for the strong potential of bifacial PV to reduce the LCOE of PV-generated electricity. When assuming equal technical and financial conditions as for monofacial systems, a 10% price premium (on system CAPEX) requires the bifacial energy yield gain to exceed 10% in order to reduce the bifacial LCOE below the level of the monofacial one. Higher bifacial gains – while maintaining a constant system CAPEX will lead to further reductions of the LCOE (see Figure 6.2(a)–(c)).

The WACC has a significant effect on the LCOE of a bifacial PV system. The benefits of bifacial PV in terms of reduced LCOE are subject to a technological risk evaluation by the lenders (see Chapter 7). Once bifacial systems will be well established on the market thanks to a significant track record of large-scale bifacial

PV systems and thanks to the availability of validated and accurate energy yield prediction software tools, the related financial risk should not be significantly higher than the risk attributed to established monofacial PV technologies. Accordingly, it is expected that in the future the WACC to be applied for LCOE calculations for bifacial PV systems will reach the same values as for monofacial PV systems. However, at the time of writing, a certain increase in WACC should still be assumed. As shown in this chapter, when applying a WACC of 5% to a bifacial PV systems while a monofacial system at the same location would have WACC of only 4%, an additional bifacial gain of 10% abs. (e.g. 15% gain instead of 5%) is required to achieve a lower LCOE with the bifacial system. The results of this chapter show that, in particular for large ground-mounted PV systems, bifacial PV technology has a strong potential to significantly reduce the LCOE of PV-generated electricity. An interesting aspect regarding the ground cover ratio of monofacial and bifacial PV systems has been investigated in the final part of this chapter: in a typical scenario, the use of bifacial modules (instead of monofacial modules) requires a larger row-to-row spacing in order to exploit the benefits of bifacial PV in terms of energy yield. Depending on the actual cost for land preparation, it is shown that different optimum values for the ground cover ratio exist for which the LCOE of the bifacial system is minimized and lower than the minimum LCOE that is achievable for the equivalent monofacial PV system.

This potential of bifacial PV for achieving the lowest values of LCOE for PV systems is the motivation for many industrial and academic research groups to work on the further increase of the achievable module efficiencies and bifacial energy yield gains as well as for industrial companies to implement large-scale PV plants using also bifacial modules in order to improve the track record of bifacial PV technology and – in this way – to fully exploit its potential.

References

[1] National Renewable Energy Laboratory. Golden, CO. "System Advisor Model Version 2017.1.17 (SAM 2017.1.17)" [Accessed 22 May 2017].

[2] Seth B. Darling, F. You, T. Veselkad, and A. Velosae. Assumptions and the levelized cost of energy for photovoltaics, Energy Environ. Sci., 2011, 4, 3133.

[3] K. Branker, M. J. M. Pathak, and J. M. Pearce. A review of solar photovoltaic levelized cost of electricity. Ren. Sustain. Energy Rev., 2011, 15(9), pp. 4470–4482.

Chapter 7
Importance of bankability for market introduction of new PV technologies

André Richter[1]

In previous chapters we have seen lot of technical discussions about PV technology and bifacial PV applications. If one starts to implement such a project, in most cases a bank has to be approached in order to cover a part of the financing.

The bank verifies the project and the fact that it provides some financing means that the project is bankable. There is no clear definition of this word because it depends on a lot of conditions. Here are some examples.

Considering a big multi-international company with total turnover of 50 billion dollar and with a small PV subsidiary. The community expects that this company has lot of production expertise and in case of product failures the company can cover it by its own financial strength. In addition, this kind of company is rated by the known big three rating agencies Standard & Poor's (S&P), Moody's and Fitch. Usually it can be expected that all products and projects of such a company are bankable. This is the same for government backed companies and governmental projects. The rating is most likely the rating of the country. Mid-size, only PV business–related companies with well-known names like big solar manufacturers or big EPC companies: this case is not clear and the risk has to be investigated. This risk is mostly non-technical and depends mainly on the financial strength of this company and the overall market development. We will later analyze these risks.

Small companies or "no names" like startups have created a new product to attain specific market segments and most likely they have not the financial strength to cover bigger claims. There is a very high risk that the company will disappear in the next 20–30 years. To reduce the risk exposure for the company and the clients, these companies use instruments like insurances and re-insurances to get a rating which is just sufficient to participate in bigger projects.

The outcome of the three examples is that bankability can be achieved, if the overall risk involving this company in the considered project is low enough to prevent a default and to ensure return rates that are high enough to cover at least the costs for the bank.

[1]Meyer Burger Technology Ltd, Switzerland

Bankability is not limited to the suppliers in the project; it is also related to the risk exposure of the project owner. The bank will investigate the full project, all participants (suppliers of modules, supplier of mounting systems and inverters, project developer, EPC, country and support schemes like feed-in tariffs) and the project owner: e.g. how much money the project owner requires as a debt, the reimbursement period, contracts with EPC, O&M companies, and service providers. A profound investigation includes the project schedule and a supervision of each milestone and additional tests during the project, like checks of the delivery and control of the project progress.

Summarized only 25% of the risk are related to the solar module, 25% to the project execution and 50% to the partners, country, etc. Every project is different. Of course there are existing processes to rate the risks. Projects with non-known factors, such as new technologies, new companies, sizes of project with the partners, have no reference for and every other new element in such a project can cause a premium to the debt (i.e. increase of requested interest rate) to cover this additional risks. Of course bifacial modules have only a very small market share at time of writing of this book and they are still considered as "new".

In the next sections, these risks are discussed including guidelines of many available rating schemes and best practice, such as solarbankability.org and S&P. This investigation is more non-technical, but it is focused on PV systems and when possible, the differences of standard PV systems to bifacial systems are mentioned. If risks are not investigated carefully, the resulting analysis is most likely flawed and will cause surprises of the stakeholders in future. Out of general observations one cannot conclude specific decisions like: Is the cheap module better, because lower CAPEX or should the more expensive one, with potentially smaller issues, be considered?

7.1 Value chain and cost types

In this chapter, we are looking only from the commissioning of the PV system up to the energy market.

Along the PV value chain the measures of the market change, e.g. the silicon is traded in USD/kg the wafer in piece and the solar cell and module in USD/Wp. In the end the investor is interested in lowest cost per kWh (Figure 7.1).

This shows a first conflict of the participants of the market: the optimization of a production of wafers where you are paid per piece is different to an optimization of a cell prices per Wp. The cell relies on a good wafer quality but this is not directly reflected in the price per piece of wafer. Of course there are second-order properties and different qualities of wafer.

A second example is the energy generation of a solar system: the Wp value is an indication of the possible final value of generated energy, but it depends on many more external conditions: the irradiation at the site where the considered PV system is located, the irradiation and temperature which cause, e.g. degradation and soiling effects and the longevity which is caused partly by the quality of construction and many more: how to rate a system which depends on so many influences?

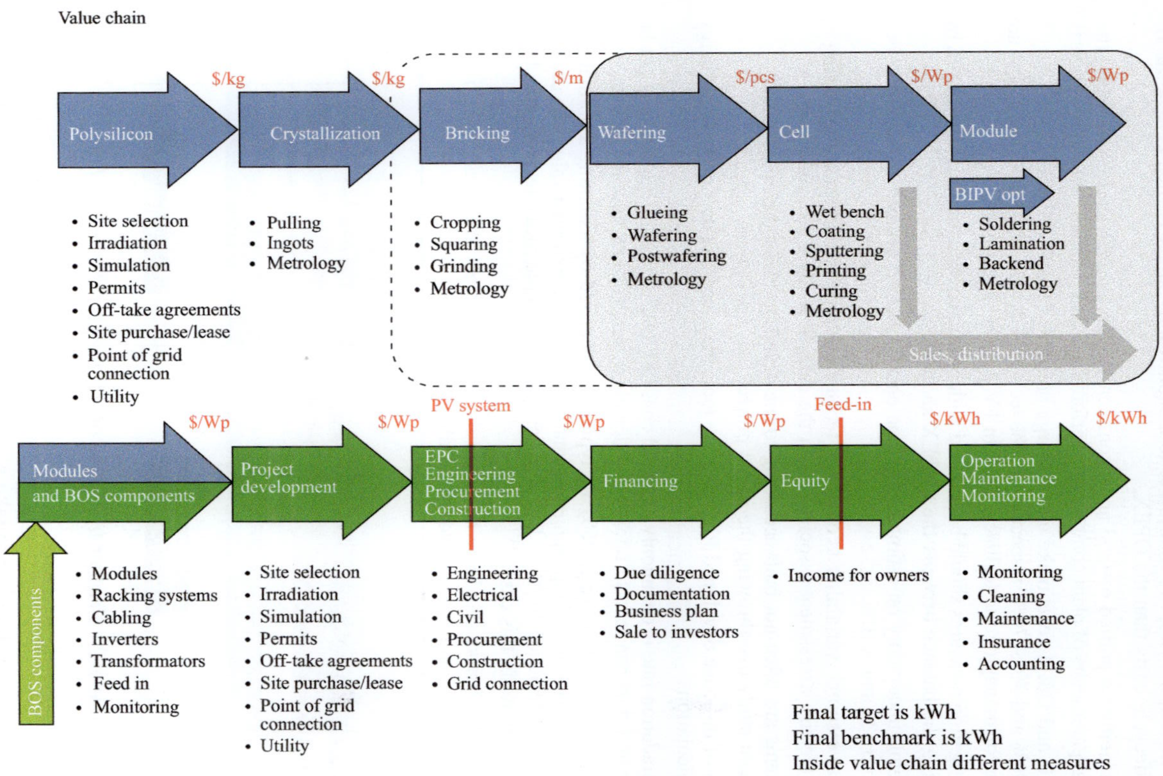

Figure 7.1 Value chain of the PV industry

Table 7.1 Extract of Solar Bankability summary of technical assumptions [1]

Summary of technical assumptions in present-day financial models for PV
1. For PV LCOE, the CAPEX contributes to a significantly larger portion (~75%–90%) to the lifecycle costs than the OPEX.
2. There is neither a unified method nor a commonly accepted practice for translating the technical parameters of plant components, performance and reliability into lifecycle costs.
3. The EPC and O&M costs make up to a large portion of the CAPEX and OPEX (70%–90% and 30%–70%, respectively); the technical details in the EPC and O&M are decisive for managing the technical risks in PV project investment.
4. Risk mitigation measures should be selected with an objective to minimize the LCOE by optimizing the balance between the CAPEX and OPEX.
5. The overall uncertainty on estimated lifetime energy yield is typically assumed to be between ±5% and ±10%.
6. The solar resource variability is one main technical source of uncertainty impacting mainly the risk assessment associated with the cash flow during a single year.
7. PV systems are often not built according to the design used for the initial yield assessment study overthrowing the initial project risk assessment.
8. The use of in-house developed PV modeling tools may lead to flaws in lifetime energy yield calculations.
9. The degradation rate is commonly assumed constant over time although this may not be the case and thus can lead to unexpected deviation in cash flow over the years.
10. Exceedance probabilities (e.g. P60) are typically calculated by assuming a normal probability distribution of e.g. annual irradiation around the expected value; the use of a cumulative distribution function based on long-term resource measurements may be more appropriate in this case.
11. Not all technical risks should be mitigated thorough technical measures. Financial or legal mitigations should be considered as alternatives.

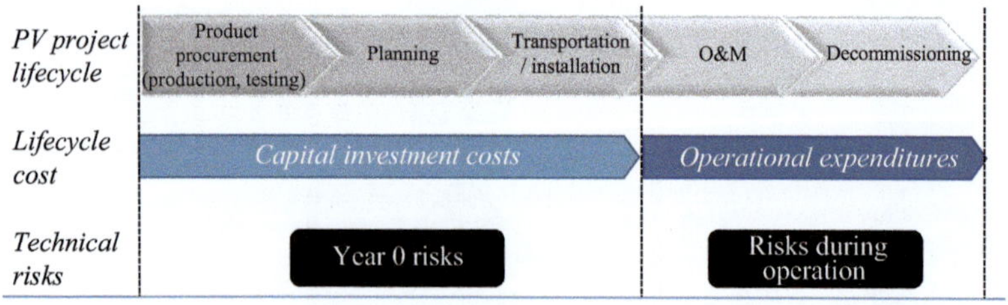

Figure 7.2 Project life cycle of PV project, based on [1]

The rating of a system has to take all these parameters into account. Risks are based on deviations of the initial project assumptions. A good summary is done by the Solar Bankability project (see Table 7.1).

And, as shown in Figure 7.2, the risks can be related to the value chain and project realization.

Very specific to PV system projects is that 80%–90% of all costs are related to the CAPEX of a PV project and thus, have to paid up-front, before any revenue is generated from the sales of electricity. This circumstance increases the amount of debt and therefore, the absolute risk for a bank or any other stakeholder.

Risk mitigation is very important. A simple approach is just to limit the involvement of a stakeholder to a certain amount, e.g. 50% of the total value. A second possibility is to control the project commissioning, the condition in year 0 and condition in later years. Monitoring systems are helpful for this purpose.

For bigger projects a third approach is considered: quality supervision of the upstream process, particular the production of module and solar cell to ensure the quality of the initial PV system. Check of the bill of materials of the certification, check of the used material quality and of course check of the final product with electroluminescence or thermal images (see Figure 7.3).

Very important is to minimize the risks at the beginning of the project; costs to influence the project at later stage are increasing with the progress of the project (Figures 7.4 and 7.5).

The Solar Bankability project summarized most important parts of the project until the initial phase.

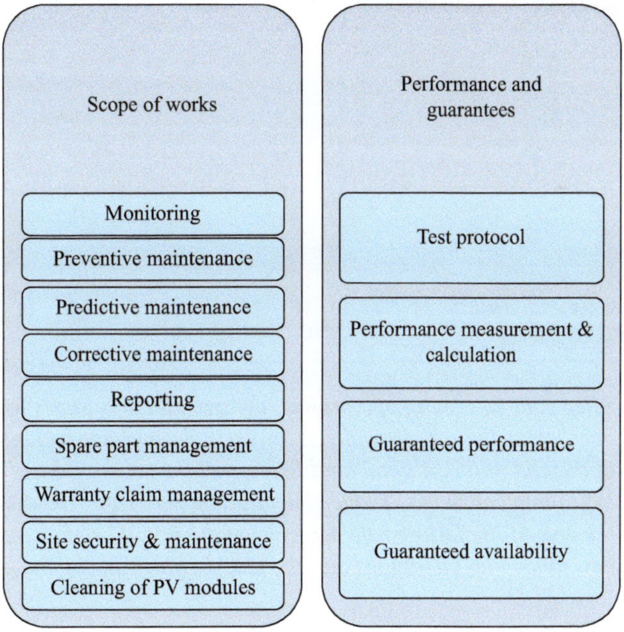

Figure 7.3 Lists of measures for risk mitigations of EPC. Partly other stakeholders like banks will add their own supervision procedures. (Based on [1].)

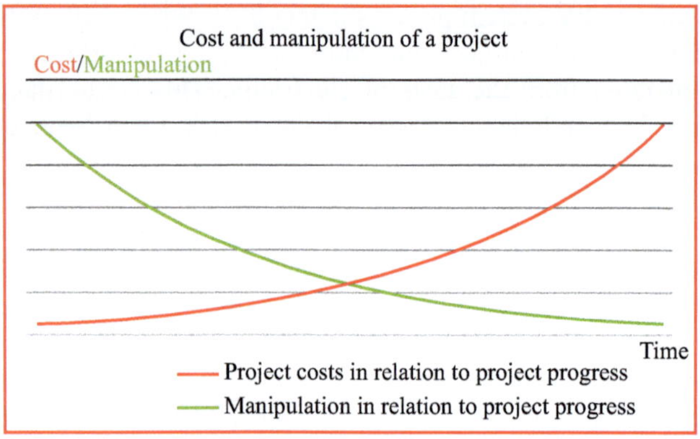

Figure 7.4 Typical cost development during the project. In later stage of a project the costs to make changes in the project are increasing exponentially. This leads to the recommendation to plan carefully the project and not to change it during the execution

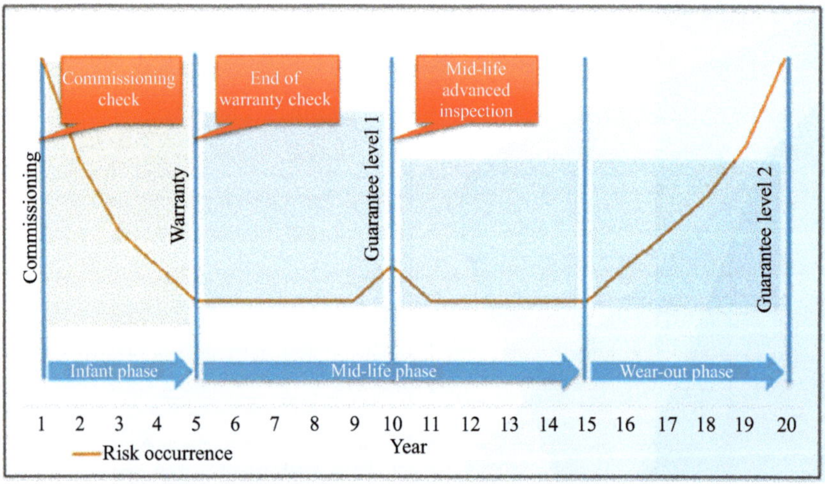

Figure 7.5 After commissioning of a project, the typical curve starts with a decreasing failure rate which is lowest in the mid-age and increase again at the end of life. A good qualification of the used products at the beginning can help to reduce the initial failure rate. The development of the end-of-life failure rate depends on the initial product quality, the environmental conditions and the O&M efforts during the project [1]

The following factors are described in [1] (Figure 7.6):

- **Year-0 parameters**
  - **Plant parameters:** Location and type of PV system (roof-top/ground mounted, etc.), nominal capacity, annual yield, annual degradation, start of operation, project duration.

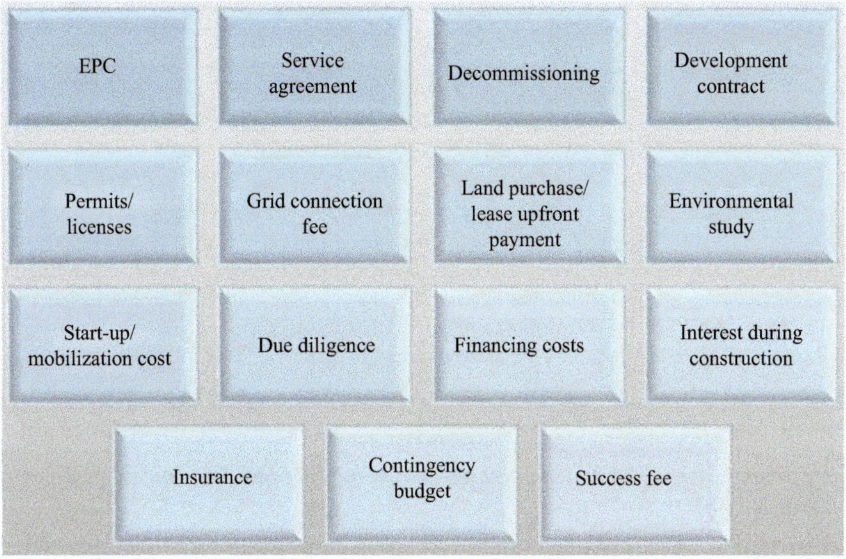

Figure 7.6 Main factors for the initial phase until year 0 [1]

  - **CAPEX:** Encompass total investment costs including project development, land purchase, EPC, due diligence and financing.
  - **Financing:** Equity capital, debt capital and conditions of credit, including term, interest rate and redemption.
  - **Legal/tax:** Legal and ownership structure with respective income tax and depreciation rates.
  - **Electricity tariff/business model:** The electricity tariff will depend on the type of, the nominal capacity and the start of operation of the PV system as well as the underlying business model, i.e. feed-in tariff, net-metering, self-consumption or power purchase agreement.
- **Parameters during operation**
  - **Revenues:** Revenues depend on the electrical yield of the PV-system, its annual degradation and the respective electricity tariffs under a feed-in-Tariff (FiT), net metering or power purchase scheme. In case of self-consumption the relevant demand curve and the electricity tariff for self-consumption have also to be considered.
  - **OPEX:** Encompass all expenses to operate and maintain the PV plant during the operational years, including costs for operations and maintenance, land lease, debt service, insurance and tax (Figure 7.7).
  - **Reserves:** Different reserves are included to reflect seasonal fluctuations and single events. Most common is the repair and maintenance reserve to cover ongoing repair and maintenance including the replacement of inverters at the end of service life. In case of debt financing, banks will ask for a debt service reserve account to ensure complete and on-time payments. A decommissioning reserve can accumulate the costs of dismantling the PV system at the end of its service life.

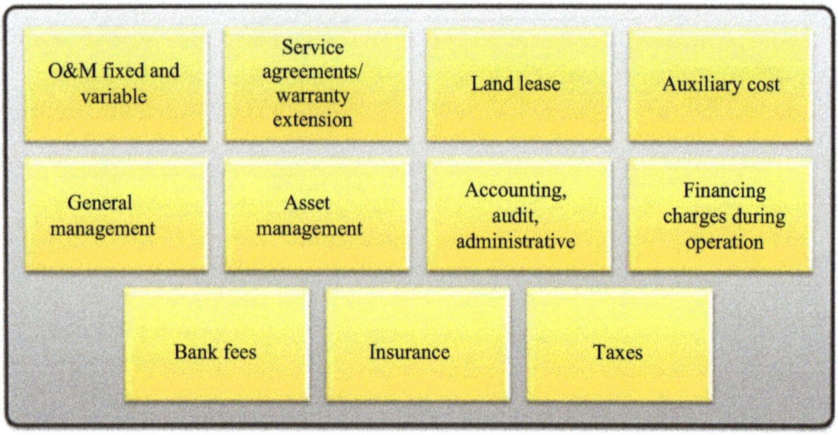

Figure 7.7 OPEX items to consider in most of the PV-projects [1]

In this stage the commissioning of the PV project is finished and the operational phase is taking place with the following main factors:

- administration costs (meter readings);
- maintenance costs;
- repair costs of inverter, installation, mounting system;
- costs for protection and night watch.

In the financial world the project costs are translated in the following terms (based on [1]):

- **Cash flow/cumulative cash flow**: Cash flow is the net amount of cash moving in and out of the PV-project. The cumulative cash flow indicates the sum of all cash flows over the course of the PV project.
- **FiTs**: Guarantees the reimbursement per energy unit over a certain time period, e.g. 15 or 20 years. FiTs reduce the risk to the technical risk of forecasting the energy yield over this period of time.
- **Liquidity**: Reflects the cash flow plus reserves, excluding depreciation. It indicates if an investor is able to pay out dividends or might need to inject fresh capital. Good projects never running out of liquidity.
- **Payback time/breakeven**: Indicates the point in time when cumulative revenues equal cumulative costs, that means from this point onwards profits begin to accumulate and the project becomes financially viable. Each party can have a different payback definition of a project: a bank with e.g. front-ranking assets, will consider the payback when they get paid the debt including interests. The owner might have a longer payback period.
- **Internal rate of return (IRR)**: The IRR is the discount rate at which the net present value of all cash flows from the project equals zero. The higher the IRR the more attractive it is for the investor to engage in the PV project. A distinction is being made between the equity IRR based on 100% equity financing and the project IRR based on partial debt financing.

- **Debt Service Coverage Ratio (DSCR)**: The DSCR is a measure of the cash flow available to pay current debt obligations. The ratio states the net operating income as a multiple of debt obligations due within one year. The DSCR should always be higher than 100%, typical good values are 120%.

Looking at the bunch of different fields that need to be covered in a PV project some main parts should be highlighted:

- The used technology is of minor importance. This means the project is not done because of a technology, but it's realized because of the economic values. Of course these values are driven partly by the technology. For example bifacial technology will gain 5%–30% more energy with the same front side module power installed. The higher energy yield will help to increase margins, etc.
- Risk mitigation is the highest interests of all parties: PV projects need a relative long pay-back compared to other financial involvements for the owner of the system: until the payback time, the project is cash-flow negative, after this period is takes again some years to reach a certain return rate. All occurring problems during this time will affect most likely the financial sheet.
- Regarding bifacial systems there are some good points: the installed area is smaller and O&M costs, depending mostly on area, are reduced by this technology. The higher energy yield can help to decrease pack-back times and reduce total paid interests.

It has to be noted that the revenues of the PV system are not only the sold electric energy, but also the residual value of the installation has to be taken into account and in special situations, grid stabilization purposes can represent an additional economic value.

7.2 Measures to calculate PV systems

A PV system generates costs and revenues. This leads direct to the cash-flow calculation (Figure 7.8).

In the cash-flow, bifacial systems with a typical ground albedo of 10%–25% consists in superior cash-flow curves. This of course results in short payback times and very high returns at end of life like Figure 7.6 shows. Measurements of monofacial and bifacial show an energy gain of 5%–35% for the bifacial modules.

On a financial view the bifacial gain is not guaranteed and there are very few installations with a good track record of a third party certified energy yield of bifacial systems. At the time of writing of this chapter, most stakeholders might have no confidence on this additional gain, so this feature will not help direct in calculation of cash-flow curves. Banks will more likely calculate the well-known monofacial case and rate the PV system according these settings. The additional energy yield due to the bifaciality gets relevant after few years of successful operation of the system. In a later stage this gain can be used to rate the system and to get better financial conditions.

Because of the higher energy yield, possible degradation effects are compensated and the residual value of the system after 2 or 5 years might be valued higher

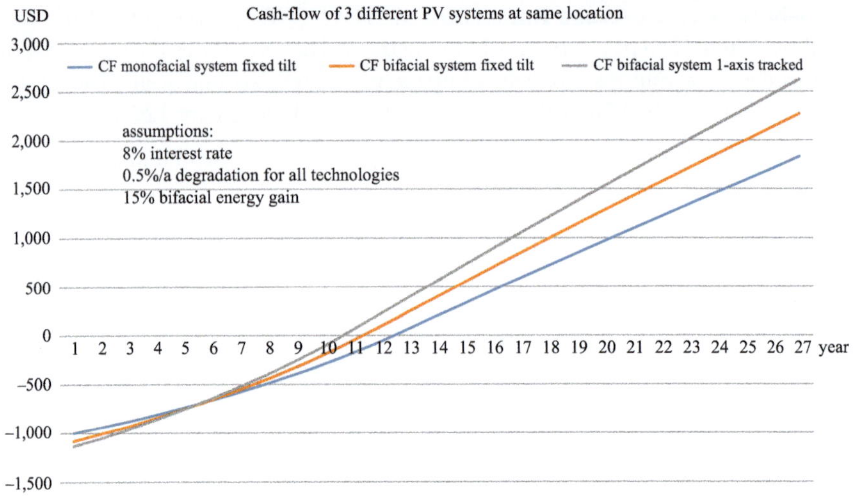

Figure 7.8 Example of three different PV systems in the same location. The monofacial system is a standard fixed tilt and the bifacial system is fixed tilt with optimized mounting structure. This will result in slightly higher starting costs, but will pay-off after few years in faster payback time

in the market. These advantages are not yet so obvious, because there are not so many bifacial projects traded.

In a few years the market will have more experience with the bifacial systems and will create benchmarks for this kind of systems to ease financial ratings for all market participants.

The points of Table 7.2 can be understood as an overall checklist of a PV project. All this points are following the major implications:

- wealthy of project partners: reduce the risk if one partner gets bankrupt, etc.;
- payback of the project: secured PPA rights and simulated energy yield of this location and technology;
- delayed timing and deadlines can cause a project default;
- monitoring: identify as soon as possible a problem, increase the residual value.

To compare the costs of electricity of PV systems to other PV systems and to any other electricity generation systems such as wind energy, coal energy or nuclear energy, the levelized cost of electricity (LCOE) generation method is used. Details about this method are given in this chapter. The relevant point regarding the question of bankability of bifacial PV system is the fact that for calculating the LCOE, all future costs and all future revenues are depreciated by a discount rate and usually, the weighted average capital costs (WACCs) are taken for this discount rate. The WACC is calculated for every project separately and considers many aspect such as the country risks, the currency risks and the capital rate (see Figure 7.9).

A very similar approach is the calculation of the maximal possible discount rate the net present value of zero. With this method the depreciated cash-flow over

Table 7.2 Risk mitigation can be done in added costs during the lifespan [1]

Risk	Phase/field		Identified critical technical gaps
Year 0	Procurement/product selection and testing	1	Insufficient EPC technical specifications to ensure that selected components are suitable for use in the specific PV plant environment of application
		2	Inadequate component testing to check for product manufacturing deviations
		3	Absence of adequate independent product acceptance test and criteria
	Planning/lifetime energy yield estimation	4	The effect of long-term trends in the solar resource is not fully accounted for
		5	Exceedance probabilities (e.g. P90) are often calculated for risk assessment assuming a normal distribution for all elements contributing to the overall uncertainly
		6	Incorrect degradation rate and behavior overtime assumed in the yield estimation
		7	Incorrect availability assumption to calculate the initial yield for project investment financial model (vs. O&M plant availability guarantee)
	Transportation	8	Absence of standardized transportation and handling protocol
	Installation/construction	9	Inadequate quality procedures in component un-packing and handing during construction by workers
		10	Missing intermediate construction monitoring
	Installation/provisional and final acceptance	11	Inadequate protocol or equipment for plant acceptance visual inspection
		12	Missing short-term performance (e.g. PR) check at provisional acceptance test includes proper correction for temperature and other losses
		13	Missing final performance check and guaranteed performance
		14	Incorrect or missing specification for collecting data for PR or availability evaluations: incorrect measurement sensor specification, incorrect irradiance threshold to define time window of PV operation for PR/availability calculation
Risks during operation	Operation	15	Selected monitoring system is not capable of advanced fault detection and identification
		16	Inadequate or absence of devices for visual inspection to catch invisible defects/faults
		17	Missing guaranteed key performance indicators (PR, availability or energy yield)
		18	Incorrect or missing specification for collecting data for PR or availability evaluations: incorrect measurement sensor specification, incorrect irradiance threshold to define time window of PV operation for PV/availability calculation
	Maintenance	19	Missing or inadequate maintenance of the monitoring system
		20	Module cleaning missing or frequency too low

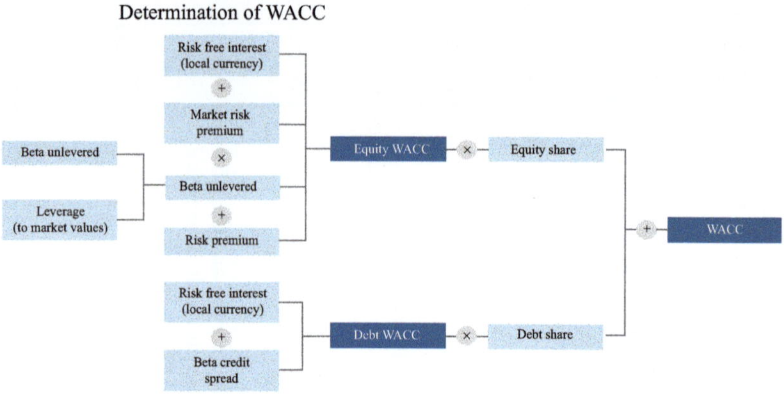

Figure 7.9 This calculation scheme for the WACC is typical concept to calculate and weigh the risks in industry [2]

a specific period is set to zero. With approximation procedures the depreciation rate is calculated:

$$\text{NPV} = \sum_{n=0}^{N} \frac{C_n}{(1+r)^n} = 0 \qquad (7.1)$$

"r" describes the depreciation rate to a value down to zero in a specific period. This rate is named IRR.

In a rule of thumb the IRR of a project should cover the country risks, currency risks, technical risks like new technology and project risks like a deadline of fit-in-tariff. The remaining part of IRR is good for the margin of debt or equity.

7.3 Energy yield simulation

While the energy yield simulation is already discussed in detail in Chapter 4, in this section some quantitative aspects regarding the accuracy of energy yield forecasts are presented. Because more than 80% of the PV lifetime costs are generated until year 0, a very good estimation of the generated energy is indispensable. This simulation delivered the generated energy and therefore the earnings of the system.

Energy yield prediction by simulation is not yet established for bifacial systems, so while technicians might see the advantage, this bifacial gain will not be valued from banks. In 2017 first commercial systems like PV system are available for these specific modules, but not yet validated by experimental data. The simulation has not only taken into account the front side shading and illumination but as well the rear side illumination and self-shading. Because of this the simulation need more detailed information of the PV system itself. In praxis every input parameter has a tolerance or has to be estimated and all irradiation based parameters have by nature a higher tolerance.

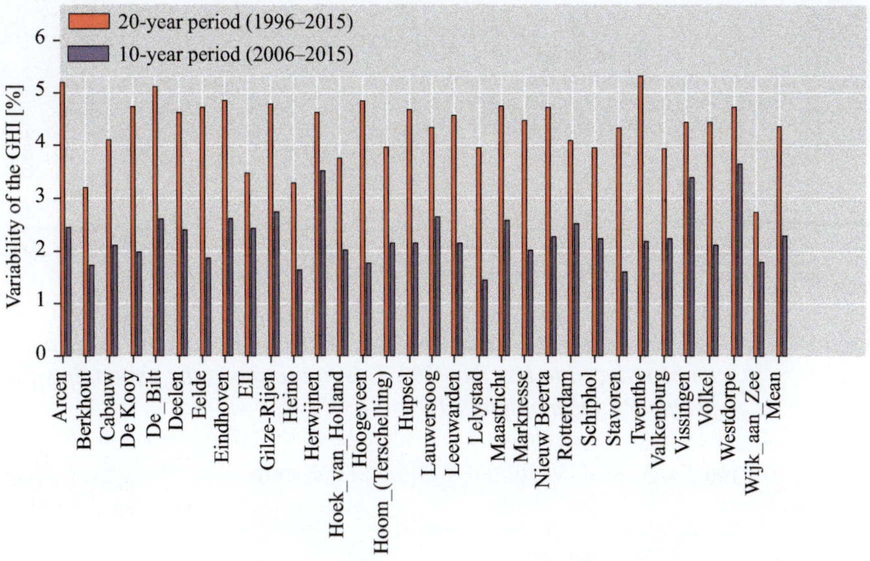

Figure 7.10 *Annual variability of the GHI in the Netherlands [1]: this is the primary input of each PV system simulation. Even this parameter has a variability of 1%–3.5% for a 10-year period and 3%–5% for a 20-year period in the Netherlands for specific investigated stations. This is the basic unknown factor in forecasting and simulation of PV systems. A local monitoring system can measure the PV performance relative to the local irradiation and temperature. Then remaining uncertainties are only differences of temperature coefficients, spectral effects which can be considered as second-order magnitude. Other effects like shading effects, etc. are not considered here because this effect can be considered already at the design phase [1]*

In the best case, the energy yield simulation is based on 10 years average weather data of a specific location and assumptions about degradation and soiling. There are possible distributions for other climates and e.g. degradation rates. This results in different energy yields (Figures 7.10 and 7.11).

Because of this uncertainty the energy yield forecasting quality is classified in certain classes: a P90 energy yield value will cover 90% of all cases and the possibility of not reaching this value is only 10%. A P50 value gives a 50% possibility not to reach the value (Figure 7.12).

In Figure 7.12, 41 different systems are evaluated and the real energy yield in year 1 was measured.

While the measurement of the energy yield is very accurate and better than 0.5%, the measurement of weather data, in particular irradiance data, has significant tolerances (Table 7.3).

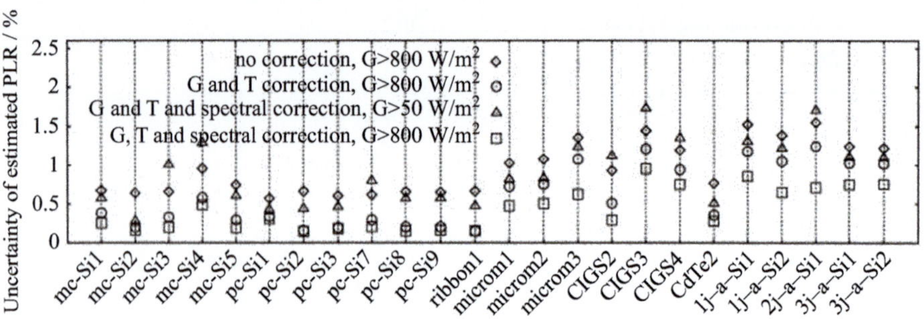

Figure 7.11 Considering more local parameters in the simulation, the uncertainty can be reduced to less than 2%. This figure shows the uncertainty for specific PV technologies [1]

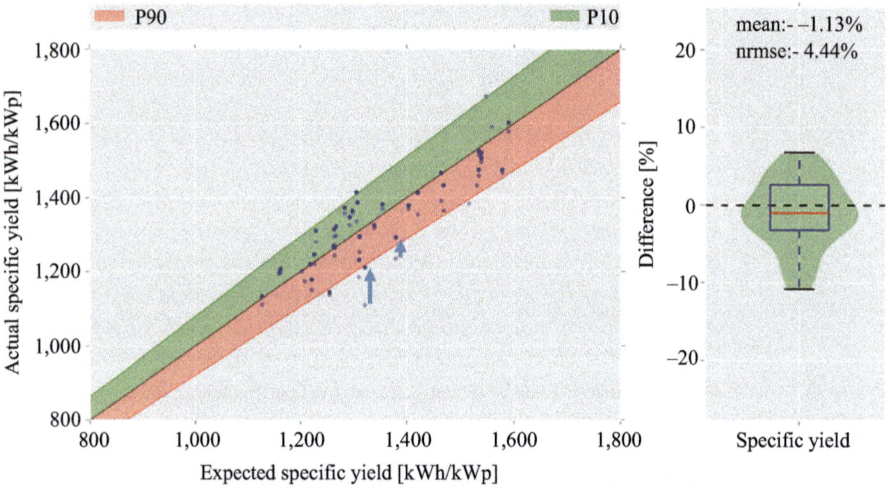

Figure 7.12 Forty-one analyzed PV systems after 1 year of operation show that most systems perform on the given P90 limit (red) or even better. Only 5 systems of 41 are performing slightly less than the P90 forecast [1]

Table 7.3 Typical uncertainties of sensors for irradiance measurement [1]. The uncertainty for a specific sensor in a specific use can be greater. Best values are 50% of the shown data

Source	**Uncertainty (ca.)**
Secondary standard pyranometer	±2%
First-class pyranometer	±5%
Second-class pyranometer	±10
Silicon sensor	±5%–8%
Satellite-derived data	±2.5%–5%

7.4 Risk—the key factor in a project

Risk assessment is essential for a project and for an investor. Higher amount of risk might be mitigated by higher return rates or the risk is shared between all parties. Beside all technological details, to evaluate the risk is always a key factor in projects.

Everyone knows what is meant with "risk," but it is very hard to quantify it for a specific project.

There are some ways to measure it, a good overview with extended literature is published in [3], but in most cases the investors will define a risk of a part of a project in relation to another known project. Experience can make the difference to get a project (avoiding too high risk premium) or to run a project successful (not to tap into unknown costs during the project).

Bifacial photovoltaic technology is in first sight very similar to standard photovoltaic because most of the components are very similar or the same to monofacial PV systems. The main difference is the system design to maximize the rear side illumination and to expect in operation higher currents than the Impp measured at STC conditions. Almost every existing cell technology can be converted in a bifacial cell: PERC or by design bifacial technologies like PERT and HJT. Taken these cells in a module with two transparent sides, like glass/glass, the bifacial module is ready. The advantages of glass/glass modules are very well known from monofacial thin film technologies. In summary there should be no big surprise when using bifacial technologies instead of monofacial technologies. The investors have a good right to take care about every change in a project; the further chapter discusses the influence of the different kind of risks particular to bifacial PV technology to reduce this asymmetry of information and to increase investor's confidence.

In general there are three main risk categories: country risk, market risk and technology risk. Usually all three groups are rated separately and added in a project specific way.

A PV system has a specific location. The jurisdiction and other rules are defined by the country hosting this location. The energy of the system is sold into a grid or is consumed by a client directly. Investors covering this with a market risk assessment. Sometimes the "market risk" includes the risk to do business with a certain partner. The last category is the technology risk with main focus of the reliability toward an expected energy yield over a long time.

- **Country risk**

The country risk represents the security to do business in a country: independent justice, stable government with seamless laws and a good infrastructure driving this risk lower. A higher country interest rate reflects a higher risk in general for business in a country. An AAA-rated country with very low interests for government bonds is preferable.

Solar projects are selected by the main revenue driver: Irradiation! The location plays the most significant role to calculate the revenue. Of course there are other parameters like AC connection costs, land commissioning costs, etc., but all these parameters are already second order compare to the irradiation. The place of a project will define the country and its infrastructure such as justice, transportation and regulation.

For specific business areas, specific indices exist: the RECAI (Renewable Country Attractiveness Index) [4] computed and published since 2003 by advisory company Ernest & Young is one of the most famous (Figure 7.13(a) and (b)).

Around the most important 40 countries for renewable energy are rated for the categories: Onshore and offshore wind, PV and CSP solar, biomass, geothermal, hydro and marine (waves, tide, etc.). Every country has different potentials, so the renewable index has to pay attention for this difference (Table 7.4).

The RECAI offers beside the attractiveness number also in depth information about actual topics and it is distributed quarterly.

It includes two specific information for each renewable sector:

A **relative ranking** of the countries:

Investors familiar with one country can estimate if the risk of a project in another country may be higher or lower to their known one.

This information is in many cases already sufficient for investors, if the project is in a higher ranked country, the investors can used worst case the known risk. If the country is rated lower, the investor has to take attention for the country risk and has probably to start an assessment to discover the additional risks here.

An **absolute number**: This absolute number an investor can compute in its own balance sheet to get an automated calculation for the risk in his project. An example is shown in [3].

Beside this figure the RECAI explains relative changes in the view of investors. This background information specifies why a specific country is rated different this time.

- **Market risk**

Market risks have lot of connections to the country, its infrastructure and the technology.

In principle in every country or market occur expected or unexpected changes.

A very actual example for market risk is the new US tariff for imports of solar modules and solar cells [5]. Imports over the 2.5 GW will be punished with a 30% tariff in 2018 with is melting down 5% a year and it will remain 4 years active. The United States is a big market for PV modules and 30% added cost for the main part of a solar system—the module—will have a big impact on every project balance sheet.

Another example for (unexpected) market risk is Spain with the retroactive reduction of the FiT 2013 [6]. Of course this FiT reduction gives a disastrous signal to the markets and the trust of investors was gone for years. This example shows the difficulties to introduce renewable energy.

Today we are more sensitive about costs and external costs for every investment. This makes things more complex. In the past the governments subsidized nuclear and fossil technologies. For example almost no external cost is paid by the "old style" technologies. Look at the nuclear waste or decommissioning of power plants or have a look of the very low carbon dioxide prices for fossil power plants. The actual developments in Germany and other countries show that the public will

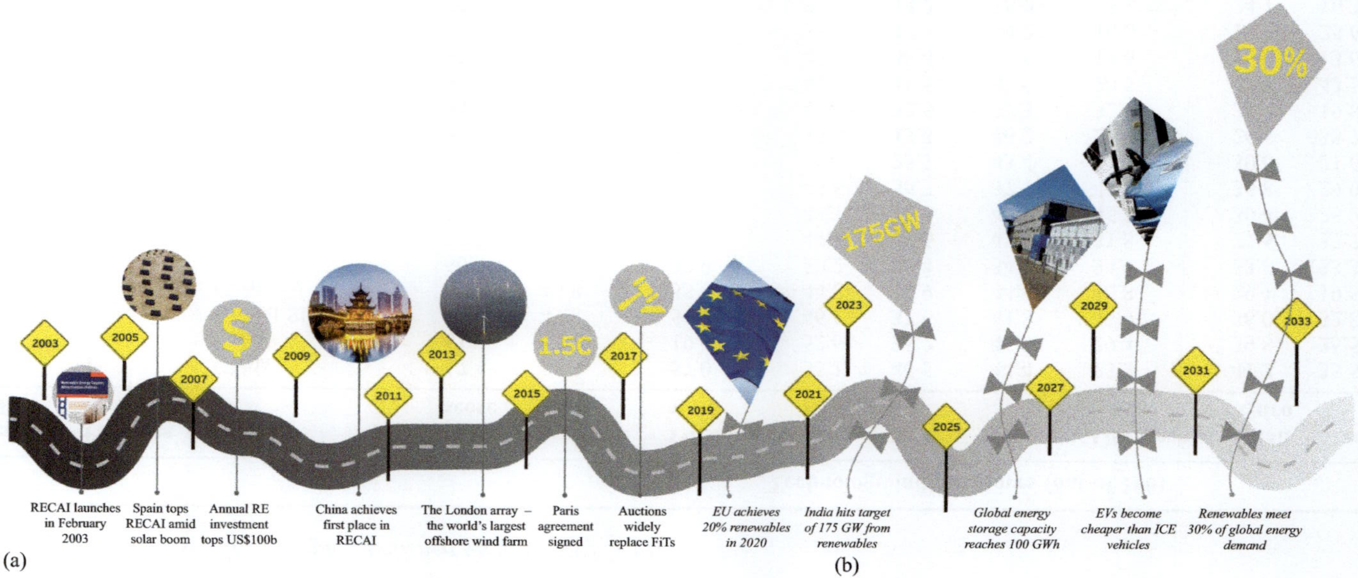

Figure 7.13 *(a) RECAI [5] evolution in the past and (b) the expectation for the next years. This underlines the intention of RECAI to give a clear picture to investors [4]*

Table 7.4 RECAI No 50 (example published October 2017) [4]

				Technology indices scores (out of 100)							
Overall rank	**Pervious rank**	**Country**	**RECAI score**	**Onshore wind**	**Offshore wind**	**Solar PV**	**Solar CSP**	**Biomass**	**Geothermal**	**Small hydro**	**Marine**
1	1	China	67.4	51.3	57.0	55.1	40.5	44.4	23.2	41.2	35.5
2	2	India	61.9	49.2	19.0	52.6	38.2	45.3	29.4	39.8	25.5
3	3	United States	61.8	49.8	51.6	46.1	37.6	41.8	43.9	36.0	32.8
4	4	Germany	60.7	45.8	55.3	44.8	16.9	44.4	36.8	29.1	19.5
5	5	Australia	60.5	45.9	32.9	50.2	38.4	34.8	24.9	33.8	33.3
6	8	France	57.2	43.5	39.0	44.4	22.6	45.5	31.8	27.5	37.7
7	7	Japan	56.7	41.3	45.4	43.7	18.0	47.9	45.7	30.4	25.2
8	6	Chile	56.1	43.2	20.2	45.6	36.7	37.7	41.2	36.8	29.0
9	9	Mexico	55.8	42.6	19.5	48.8	25.1	43.4	43.3	30.6	21.6
10	10	United Kingdom	54.6	42.8	57.3	36.6	13.3	46.2	25.7	26.8	38.7
11	12	Argentina	54.1	44.3	20.8	45.4	32.5	37.3	32.3	34.1	19.8
12	11	Canada	53.5	44.6	28.8	41.0	18.5	37.9	20.5	41.7	42.5
13	14	Morocco	53.3	41.2	17.1	45.3	38.4	6.6	13.6	16.9	13.6
14	13	Denmark	53.2	43.5	47.3	35.3	17.2	44.2	16.6	18.9	25.6
15	16	Netherlands	52.5	41.6	45.6	36.2	14.2	36.0	25.2	24.1	16.3
16	17	Turkey	52.2	43.4	18.9	42.1	24.6	35.9	40.1	38.3	16.4
17	15	Brazil	52.0	44.2	22.8	43.5	21.5	49.3	14.9	42.9	17.8
18	18	Italy	50.5	39.5	31.3	37.2	30.5	45.6	37.4	39.9	19.6

(Continues)

Table 7.4 (Continued)

19	27	Egypt	50.5	42.1	14.1	45.7	39.0	12.8	11.6	14.8	11.6
20	22	Portugal	50.3	34.9	32.0	38.8	25.7	36.9	23.7	30.5	26.7
21	23	Philippines	50.2	38.1	15.6	42.1	18.1	42.9	42.0	38.6	26.3
22	24	Belgium	50.1	40.9	43.6	33.3	13.8	41.0	19.8	22.8	14.2
23	19	South Africa	49.8	39.7	17.8	42.5	37.1	32.8	13.9	29.0	22.9
24	21	Jordan	49.6	37.9	13.6	42.6	29.3	20.9	13.1	16.6	13.1
25	20	Sweden	49.1	42.2	32.2	32.0	14.0	40.0	20.1	35.0	28.7
26	26	Pakistan	48.9	39.2	12.8	42.9	21.4	20.3	18.5	34.5	16.8
27	29	Spain	48.8	39.5	21.9	36.9	24.9	37.4	17.8	26.2	23.1
28	28	Peru	48.5	37.0	14.5	40.2	23.0	32.9	23.6	36.2	18.4
29	33	South Korea	48.2	28.2	29.7	39.9	18.5	32.1	18.9	26.2	39.1
30	25	Israel	48.0	30.9	13.7	43.2	31.7	21.2	14.3	21.6	17.7
31	40	Ireland	47.6	41.8	26.7	31.9	13.6	31.5	22.6	25.1	29.1
32	30	Finland	47.5	43.8	41.2	24.3	14.5	48.1	16.9	27.8	14.5
33	32	Greece	47.4	37.4	24.1	37.1	28.7	18.8	22.3	25.5	12.8
34	31	Thailand	47.1	34.9	15.0	38.6	21.7	40.4	16.1	27.3	17.6
35	36	Uruguay	46.1	39.0	16.6	36.7	17.6	34.9	14.2	23.5	18.2
36	–	Vietnam	45.7	36.8	20.8	36.5	16.7	38.9	16.7	34.4	17.1
37	35	Kenya	45.7	37.3	13.7	38.9	21.6	27.1	45.9	30.5	11.7
38	–	Algeria	45.6	33.5	14.2	42.8	32.9	17.3	11.2	17.7	11.2
39	39	Dominican Republic	45.5	34.6	14.1	37.6	19.8	20.0	14.4	31.6	12.8
40	37	Kazakhstan	45.4	36.4	12.7	38.8	16.6	13.4	12.2	25.6	12.2

carry the most of the costs in future, so the market risk for existing power plants of the "old style" is still low.

New projects, and this includes all power generation technologies, have not the same luxury any more: In developed countries all technologies have to consider more and more external costs, like CO_2 taxes or decommissioning fees.

A way out of this is to limit risk by the time of investment: projects with a very high return rate with a fast payback are preferred. High efficient PV, particular bifacial systems, with short payback times of few years, are already interesting for investors: fossil fuel prices had been low last years, but there are very volatile and the payback of "old style" investments is in the same level of high efficient PV systems (refer to Figure 7.8).

For PV systems other "market risks" are more important, here are some examples:

- Will the price of solar modules drop in the next 2 to 12 months?

In the view of 2018 the market price for monofacial solar modules is already in the level of the square meter price of the materials. The market price is fluctuating on the actual level. It's expected that bifacial modules will have a fast learning curve to team with the monofacial PV modules. Many manufacturers are increasing the production capacity already and this will relax the prices.

Second important point is the new IEC 61251 standard for bifacial modules to rate the power. This standard is expected in the first half of 2018 and will reduce market risk due to measurement significantly, because manufacturers and clients are a same base to negotiate contracts.

The "new" bifacial modules will need a specific time to equalize with the market price. Manufacturing costs of a 60 or 72 cells solar module in bifacial are the almost the same compared to monofacial modules, so there is no reason to sell the modules more expensive than existing PV monofacial modules.

- Will I have a replacement of my bifacial module in the next 5 years?

This question is the same than for all other solar systems and solar modules. The main challenge is the mechanical dimensions of a panel. This is given by the amount of cells (today 60 or 72), the size of the wafer (today M0 and M2) and the insulation class (today 100 Vdc, but going to 1,500 Vdc). Bifacial modules will have the same dimensions than standard modules of the same conditions, so mechanical issues are the same and of course of same risks.

Market risks due to regulations like FiTs or project partners are the same compared to monofacial PV.

The supply of a specific PV module for a big PV project remains crucial for all projects, monofacial and bifacial. Because more manufactures are switching to bifacial today and the small premium paid for bifacial modules will add security to projects and might cause problems to very low cost monofacial projects.

- **Technology risks**

Bifacial technology is discussed throughout this book. Like the country risk we can try to evaluate the technology risk of bifacial technology in comparison to the known monofacial technology, which has been installed around 100 GW in 2017 (Table 7.5).

Table 7.5 Listing of technology risks relative to monofacial PV in the view of an investor

Technology risk table—comparison with monofacial PV			
0: no change, + better, – worse			
	Monofacial	**Bifacial**	**Remark**
Cell stability	0	0	Same cell technology
Module materials	0	0	Same materials
Cell interconnection	0	0	Same interconnection
Mounting systems	0	0	Same type of mounting systems and mounting
System design	0	(–)	Bifacial systems need little more design knowledge
Power measurement	0	(–)	New IEC standard for bifacial modules published soon
Risk mitigation—payback time	0	++	Faster payback time with same energy rates
Risk mitigation—soiling	0	++	Less sensitive to soiling
Risk mitigation—stability	0	0/+	Same quality of module will result in same stability bifacial cells have better passivation, potential advantage
Risk mitigation—O&M	0	0	Same procedures than standard PV
Risk mitigation—experience	0	–	Bifacial systems are not yet so common, this leads to a perceived risk for bifacial today

Negative points are based on the lack of experience; these risks are perceived and not based on negative cases. On the other hand the positive aspects are based on physics and are evident.

In most cases the monofacial and bifacial PV technologies differ not too much in the risk exposure. The technology risk for c-Si monofacial PV is already rated as low [3], so the risk exposure for bifacial PV will be not higher as medium for worst case.

- **Risk depreciation**

  Already discussed is the LCOE calculation method in Chapter 6.

  Most of the risks can be mitigated by shorter payback times. In LCOE is the term

$$\sum_t \frac{1}{(1 + WACC)^t} \qquad \text{(term1)}$$

used to depreciate either costs and generated energy. This term is displayed in Figure 7.6 for some main conditions occurring in the PV. After commissioning (year 0), in the energy generation phase, PV systems have most likely the same yearly O&M costs and the same yearly energy yield. In LCOE calculation term1 causes the variation by time and interest rate. In a general view we can look only at this term for typical conditions like 10, 20 and 40 years project life and 1%–20% WACC. Term1 is multiplied with the constant O&M costs or with the constant energy yield. The LCOE is then simplified and reduced to

$$\begin{aligned} \text{LCOE} &= (\text{CAPEX} + \text{O\&M} \times \text{Term1})/(\text{energy\_yield} \times \text{Term1}) \quad \text{or} \\ \text{LCOE} &= \text{CAPEX}/(\text{energy yield} \times \text{Term1}) + \text{O\&M}/\text{energy\_yield} \end{aligned} \tag{term2}$$

High values of term1 lead to lower LCOE value mainly because of the term CAPEX/term1. The operative LCOE stays most likely constant with O&M/energy_yield. For the interesting range of 8%–15%, WACC term1 will not differ so much by project lifetime.

With high interest rates for equity financed projects of WACC >15%, the influence of the lifetime of term1 is very small. One interpretation of term1 can be assumed like depreciated years or effective years: in a similar way like any investment term2 shows that the CAPEX is depreciated by term1 over a certain lifetime. The example in Figure 7.6 covers most settings of WACC and lifetime for PV projects: according to this, the interpretation of term1 with values in the range of 5–20 represents an effective distribution of CAPEX over 5–20 years. Therefore, a PV system with generates high returns covers a high WACC is equivalent with a short depreciation time. The project lifetime is less important, but the CAPEX and/or the O&M/energy_yield should be low to get a low LCOE.

Figure 7.14 shows the added factors for depreciation. A high depreciation rate will cause lower numbers. Interpretation of this factor can be done as effective

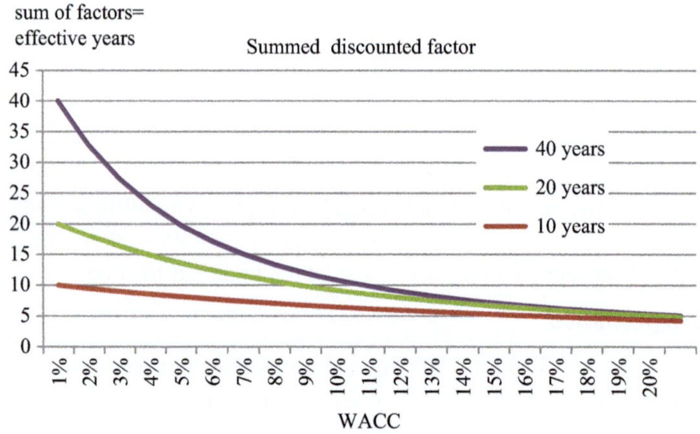

Figure 7.14 The discount function calculated only: a WACC of already 10% will depreciate every value so strong that a project extension by four times will cause no big variation for this factor. Most investors asking for revenues bigger than 10%, this will not support longer lifetimes of systems, but this supports bigger revenues: Bifacial systems can be build-up to almost same costs but will have typical 10%–20% higher generation in every year. This higher generation is significant compared to monofacial PV and will help for faster payback of the system

years of view of an investor. A 10% WACC reduced the effective years down to 6–10 years and all WACC above 20% will reduce the time of 5 years, no matter how long the system will last.

Bifacial systems have not significant different CAPEX to monofacial PV systems, but the energy yield is significant higher. The O&M costs are most likely the same for both PV systems. Therefore, the LOCE must be lower for bifacial PV systems in general compared to monofacial PV systems. The lower payback time of bifacial systems support the WACC requirements which lead to lower effective years (Figure 7.15).

This example shows that investors might like good quality long lasting systems, but higher revenue streams are much more important. Bifacial systems have a big potential to generate higher revenue streams from very beginning and will help to mitigate risks in this way.

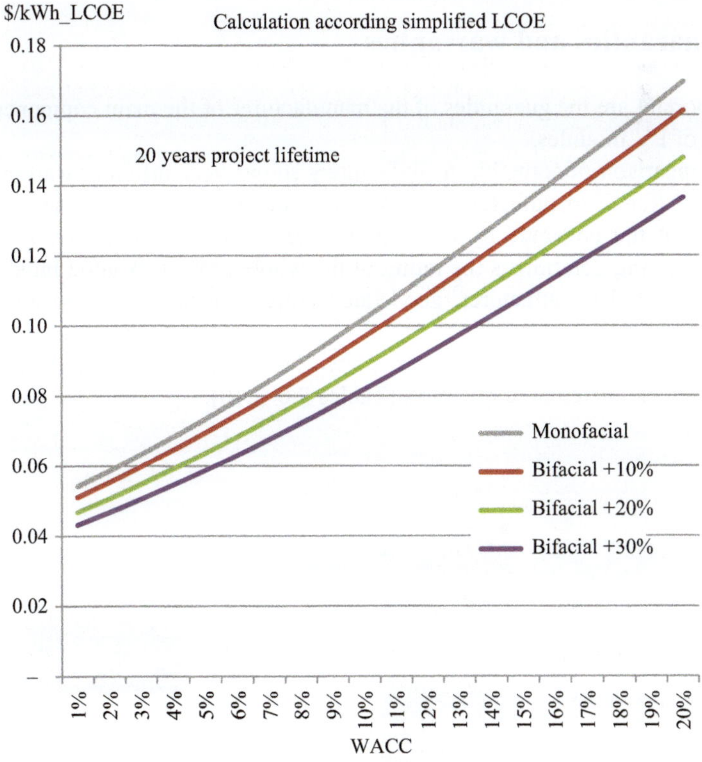

Figure 7.15 Calculation of simplified LCOE according to term2: specific CAPEX of bifacial system assumed +5% higher than monofacial system, specific O&M costs ($/Wp front surface) the same and energy yield 1,300 kWh/kWp for monofacial and +10%, +20% and +30%. CAPEX of monofacial system 1,000$/kWp, O&M of monofacial system 15$/a/kWp

7.5 Risk assessment

We already listed and categorized the points for risk mitigation until year 0 and after year 0.

In the view of a bank each phase of a project is considered with its specific risks. The risk categories are mentioned, but the specific questions to get enough answers to each category are collected by experts in checklists.

The rating agency S&P is taken as an example because many companies and industrial projects are rated through this scheme.

Figure 7.16 shows one example for project financing—PV projects belonging to this category.

After evaluation of the detailed checklists, the different categories are rated with the S&P scheme (see Table 7.4 and Figure 7.17).

Additional aspects like the maturity of a technology are rated according to Table 7.5.

7.6 Guaranties and warranties

Very important are the guaranties of the manufacturer of the main components like inverters of PV modules.

A comparison of some big manufacturers shows very different warranty conditions of different manufactures. Banks are evaluating these conditions in details to figure out the coverage by the manufacturer. If the manufacturer has a low rating, this rating determines the rating of the whole project. A good rated project needs good rated manufacturer, good rated project management and good rated material suppliers.

In the part "rating schemes" are shown how to proceed after decoding the numbers for the different single risks (Tables 7.6–7.10).

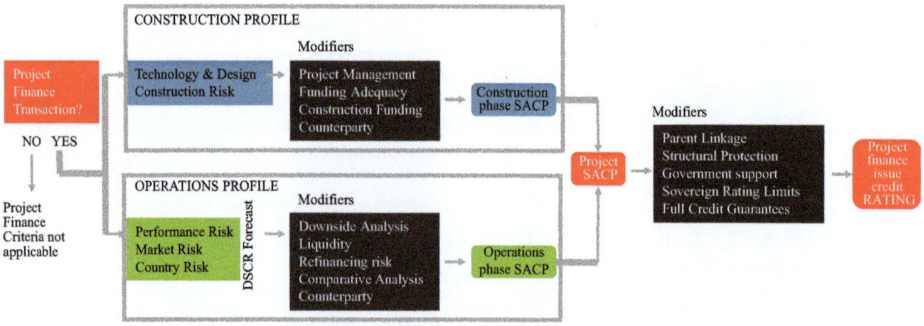

Figure 7.16 Part of general decision matrix of S&P guide: despite the initial calculation of the project, the main target is to evaluate the downsides of a project: what are realistic occurrences to increase costs, to delay payments or to reduce payments [7]

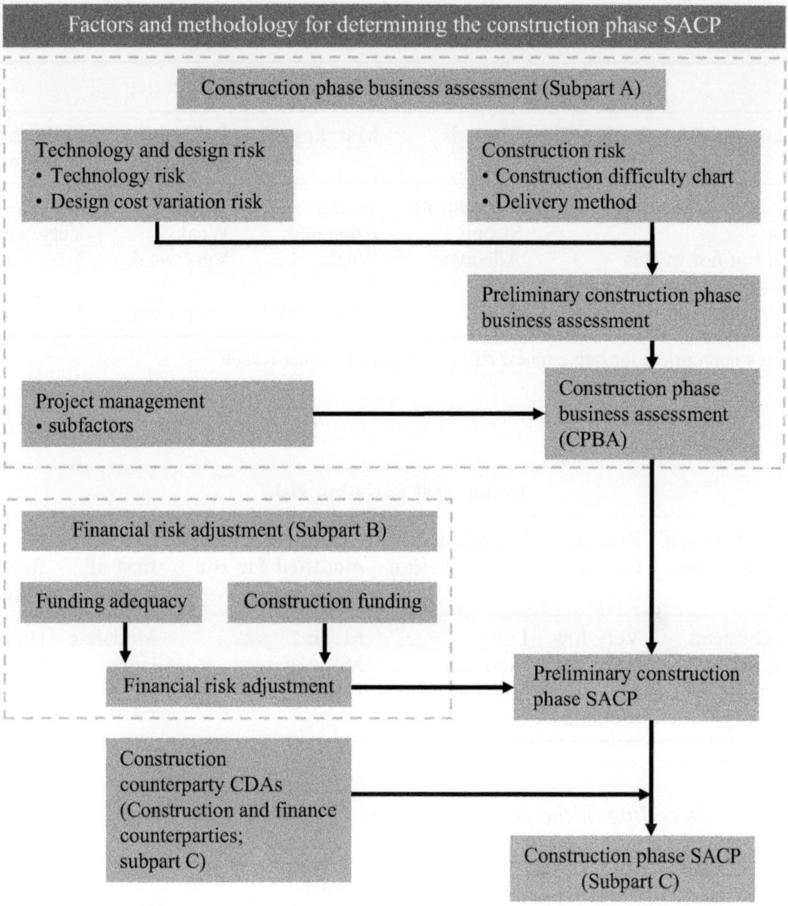

Figure 7.17 More detailed S&P decision list for the construction phase of a project. Two risk tables for technology risks and construction risks. These tables are shown in Tables 7.6–7.9 in detail [7]

Table 7.6 Figures of technology and design risks part 1 of 2 (based on [7]). A low number represents projects with small risks. Higher numbers are for projects with higher risks

	Technology and design risks				
Design cost variation risk	**Very strong**	**Strong**	**Adequate**	**Weak**	**Very weak**
Very low	1	2	3	4	5
Low	2	2	3	4	5
Modest	2	3	4	5	5
Moderate	3	4	5	*	*
High	4	5	*	*	*

*Projects with very weak technology and high design costs will block any further investigations, they have no rating number.

Table 7.7 Decoding of the technological risk definition (based on [7])

Technological risk				
Technological track record	**Exceeds**	**Matches all**	**Falls short or minor**	**Falls short of materials**
Commercially proven	Very strong	Strong	Weak	Very weak
Proven	Strong	Adequate	Weak	Very weak
Proven but not in this application or arrangement	Adequate	Weak	Very weak	*
New or unproven technology	Weak	Very weak	Very weak	*

Note: *not appropriate for rating unless risk is mitigated by other reason.

Table 7.8 Decoding of the design cost variation risk definition (based on [7])

Design cost variation risk					
Degree of design completion and costing	**Proven design**	**Modified proven design**	**Established design modified for site conditions**	**Simple first of a kind**	**Simplex first of a kind**
Very advanced	Very low	Low	Modest	Moderate	High
Advanced	Very low	Modest	Moderate	High	High
Moderate	Low	Moderate	High	High	*
Preliminary	Moderate	High	*	*	*

Note: *not appropriate for rating unless risk is mitigated by other reason.

Table 7.9 Decoding of the construction risk definition (based on [7])

Construction risks					
Delivery method	**Simple building task**	**Moderately complex building or simple civil engineering task**	**Civil or heavy engineering task**	**Heavy engineering-to-industrial task**	**Industrial task simplex building task**
Very strong	1	1	1	2	3
Strong	1	2	3	4	4
Adequate	2	3	4	5	*
Weak	4	4	5	*	*
Very weak	5	5	*	*	*

Note: *not appropriate for rating unless risk is mitigated by other reason.

7.7 Rating schemes

Each rating company has a final rating of the project. According to this scheme the whole project gets a rating. This rating defines the interest rate and additional conditions (Tables 7.11 and 7.12).

Table 7.10 Example of warranty conditions of selected companies: in some parts there is a big difference in this conditions (based on [8])

Warranty power	Product guarantee		Power guarantee years	Exclusion	Compensation
	Time (years)	Valid for			
+5W/−0	12	Defective materials	Linear 12 years down to 97% Linear to year 25 years down to 83%	Wrong installation, wrong maintenance % operation of not authorized persons, accident, modifications of product	Replacement of modules refurbishment, max to actual market price
±3%	10	Defective materials	97% (1st year), 80.2% linear to year 25	Wrong installation, damages by animals, external influences	Replacement of modules refurbishment, max to actual market price
N/A	10	Defective materials	mc-Si: 95.5% (year 1), 80.7% linear to year 25 cz-Si: 97% (year 1), 80.68% linear to year 25	External occurrences, wrong installation, wrong components	Replacement of modules refurbishment, max to actual market price
N/A	10	Defective materials	98% (1st year), 82.4% linear to year 25	External occurrences, wrong installation, wrong components, microcracks	Replacement of modules refurbishment, max to actual market price

Table 7.11 Overall rating of a project in construction phase (based on [7]). The figures representing the rating result of this project for this phase

	Preliminary construction phase business assessment (CPBA)				
Technology and design risk	**1**	**2**	**3**	**4**	**5**
1	a+	a	a−	bbb+	bbb−
2	a	a−	bbb+	bbb	bb+
3	a−	bbb+	bbb	bbb−	Bb
	bbb+	bbb	bbb−	bb+	bb−
	bbb−	bb+	bb	bb−	b+

In the following the rating of S&P after the risk assessment and the tables of the chapter before.

Stakeholders are very interested to avoid any kind of default. PV systems in operation after 2 years are less prone to default, but during system commissioning the risk might be higher.

Table 7.12 DSCR of a project in the operational phase. OPBA (operation-based business assessment is the S&P assessment to project's performance, market and country risks during the operations phase) (based on [7])

Preliminary operations phase SACP					
Minimum DSCR ranges					
OPBA	**aa**	**a**	**bbb**	**bb**	**b**
1–2	>1.75	1.75–1.20	1.20–1.10	<1.10	<1.10
3–4	N/A	>1.40	1.40–1.20	1.20–1.10	<1.10
5–6	N/A	>2.00	2.00–1.40	1.40–1.20	<1.20
7–8	N/A	>2.50	2.50–1.75	1.75–1.40	<1.40
9–10	N/A	>5.00	5.00–2.50	2.50–1.50	<1.50
11–12	N/A	N/A	N/A	>3.00	<3.00

7.8 Summary "bankability"

Whether a project is bankable or not, depends not only of the project, but depends also on the partners and how the partners are rated. The evaluation sheets and check lists of rating companies and banks have been introduced shortly within this chapter. New technologies with not existing deep experience, like bifacial energy gain, cannot get rated all its upside potential. For new technology a sound cooperation between technicians, project developers, engineers for simulation and banks are very important in order to make such a project feasible from the financing point of view.

At the time of writing of the present book, bifacial system has a small market share and the value chain is not yet fully developed: energy yield simulation, demonstration plants and technical descriptions are not yet at a mature stage or no relevant track record exists. In the near future, bifacial PV systems are expected to gain more market share and, with improving track record, the technological risk perceived by banks and rating companies is expected to approach more and more the rating of comparable standard (monofacial) PV systems.

The market is moving away from protected schemes like FiT to the common energy market. In the fierce competition, every financial and technical advantage plays an important role. Bifacial PV technology has some very strong advantages. It is sure that this technology will come in focus in the very next years.

References

[1] Available from www.solarbankability.eu [Accessed 26 Jan 2017].
[2] Risikogerechte Entschädigung für Netzbetreiber im schweizerischen Elektrizitätsmarkt, IFBC AG, 25 July 2012.
[3] Mazzucato S. Financing renewable energy: Who is financing what and why it matters. Technology Forecasting & Social Change 127 (2018) 8–22, doi:10.1016/j.techfore.2017.05.021.

[4] RECAI. Available from http://www.ey.com/gl/en/industries/power—utilities/renewable-energy-country-attractiveness-index.
[5] Available from https://ustr.gov/sites/default/files/files/Press/fs/201%20Cases%20Fact%20Sheet.pdf [Accessed 5 Feb 2018].
[6] Spain decree, IET/221/2013, see more under https://www.pv-magazine.com/2013/02/21/spain-publishes-retroactive-pv-fit-cuts_100010298/ [Accessed 5 Feb 2018].
[7] Standard & Poor's Project Finance Ratings Criteria Reference Guide, 16. (2014) pp. 11, 13, 94–102. Available from https://www.spratings.com/documents/20184/86990/SPRS_Project%2BFinance%2BRatings%2BCriteria%2BReference%2BGuide_FINAL/cdfde690-57d1-4ff4-a87f-986527603c22 [Accessed 23 Jul 2018].
[8] Photon Deutschland Mai, 2017, S. 40ff.

Chapter 8

A "global" view on bifacial gain: dependence on geographic location and environmental conditions

Eric Gerritsen[1], *Gaby Janssen*[2], *and Chris Deline*[3]

8.1 Introduction

It may be straightforward to get an increased yield from a bifacial system compared to a monofacial system in the same location and having the same orientation. But how much will that extra gain be, and in which geographical regions can we expect bifacial systems to be most advantageous? And where does it make sense to deploy vertical bifacial systems?

Experimental data, reviewed in Chapter 5, are still scarce, but predictions can be made based on simulations (see Chapter 4 for descriptions of simulation methods). These methods are still being developed, but recently reported results give good insight into the mechanisms of bifacial gain and how the geographical and environmental factors affect gain.

Similar to the yield of monofacial modules and systems, bifacial gain is determined primarily by irradiance. Bifacial gain depends strongly on the ratio G_{rear}/G_{front}. We first discuss location-specific factors of the irradiance that determine the G_{rear}, such as the ground-reflection coefficient or albedo and the amount of diffuse irradiance, which is related to the clearness index. In addition, we discuss design factors for stand-alone systems that affect the bifacial gain such as ground clearance (elevation), tilt angle, and module transparency.

In recent years, results have been published not only for stand-alone panels but also for systems consisting of many sequential rows. South/North (S/N) systems—where the front side is oriented with respect to the Equator—with optimized tilt angle have been studied. We will focus on the expected annual gain of bifacial systems compared to their conventional monofacial counterparts, at varying irradiance conditions and latitudes, and also taking design parameters into account. The few studies that address single-axis tracking will also be considered here.

[1]CEA-INES – Institut National de l'Energie Solaire, France
[2]ECN part of TNO – Solar Energy, The Netherlands
[3]National Renewable Energy Laboratory, USA

Bifacial gain can be increased by improving the irradiance collected at the rear side, such as by increasing the ground albedo. Moreover, bifacial gain can be high in situations where the front-side system output is not optimized. Vertically placed panels are a good example, and several recent studies have been devoted to these systems when they are at an East/West orientation.

Other aspects of bifacial operation are temperature effects and electrical aspects, such as inhomogeneous rear irradiance. Many studies predict the yield by multiplying the irradiance on the front and rear by a front- and rear-side standard efficiency. Other studies include a thermal and electrical model, which enables taking into account the effect of operating temperature, low light conditions, ohmic losses, and shading losses (which are also relevant for monofacial systems), as well as the effect of inhomogeneous irradiance on the rear of the panels (which is typical for bifacial systems).

From the review of simulation results available at this time and presented in this chapter, we have derived several "rules of thumb" on the relation between the location and system layout of bifacial photovoltaic (PV) installations and their expected yield. These rules will be presented first, with references to locations later in this chapter.

8.2 Some design rules (of thumb) for bifacial PV installations–as presented in the indicated sections of this chapter

1. Bifacial performance increases linearly with ground reflectance (albedo). The bifacial gain responds more strongly to albedo for systems with low tilt angle (Section 8.3.2—Figure 8.1).
2. At similar latitude, installation sites with high cloudiness (low clearness index K_T) will give a 2%–5% higher bifacial gain, compared to high direct normal irradiance (DNI) sites (Section 8.3.3 and Figure 8.4).
3. Installation sites at lower latitude and/or higher albedo benefit more from increased module elevation above ground (Section 8.4.1).
4. Module transparency, by spacing out the cells, helps increase rear-side irradiance. The increase is most pronounced for lower (<0.5 m) module elevation above ground (Section 8.4.3—Figure 8.8).
5. Bifacial modules favor a larger tilt angle, particularly with high albedo conditions. At all latitudes, the optimum tilt angle for bifacial panels is close to the latitude tilt—substantially larger (from 10° to 15° larger, increasing with albedo) than for monofacial panels (Section 8.5.2—Figure 8.10).
6. Bifacial panels at the end of a row have less self-shading than those in the center. Harvesting this extra energy depends on the electrical layout of the system (Section 8.5).
7. At high latitudes, the sun-path is more tilted, resulting in larger optimum row-distance to avoid mutual shading between adjacent panels. At the same latitude, locations with more diffuse insolation (low sky clearness) will have a larger panel row-distance, and thus a lower ground-coverage ratio (Section 8.5.5).

8. High-latitude, snowy climates are particularly well suited for bifacial systems due to large tilt angles and wide row spacing. Moreover, the low light conditions at these latitudes will lead to higher efficiency for bifacial modules compared to monofacial ones (Section 8.5.5—Figure 8.12).
9. The combination of bifacial PV and single-axis tracking always gives a higher energy output than a monofacial tracked system or a bifacial system with fixed tilt angle. The bifacial gain of tracked systems is 1%–2% lower than that of fixed-tilt bifacial systems (Section 8.6).
10. A vertical, E/W-oriented, bifacial ground-mounted configuration outperforms monofacial S/N tilt-optimized configurations in high latitude regions of the world, and areas with naturally high albedo (Section 8.7.1—Figure 8.15).
11. At specific conditions (low latitude, zero ground clearance, and low albedo) or very high latitudes, E/W-oriented systems panels may even be more advantageous than bifacial S/N systems (Section 8.7.1— Figures 8.19 and 8.20).
12. The nonuniformity of the rear irradiance, caused by self-shading and rear shading due to mounting structures, increases with increasing albedo (to about 5% relative standard deviation [RSD]) but will remain below the threshold for activation of protective bypass diodes (Section 8.8.2—Figure 8.23).

8.3 Location-specific factors

8.3.1 Albedo

Albedo or ground reflectance values for a site can change seasonally as snow and local foliage changes occur. In general, arid environments without foliage will have constant high albedo because vegetation absorbs sunlight. Desert locations, in particular, can have consistent high albedo. At high latitudes, winter conditions can introduce seasonal bifacial improvements as high albedos increase irradiance reflection. Field tests have also shown bifacial modules to more quickly shed snow following a storm because of rear irradiance collection.

Monthly albedo values can be extracted from data available online from NASA [2]. Figure 8.1 gives an example of NASA data during the month of April 2002; it shows the high albedo values in desert areas such as Northern Africa, the Middle East, Australia, and China.

Although snow and vegetation have a large seasonal impact, variations over the day must also be expected, e.g., during and after periods of rain. Chiodetti *et al.* [3,4] showed that the albedo can change substantially over the day, and they therefore proposed an albedo model that depends not only on the zenith angle [5] but also on the fraction of diffuse light.

The ground albedo of a given site has a linear effect on G_{rear}. The values shown in Figure 8.2 are based on ray-tracing simulations for a single module under 1-sun irradiance conditions. This figure indicates that when ground reflectance increases by a factor 4, the resulting G_{rear} increases by a factor 3. Although control over the ground albedo surface may be limited for certain ground-mount systems—as in the case of PV deployments in the built environment—many high-albedo

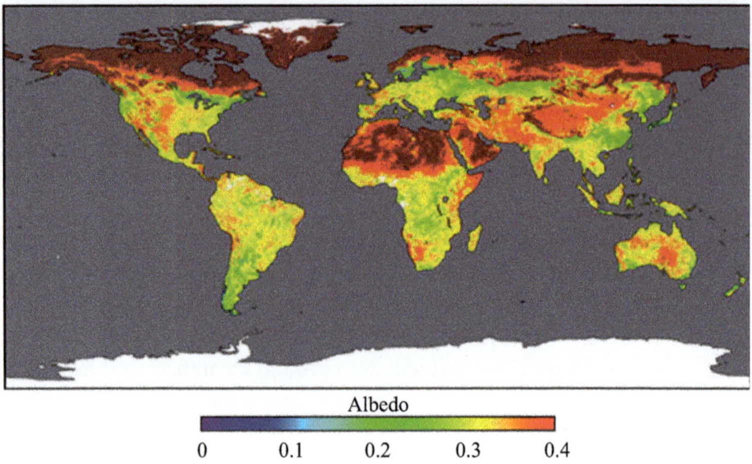

Figure 8.1 Global albedo map, April 7–22, 2002. NASA MODIS [1]

surface options exist. High reflective cool-roof coatings use white paint or materials to reduce thermal gain to buildings in hot climates. In addition to providing a lower ambient-temperature environment, these coatings have high solar reflectance, typically between 0.55 and 0.85. As such, they are ideally suited for bifacial rooftop applications. Note that reflective coatings typically weather with time and will exhibit a degraded ground albedo and bifacial gain unless cleaned and rehabilitated periodically, depending on the regional climate condition.

In 1993, Chieng and Green had already published simulations and experimental results for bifacial modules that emphasized the importance of the albedo, and they recommended that for bifacial applications the background should be carefully chosen [6]. Estimates by Dupeyrat *et al.* suggest that with albedo values in the order of 0.7, bifacial gains up to 50% would be achievable at locations in Europe [7].

8.3.2 Latitude

With increasing the latitude PV panels are usually deployed at a larger tilt angle to enhance the capture of direct light by the front of the panel. This angle also determines how much of the reflected and diffuse light can be captured. The plot in Figure 8.2 shows clearly that rear irradiance and bifacial gain increase linearly with albedo. However, the exact slope of this gain vs albedo plot depends on the tilt angle the installation in question, and therefore on the site latitude. Some of the first detailed simulations of annual bifacial gain were reported by Yusufoglu *et al.* for single modules [8,9]. In those papers, the combined effects of latitude, tilt angle, and diffuse climate were studied by comparing a northern location with a large fraction of diffuse irradiance (Oslo, Norway) to a low-latitude location with a high amount of direct light (Cairo, Egypt), over different albedo surfaces. It was found that bifacial gain increases with ground reflectance as described above, but this

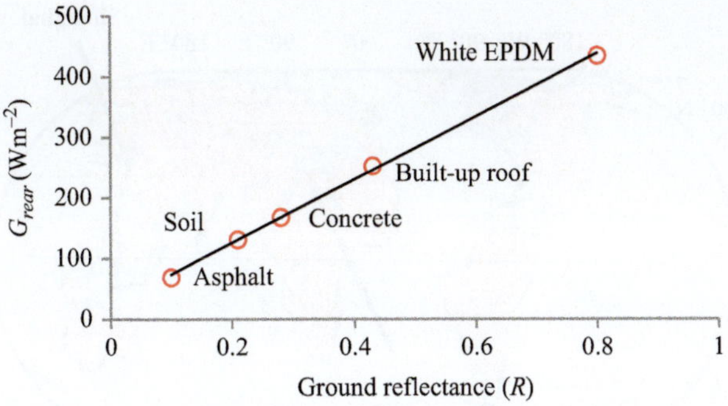

Figure 8.2 Effect of ground material type on G_{rear} for a single bifacial module deployed at 1-m height

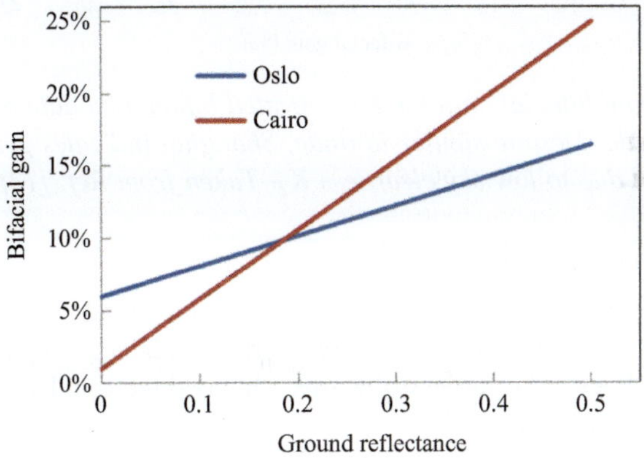

Figure 8.3 In places closer to the equator (Cairo), the bifacial gain increases more strongly with the ground reflectance (albedo) coefficient than at high-latitude locations (Oslo). However, due to the high tilt angle the gain in diffuse irradiance is higher in Oslo, resulting in a larger bifacial gain at low albedo. After Ref. [9]

effect is much stronger in Cairo than in Oslo, due primarily to the higher tilt angle for the Oslo system (Figure 8.3). In Cairo, larger bifacial gains are obtained for albedos >0.2 because the low tilt angle and higher direct irradiance allows more ground-reflected irradiance to be captured. Interestingly, high bifacial gains are still obtained in Oslo at low albedo (<0.2) because a large tilt angle captures more of the available diffuse irradiance, relative to the Cairo site. Using a similar approach, but using a slighter higher bifaciality factor (90% instead of 80%), Shoukry *et al.* obtained somewhat larger gains, i.e., 13% and 15% at 0.2 albedo in El-Gouna and Konstanz, respectively [10]. Also, in this study, the higher tilt and diffuse component at Konstanz caused the higher bifacial gain at this low albedo. At 0.5 albedo, a gain of about 34% (Konstanz) and 36% (El Gouna) was predicted.

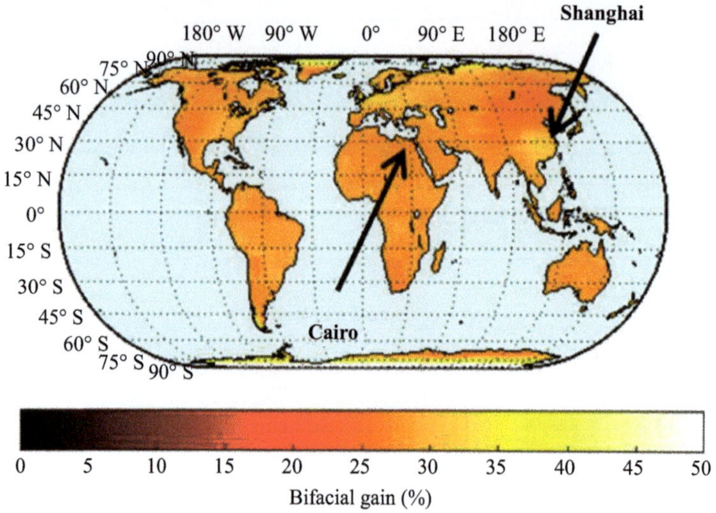

Figure 8.4 Global bifacial gain for 1-m elevated bifacial module over 0.5 ground albedo. Despite similar latitude, Shanghai indicates greater bifacial gain due to low sky clearness K_T. Taken from Ref. [12]

8.3.3 Clearness index

A site's clearness index K_T is a calculation of measured global horizontal irradiance relative to the available (clear-sky) extraterrestrial irradiance [11], and it indicates the relative cloudiness or clearness of a site. This is directly related to the diffuse irradiance ratio of the climate.

Sun *et al.* have investigated the effect of clearness index on bifacial gain in their global analysis [12]. Figure 8.4 shows that at the same albedo and latitude, significant variations in bifacial gain can occur. A particular comparison was done between Cairo, Egypt (clear conditions, annual $K_T \approx 0.7$, latitude 30°) and Shanghai, China (cloudy, annual $K_T \approx 0.35$, latitude 31°), with a fixed ground albedo of 0.5. Due primarily to the less clear conditions, the bifacial gain in Shanghai was 5% greater than in Cairo. This can be understood because low sky clearness ("cloudiness") mainly attenuates DNI, whereas sky diffuse irradiance is increased. Therefore, despite the lower total solar insolation, bifacial solar modules benefit more in Shanghai than Cairo due to the additional rear-side absorption of diffuse light. In agreement with results reported in the previous section, this study shows that bifacial modules in western and central parts of Europe also benefit from the large fraction of diffuse light.

8.4 Single-module factors

8.4.1 Single modules – ground clearance

As shown in Figure 8.5, many studies indicate that elevating single bifacial modules to 0.5 or 1 m above the ground will increase bifacial gain [8–10,12–14].

(a) (b)

Figure 8.5 Self-shading of the module for an elevation of (a) 1 m and (b) 10 cm. G_{rear} is attenuated significantly in case (b) due to the shorter distance between the shadow and the module

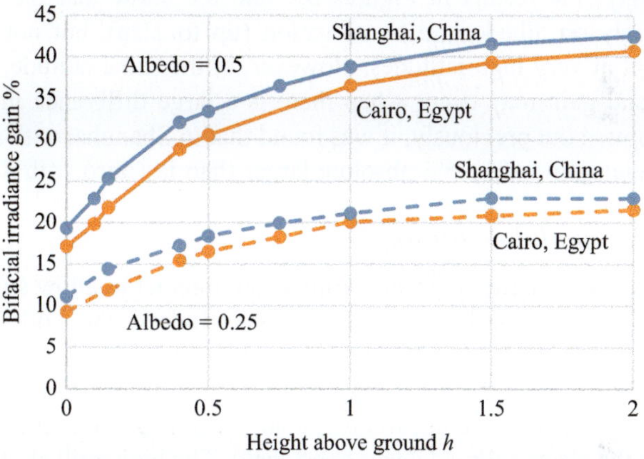

Figure 8.6 Rear irradiance fraction G_{rear}/G_{front} vs height above ground (clearance) for two climates (high K_T Cairo and low K_T Shanghai) and two albedos

The larger distance between the panel and the self-shade reduces the view factor to this shade, and also leads to reduced inhomogeneity of the rear irradiance [9].

Additional analysis is conducted for the two specified sites in Figure 8.4 to identify the impact of height h above ground at Cairo and Shanghai. Ray-tracing simulations on single modules confirm the results of Sun *et al.* [12] that bifacial yield depends on the albedo and elevation, as well as variations in the clearness index K_T (Figure 8.6). They show that for zero clearance ($h = 0$), the bifacial gain globally averages ~10% when the albedo is 0.25, whereas it increases to an average of 20% at albedo 0.5. By increasing the elevation to 1 m, the average bifacial gain increases to 30%–40%. As may be expected, the beneficial effect of elevation was found to increase at higher albedo.

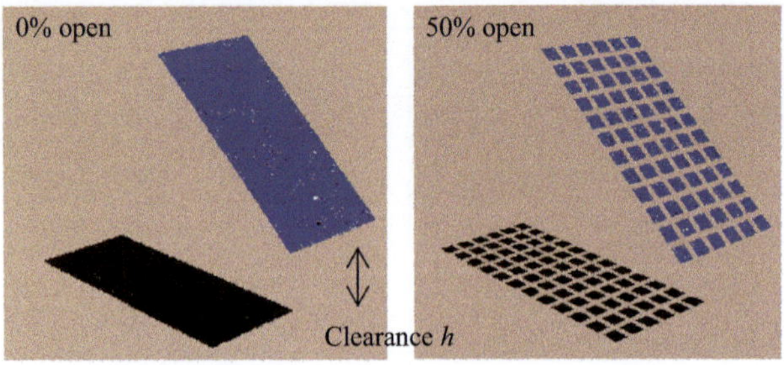

Figure 8.7 *Single-module deployment scenario representative of the IEC 60904-3 irradiance standard, namely, 1,000 W m^{-2} front irradiance, 37° tilt angle, and light-soil ground cover. Two sensitivity studies are conducted here: ground clearance h and fraction of module transparent area*

Interestingly, the results in Figures 8.4 and 8.6 show that the bifacial gain depends strongly on albedo and on elevation (up to 1 m), but not so much on latitude, except at very high latitudes. However, at the same latitude, some differences in bifacial gain may occur when there is a large difference in diffuse irradiation. As pointed out previously, it was found that in Shanghai with a K_T index of 0.35, the bifacial gain is 2%–5% absolute larger than in Cairo with K_T 0.7 [12].

8.4.2 Single module—tilt angle

As described above, the tilt angle and latitude can interact to allow solar irradiance to be collected directly on the rear of the module. Indeed, the work of Yusufoglu *et al.* showed that the optimum tilt angle depends in a complicated way on elevation, albedo, diffuse fraction, and system layout [8,9]. When only considering the irradiance, a larger albedo seemed to favor a larger tilt angle; but this was no longer the case when the electrical model was considered. The higher tilt angles resulted in more variation over the panel in distance to ground shading, and therefore, in more inhomogeneity in the reflected irradiance. Sun *et al.* found a substantial increase (order of 15° or more) in optimum tilt angle for bifacial panels compared to monofacial panels at all latitudes, and an increase with albedo and decrease with elevation [12]. However, they did not take into account the counteracting effect of increased inhomogeneous irradiance with tilt angle.

8.4.3 Spacing between cells

Module transparency increases available rear irradiance, particularly for low-clearance module deployments. Figures 8.7 and 8.8 present ray-tracing results where module cell spacing is adjusted to allow between 0% open area and 50% open module area [15]. The effect on rear irradiance depends significantly on the height of the PV module. For the $h = 1$-m ground-clearance value, the module is deployed high above the ground and has a view mainly of surrounding unshaded ground. Therefore, the rear irradiance is less sensitive to the shadow opacity

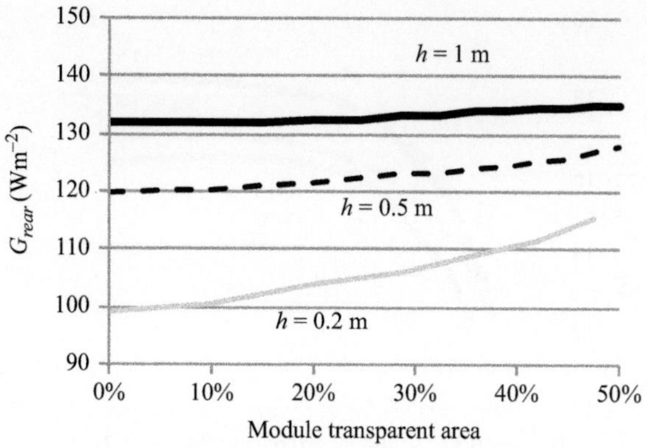

Figure 8.8 Modeled rear irradiance as a function of the fraction of the light that is transmitted through the module under 1-sun outdoor conditions, albedo = 0.2. Transparent space between cells has an impact on G_{rear} mainly at low ground clearance

beneath the module. For the $h = 0.2$-m ground-clearance module, the rear irradiance depends greatly on how opaque the shadow cast beneath it happens to be (Figures 8.7 and 8.8).

8.5 System-level configuration and effects

In large systems, the loss in ground-reflected irradiance due to self-shading of adjacent modules will be stronger [3,10]. Moreover, the view factor to both the ground and sky will be blocked by sheds (module rows) in front and behind [10]. Another factor recently included in view-factor models [16–18] is that diffuse irradiance hitting the ground is also reduced by the presence of the row itself and adjacent rows, thereby further reducing the ground-reflected irradiance. This may be a significant reduction, especially at tight row spacing.

As a result of this self-shading, the bifacial gain in large systems will be smaller than for single modules, and is dependent on the inter-row spacing [3,10,14]. In addition, the self-shading reduction seen in the center of a shed is greater than at the edge. This leads to 10%-higher rear irradiance in an end panel [3,10]. Whether all this additional irradiance can be harvested will depend on the electrical layout of the system.

8.5.1 Fixed-tilt systems—ground clearance

In the previous section, simulation results including the effect of ground clearance or elevation have been presented for single modules (Figure 8.6); in systems, however, the effect on elevation is less pronounced. At extremely low clearance values, the total available irradiance is reduced, and the rear distribution of irradiance becomes more mismatched. However, above a certain height from the

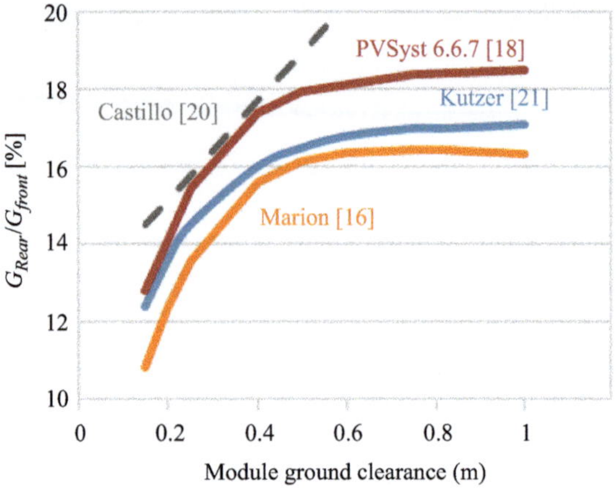

Figure 8.9 Low-tilt, high-albedo (commercial rooftop) case. 1-Up landscape over 0.62 albedo rooftop

ground, (0.5 m) the available irradiance becomes nearly constant (Figure 8.9) and the irradiance distribution becomes more uniform.

8.5.2 Fixed-tilt systems—tilt angle

In the above discussion of tilt-angle effects for single modules, bifacial modules were determined to favor a larger tilt angle, particularly with high-albedo conditions. To determine the effect of tilt angle for bifacial systems, we conducted view-factor simulations for two locations at mid- and high latitude: Albuquerque, NM, at 35° latitude, and Fairbanks, AK, at 65° latitude. In both cases, multiple rows of modules provide inter-row shading on the front of the module and reduce the available ground-reflected irradiance behind the module. Because inter-row shading depends on both row spacing and tilt angle, as the tilt angle is increased, the row spacing is also increased to maintain equal row-shading conditions. Specifically, the row pitch is chosen for each tilt angle to be shade-free at 9 a.m. on the winter solstice.

Results in Figure 8.10 indicate that the monofacial system's optimal tilt angle is slightly lower than the latitude tilt angle. This is due to the influence of inter-row shading on the front: lower tilt angles result in less row-shading loss. By including the rear irradiance contribution, optimum tilt angle increases for the bifacial system, particularly at high albedo conditions. Therefore, we confirm here that bifacial systems follow the previously reported trend for single modules: optimal tilt angle is increased with ground albedo by 10°–15°. However, the optimal tilt angle is still close to the latitude-tilt angle.

In Figure 8.10, we also note that for bifacial systems, larger tilt angles with greater row spacing result in a greater bifacial response than for small tilt angles. Two other factors were investigated here: the dependence on ground-clearance height and the effect of rear-irradiance mismatch. These two effects did not

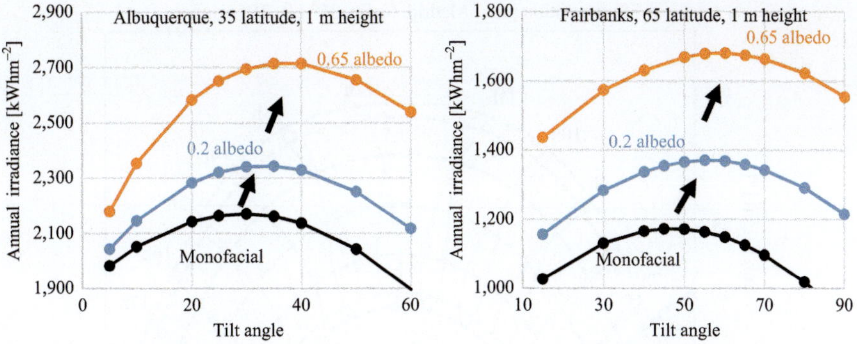

Figure 8.10 Annual front + rear irradiance for two climate conditions: Albuquerque, NM, and Fairbanks AK. Tilt-angle dependence shows an increase in optimal tilt angle for bifacial systems, particularly with high albedo

influence the optimal tilt angle of bifacial systems, even though earlier studies found some effect for single bifacial modules [9].

8.5.3 Fixed-tilt systems—latitude effects

For equator-oriented fixed-tilt bifacial modules and systems, the row-to-row shading is affected by the latitude. In general, higher latitudes require greater inter-row spacing to avoid shading in conventional systems. This can aid bifacial gain by allowing greater area between rows for reflected irradiance. Also, as was shown above, at higher latitudes, systems are more likely to be deployed at greater tilt angle. This introduces the possibility of direct incident sunlight on the rear of bifacial panels during summertime (Figure 8.11). Of course, this increases the annual rear-side irradiance collected. Recent experiments carried out in Sweden are a nice example of this effect [19].

8.5.4 Combined empirical formulae

Empirical models are simplified analytical approximations based on measurements or simulation, taking into consideration a more limited set of variables to calculate an approximate bifacial gain per year. Castillo-Aguilella *et al.* [20] developed a simple empirical expression for bifacial gain that depends on tilt angle β, ground clearance h, and albedo R:

$$BGE = 0.317 * \beta + 12.145 * h + 0.1414 * R, \tag{8.1}$$

where BGE is the additional bifacial gain in %, relative to a monofacial module; β is the module tilt angle in °; h is the module ground clearance in m; and R is the surface albedo in %. The coefficients in (8.1) were found from a fit to measured and simulated bifacial gains of single modules and single-row systems at various locations. Note that the model is calibrated for a module bifaciality of 95%;

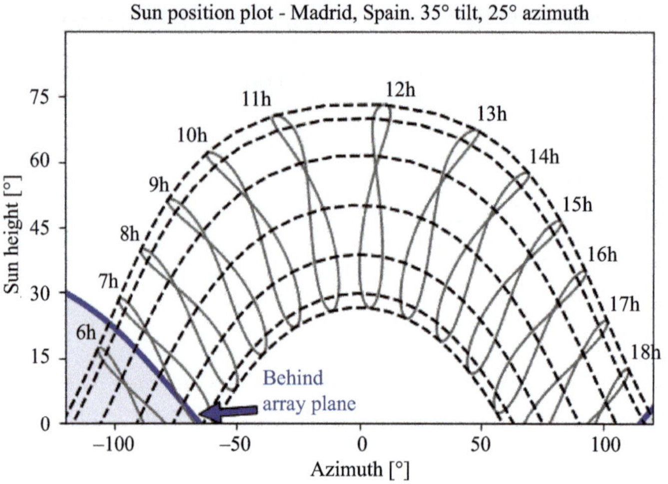

Figure 8.11 Example of tilted array geometry indicating sun paths behind the array (directly incident on the rear). For bifacial systems, these conditions increase bifacial energy gain and are increasingly present for high-tilt deployments

a different value would require linear scaling of the result. Equation (8.1) indicates a linear increase in bifacial gain with ground clearance h, which disagrees with other model results shown in Figure 8.9. The Castillo-Aguilella model therefore may be most appropriate for single-module assessments.

A second empirical model is presented here based on system-level ray-trace modeling. The Kutzer model [21] is designed for system-level approximations, and it includes row self-shading effects. The empirical model depends on two geometric factors: ground clearance h and the row-to-row pitch r. Four empirical factors are used in the equations, along with the fractional albedo and module bifaciality:

$$\text{BGE} = \textit{Albedo} * \textit{Bifaciality} * 0.95\left[1.037\left(1-\frac{1}{\sqrt{r}}\right)\left(1-e^{-\frac{8.691h}{r}}\right)+0.125\left(1-\frac{1}{r^4}\right)\right]. \qquad (8.2)$$

Again, BGE is the additional bifacial energy gain in %, relative to a monofacial module and r is the row-to-row spacing normalized by the module table chord length. The ground clearance h is also normalized by the module table chord length. Tilt-angle variations between 10° and 30° were found to be of minimal effect for the bifacial gain; therefore, the tilt angle is not included in (8.2).

Bifacial rear irradiance for this model is plotted vs height h in Figure 8.9, and it matches well with other more complete bifacial system models. Although it is not capable of capturing latitude-specific or climate-specific effects, the Kutzer model does appear to match closely with other system modeling results for mid-latitude, equator-facing systems.

8.5.5 Global combined analysis—bifacial irradiance gain for fixed-tilt systems

With the above details, we are able to conduct a global combined analysis for expected bifacial irradiance gain, based on climate, ground-surface albedo, and global latitude. From prior analysis in Figure 8.11, we identified the optimum tilt angle to be approximately equal to the site latitude. We also identified the annual average ground albedo from NASA MODIS data, shown in Figure 8.1. Therefore, we are able to conduct a survey of average fixed-tilt, ground-mount, bifacial rear-irradiance gain to identify typical rear-irradiance values available for bifacial system production. In this ray-tracing simulation, we assume a typical ground-mount system consisting of multiple sequential rows with 2-up portrait configuration. Inter-row spacing is calculated based on the site latitude and tilt angle to minimize self-shading at 9 a.m. on the winter solstice. Climate files consist of 2,865 EnergyPlus epw files that are hourly measured "typical year" climate files available for various global locations. Where ground-based epw files are unavailable, satellite climate files are used, courtesy of the European Commission's PV Geographical Information System [22].

Results of the combined global simulation are shown in Figure 8.12. Rear irradiance gain varies from 1% in equatorial regions to 40% in high latitude areas. It is clear that certain areas are particularly well suited for bifacial system deployment—namely, high-latitude, snowy climates where large tilt angles and wide row spacing enhance bifacial gain. These regions above 60° latitude enjoy

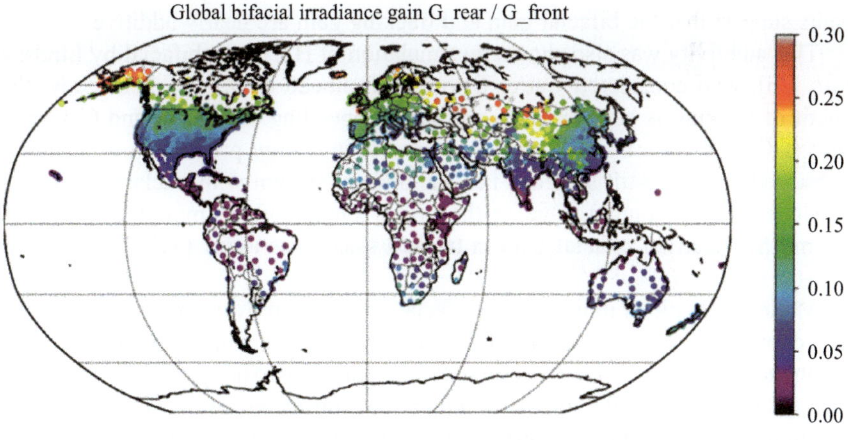

Figure 8.12 Global analysis of bifacial rear-irradiance gain for fixed-tilt ground-mount system at latitude tilt. 2-Up portrait configuration, 1 meter ground clearance, and inter-row spacing based on tilt and latitude

bifacial gains in excess of 20%. Mid-latitude regions fall between 5% bifacial gain in southeastern United States to 15%–20% in central Europe. China also displays a range of bifacial gain conditions, from 5% gain in coastal areas to 20% gain farther inland. Particularly low bifacial gain below 5% is found in equatorial regions, such as southern India, equatorial Africa, central America, and the Amazon region. In these areas, a combination of low albedo due to foliage, and high DNI conditions limit the diffuse and reflected irradiance available for bifacial ground-mount systems. Also, given the assumption here of low tilt which leads to close spacing, less ground-reflecting area is available between rows to reflect onto the rear of the module. In these low-latitude areas, alternative design choices such as increased row spacing and high-albedo synthetic ground cover may be desired to increase bifacial gain.

8.6 Single-axis tracking systems

Horizontal single-axis tracking (HSAT) can boost the output of monofacial systems by 10%–20% [23]. This technology can also be applied for bifacial systems, but so far, we have only limited information on the benefits. Bizarri recently presented results from the La Silla PV plant in Chile, where a 550-kWp single-axis bifacial module array demonstrated a 12% increase in performance with respect to a single-axis monofacial comparison, at an apparently moderate albedo [24]. However, tracking gains were not reported.

Shoukry *et al.* presented simulations on stand-alone modules with HSAT at Kasese, Uganda [10]. The benefit of tracking was 15%–18% for a monofacial module, but this did not outweigh the use of bifacial modules at fixed tilt angle, which resulted in 16.5% (albedo 0.2) and 43.8% (albedo 0.5). For the combination of bifacial modules and HSAT, predicted gains were as much as 40% (albedo 0.2) and 62% (albedo 0.5) with respect to monofacial fixed-angle modules. These results suggest that the bifacial gain and tracking gain are nearly additive.

This additivity was also shown by simulation of HSAT for bifacial by Lindsay *et al.* [25], who considered Cairo, Egypt, with albedo of 0.2. In their study, the benefit of tracking is between 21% and 27%, depending on the Ground Coverage Ratio (GCR). With bifacial panels, the increase in energy production was similar for tracked and fixed-tilt systems, i.e., the bifacial system with tracking results in the highest energy output. Due to the higher output of the monofacial tracked system, the (relative) bifacial gain in tracked systems is lower (3%–7%) than for fixed-tilt systems (5%–8%).

Model evaluations indicate that bifacial energy improvement depends on the site's climate, as well as on the GCR. Figure 8.13 shows a comparison between two hypothetical HSAT systems—one in a cloudy climate of Shanghai, China, and the other in a sunny climate of Cairo, Egypt, as considered above in Figure 8.6. Although the overall irradiance captured in the sunny climate is greater, the bifacial gain is actually higher in the low-irradiance, cloudy Shanghai climate. This can be attributed to the greater sky-diffuse irradiance captured on the back of the module, along with lower front-side DNI. The effect of reduced row spacing was also

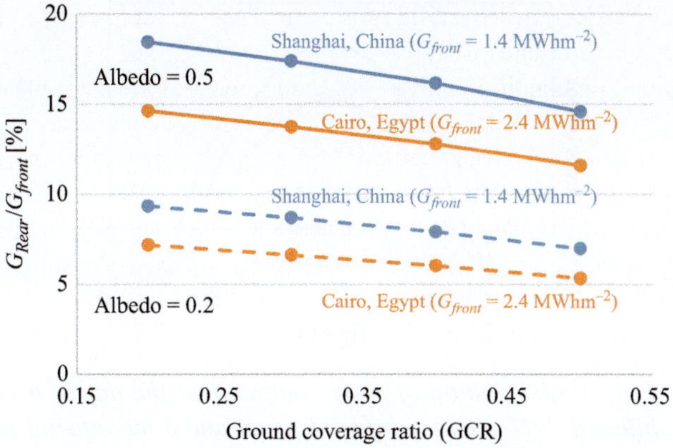

Figure 8.13 Single-axis tracking rear irradiance modeled for two locations: sunny Albuquerque, NM, and cloudy Seattle, WA. High GCR reduces rear bifacial gain (and front irradiance) due to self-shading. Assumed tracker hub height: 1 m

investigated by increasing the GCR. For this comparison, in this work doubling GCR has the effect of reducing bifacial gain by 15% relative, because increased row self-shading reduces the available ground-reflected rear irradiance.

8.7 Vertically mounted panels

8.7.1 East-west–latitude effects

The deployment of vertical bifacial modules was assessed for different global latitudes. Frequently, this has been done for a single module by comparing a latitude-tilt Equator-facing monofacial module, with a vertical E/W-facing bifacial module. As described above, the performance of the bifacial module is increased at high ground albedo, as well as at lower clearness ratio (higher diffuse irradiance).

Guo *et al.* [26] and Ito and Gerritsen [27] investigated the potential for vertically mounted, E/W-oriented systems around the world, based on irradiance and data on natural albedo. The main advantage is that E/W-facing systems have their peak output in the morning and afternoon, thus contributing to peak shaving (Figure 8.14). In addition, vertical systems are less prone to soiling, which can be a substantial advantage in desert areas, and they offer options for concurrent usage of the land. But E/W bifacial systems can, depending on latitude, collect a larger fraction of diffuse and reflected irradiance, and therefore even outperform equator-facing monofacial systems, shown in Figure 8.15.

As for S/N bifacial modules, the bifacial gain of E/W vertically mounted systems is strongly dependent on ground albedo and diffuse irradiance fraction (DHI/GHI). Figure 8.16 shows that the fraction of diffuse irradiance required to get to the breakeven point (where E/W vertical panels receive as much irradiance as

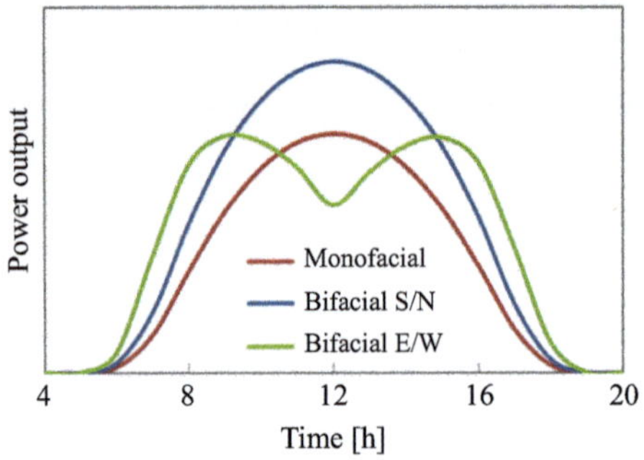

Figure 8.14 Typical distribution of power output over the day of a vertical bifacial, S/N-oriented bifacial panel and a monofacial panel

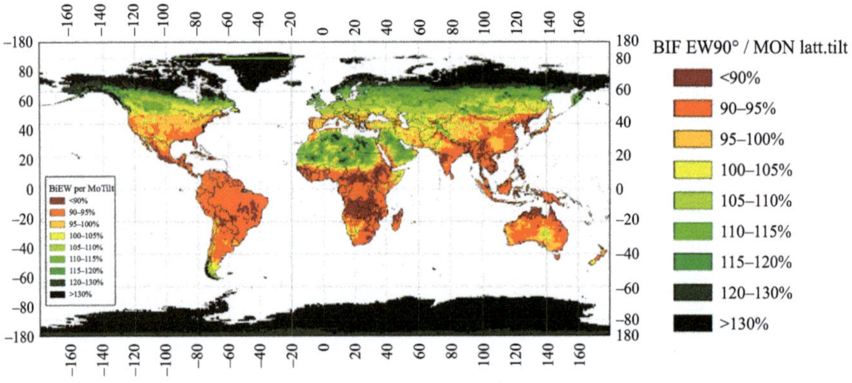

Figure 8.15 Vertical bifacial E/W-oriented compared to monofacial S/N tilted at latitude angle. From Ref. [27], based on a dataset of natural albedo values

S/N monofacial ones) decreases with latitude. This required fraction of diffuse light also decreases with increasing albedo. Taking into account the difference in clearness index K_T globally, Guo *et al.* calculated the minimum albedo required worldwide for E/W modules to outperform conventional ones [26]. Guo *et al.* [26] and Ito and Gerritsen [27] predict that E/W systems will outperform conventional PV at high latitude >45° because of the high diffuse-light fraction. In areas such as Northern Africa and the Middle East, the high natural albedo on the order >0.5 will result in E/W bifacial systems with a higher energy output than monofacial S/N systems (Figure 8.15).

In more detailed simulations, these results were corroborated for Amsterdam, Netherlands, and Doha, Qatar, by Janssen *et al.* [28], who calculated at albedo 0.2 an annual advantage of E/W (0.5-m elevation) in Amsterdam of 10%. For Doha, a loss of almost 6% was predicted at 0.2 albedo, but a gain of 17% at albedo 0.5. On the other hand, Shoukry *et al.* predicted a loss of E/W-facing systems (0.5 m) in

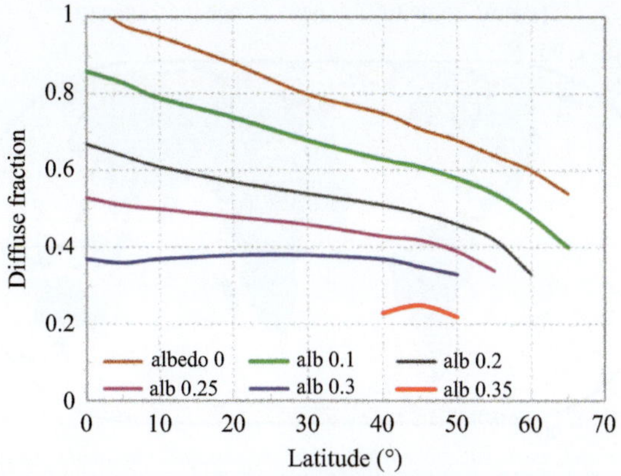

Figure 8.16 The fraction of diffuse light that is needed to reach the breakeven point for the irradiance received by E/W vertical bifacial panels compared to conventional monofacial panels. If the actual diffuse fraction of a site with a certain latitude is above its albedo line in the graph, it is preferable to use E/W vertical bifacial panels. For albedo >0.35 this is the case for nearly all latitudes. After Ref. [26]

El Gouna at both albedo 0.2 and 0.5. For Konstanz, a loss was also calculated at albedo 0.2 and a gain at albedo 0.5 [10].

Khan *et al.* [17] performed an extensive global analysis focusing on the *area-specific yield* (kWh/m^2). Their irradiance model includes direct shading by adjacent rows, self-shading losses of both the ground-reflected direct and diffuse light, as well as the fraction of diffuse light available at a specific location. To maximize area specific yield, the use of many modules spaced closely together is desired. Indeed, with vertical E/W bifacial configurations, GCR > 1 is possible which can lead to very high kWh/m^2. In their analysis, they assume vertical systems with zero ground clearance and a total installation height H and calculate the pitch-to-height ratio (p/H, for this vertical system equivalent to 1/GCR) that maximizes the specific yield. This optimal p/H is latitude- and climate-dependent, but typically falls between 0.8 in sunny climates and $p/H = 1$ in cloudy climates (Figure 8.17).

A comparison of area-specific yield of bifacial E/W and monofacial S/N systems is shown below in Figure 8.18. In low-latitude areas, monofacial tilt angles are low, GCR is high, and kWh/m^2 is high for the monofacial reference systems. The vertical E/W bifacial system specific yield is relatively poor by comparison, providing only half the energy density of a conventional system. However, in high latitude areas, the situation is reversed, showing bifacial E/W energy density 10%–20% greater than the monofacial reference. This is possible because of the higher GCR enabled by vertical bifacial module deployment.

In most studies, bifacial E/W panels and arrays have been compared to S/N monofacial panels and arrays with optimum tilt angle. But how do they compare to bifacial S/N systems? This was addressed by Appelbaum in a study for large

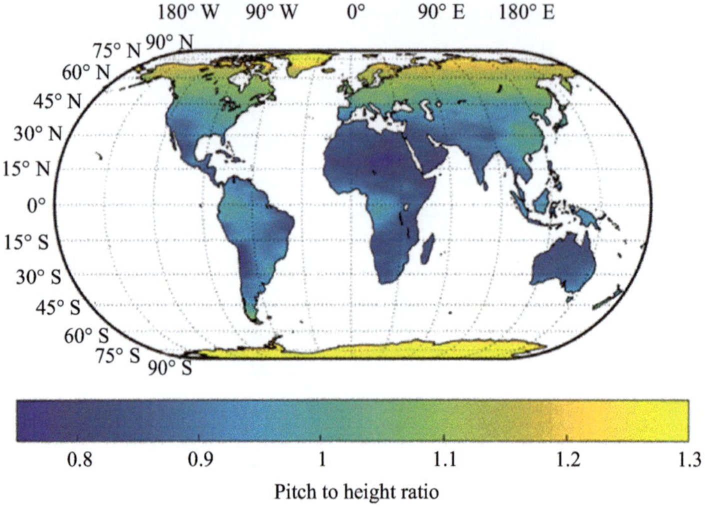

Figure 8.17 Row spacing pitch/height (p/H) for a vertical, E/W facing bifacial plant optimized for kWh/m^2. After Ref. [17]

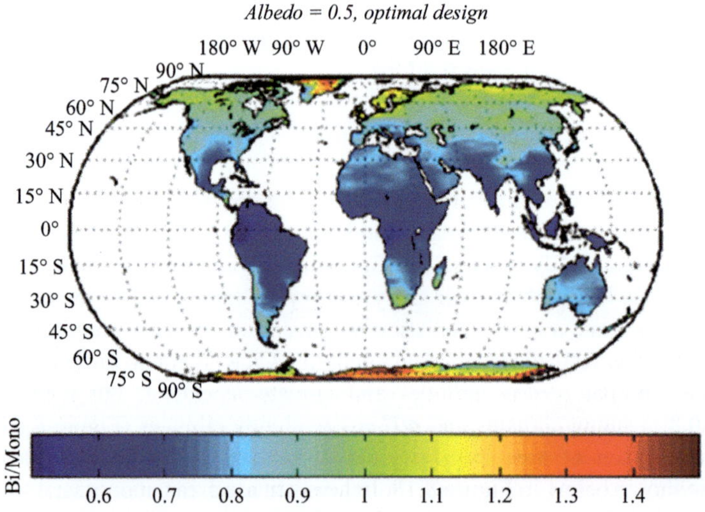

Figure 8.18 Ratio of annual kWh/m^2 of a vertical, E/W facing bifacial plant to a monofacial plant at optimum tilt angle and spacing. Zero ground clearance was assumed for all plants. Taken from Ref. [17]

systems at Tel Aviv, Israel [29]. In that work, it was calculated that the S/N system receives 30%–50% more irradiance per year; but note that this refers only to direct and diffuse irradiance, and ignores all reflected irradiance because it was assumed that self-shading would prevent any ground-reflected irradiance. The global

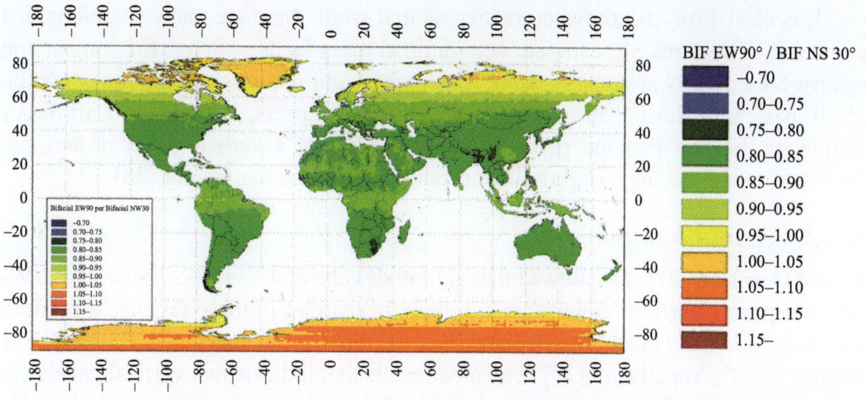

Figure 8.19 Vertical bifacial modules in E/W orientation compared to 30°-tilted, equator-oriented bifacial modules. Taken from Ref. [27]. Reduction of the ground-reflected irradiance by self-shading was ignored in this study

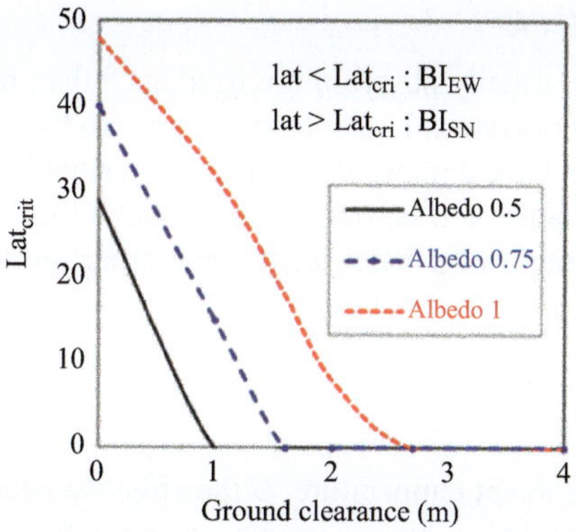

Figure 8.20 Critical latitude (Lat_{crit}, in degrees), below which BI_{EW} is more favorable than BI_{SN} and vice versa, as a function of ground clearance from 0 to 4 m and for an albedo of 0.25, 0.5, and 0.75. After Ref. [12]

comparison by Ito and Gerritsen [27], on the other hand, included ground-reflected irradiance but did not include self-shading effects. Figure 8.19 from that work suggests that E/W-facing panels will not outperform S/N-oriented panels with tilt angle of 30°—except, perhaps, at high latitudes where 30° is not the optimum tilt, but the inclusion of ground-reflected irradiance leads to smaller differences than predicted by Appelbaum.

The results of Sun *et al.* [12] show that by explicitly accounting for the self-shade correction, which will be large at low tilt angle and low clearance, the E/W configurations may have an advantage over bifacial S/N systems also at low latitudes, if the comparison is made at high albedo and low ground clearance (Figure 8.20).

It is clear from the above comparisons that while there are many situations that vertical E/W systems can compete with optimal monofacial systems, the comparison against bifacial S/N systems is much more subtle and requires careful consideration of all aspects of the PV installation. One of these aspects, in climates with heavy soiling, is the fact that the reduced cleaning costs of a vertical E/W system can readily compensate for a slightly lower, but otherwise stable, output [30].

8.7.2 Azimuth angle

Ito and Gerritsen [27] also showed that at latitudes above 40°, vertical S/N- and E/W-oriented panels perform equally well, which means that in these regions, a mixture can be used to meet electricity demand over the day ("peak shaving"). For the lower latitude of 32°, Appelbaum [29] presented results that indicate that vertical panels are actually more sensitive to the azimuth angle than panels with optimum tilt; but this was again without considering ground-reflected irradiance. As concluded above, albedo and self-shading must be taken into account to make a good comparison.

8.8 Other factors affecting the gain

8.8.1 Thermal effects

Because bifacial panels usually absorb more irradiance than monofacial panels, this may also cause them to operate at higher temperature. On the other hand, because the bifacial cells and panels have a transparent backside, a significant portion of the non-absorbed infrared irradiance will be transmitted through the cell and panel [31].

According to a simple heat balance, the operating temperature T_m of a solar module is given by:

$$T_m = T_{amb} + \frac{\alpha_f G_{front} + \alpha_r G_{rear}}{U}. \tag{8.3}$$

Here, T_{amb} is the ambient temperature, U the effective heat-transfer parameter of the module, and G_{front} and G_{rear} are the irradiance on the front and rear, respectively. The front- and rear-side absorption coefficients α_f and α_r will be different for the monofacial and bifacial modules. The heat-transfer parameter U will depend mostly on the type and thickness of the materials used in the module; glass–glass modules and glass–backsheet modules will not necessarily have the same U. In studies by Yusufoglu, normal operating cell temperature values of 45°C were quoted for monofacial modules and of 47°C for bifacial modules, indicating that those specific bifacial modules had a lower U value than the monofacial ones [8,9]. Moreover, U will depend on ambient conditions such as wind speed and humidity.

Whether a bifacial module operates at a higher temperature than a monofacial module will depend on the respective front and rear absorption coefficients, the respective U values, and the ratio between front and rear irradiance, G_{front}/G_{rear}. The increased transmittance of bifacial panels results in $\alpha_{f,b} < \alpha_{f,m}$; but because of the reflective properties of the rear backsheet, $\alpha_{r,b} > \alpha_{r,m}$ in most cases. This means that at low albedo, the bifacial module will usually be cooler than the

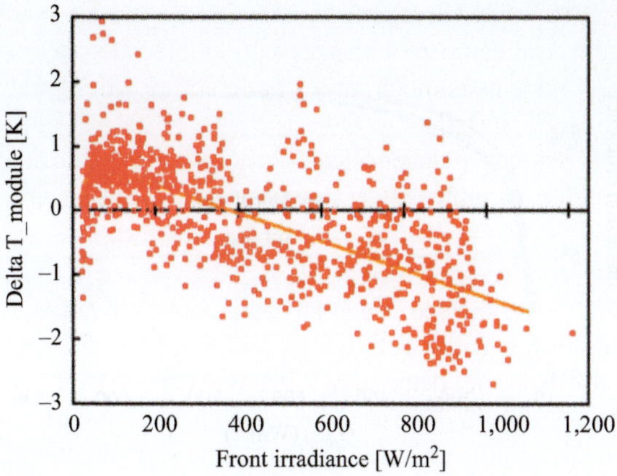

Figure 8.21 Temperature difference $\Delta T\_mod = T_{bifacial} - T_{monofacial}$ of a bifacial and a monofacial module under identical irradiance conditions. Location: Petten, The Netherlands, low albedo (concrete)

monofacial one [32]; but at high albedo, this may no longer be the case. Similarly, G_{front}/G_{rear} can also be small when the light is predominantly diffuse. In such cases, the bifacial module can also be hotter than the monofacial one [32]. This is shown in Figure 8.21.

In a study by Janssen *et al.* [28], it was found that at high albedo, differences in temperature can have an effect of about 15% relative on the calculated bifacial gain. Note that when comparing bifacial cells with conventional cells having a full-area Al rear, bifacial cells will have a more favorable temperature coefficient because of the higher open-circuit voltage (V_{OC}).

8.8.2 Electrical effects

The efficiency of a bifacial cell or module is not simply a matter of adding the front- and rear-side efficiencies. At low light conditions, the conversion efficiency of light incident on both the front and rear will be boosted, whereas at conditions with high irradiance and therefore, high currents—the series resistance of the module will reduce the efficiency (Figure 8.22). Globally, this will enhance the bifacial yield at high latitudes and decrease it at low latitudes. Moreover, vertical installations will benefit more than S/N installations.

An electrical effect that will reduce the output of the bifacial system is the inhomogeneity of the rear irradiance, which is caused by the self-shading of the ground-reflected irradiance and rear shading elements. Typically, the inhomogeneity of the rear irradiance is found to depend on orientation, elevation, and tilt angle of the panel and on irradiance conditions [8,9,33]. Because it is mainly caused by self-shading of the direct light, the inhomogeneity of the rear irradiance is largest at a high fraction of direct light on the front side [33]. However, what determines the effect on the electrical output is the inhomogeneity of the total

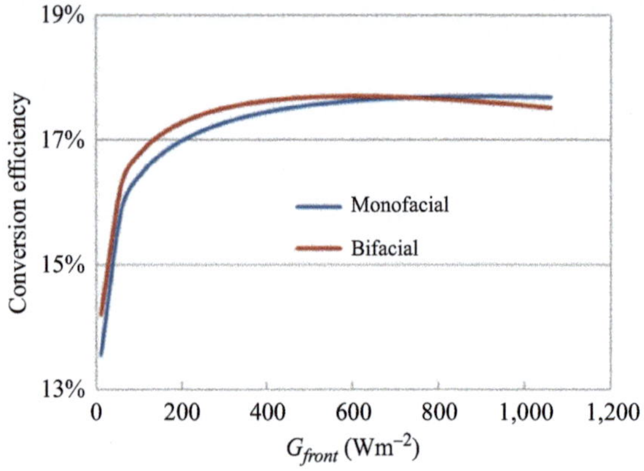

Figure 8.22 Conversion efficiency of a bifacial module with 50% additional current due to rear light compared to the conversion efficiency of a monofacial module without the additional rear light

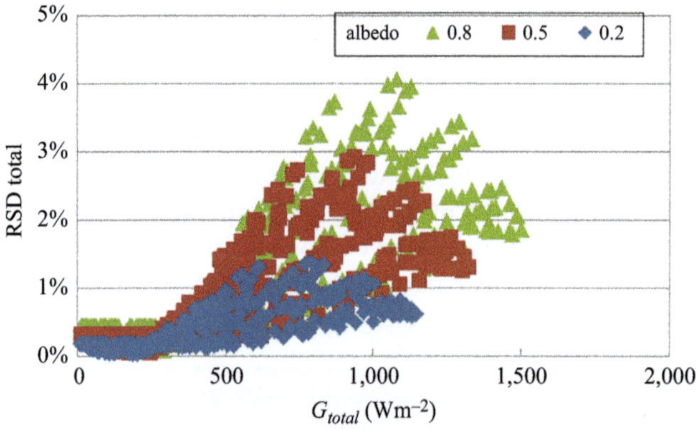

Figure 8.23 The calculated RSD of the total irradiance of a bifacial south-facing panel in Amsterdam during the month of April. Tilt angle was 38° and ground clearance was 0.5 m. G_{total} is the total irradiance received by the panel

current density. With a large amount of direct light incident on the front, the inhomogeneity of the total current density is mitigated. Simulations [33] from a stand-alone, south-facing panel with tilt angle of 38° indicated that the RSD of the total irradiance did not exceed 5%, even at an albedo as high as 0.8, and it will usually be much smaller (Figure 8.23). The same study presented LTSpice simulations indicating that an RSD < 10% in the current density of a panel will not result in activating protective bypass diodes, although one must expect an additional relative loss in the order of the RSD on top of the loss of generated current.

Globally, this means that the loss due to current inhomogeneity will be largest at locations having high albedo.

8.9 Summary and outlook

The simulation results presented in this chapter demonstrate how bifacial PV systems can bring a performance gain all around the globe, resolving a certain misbelief that they may only form a niche market restricted to specific locations or installations. Global predictions, based on irradiance, indicate bifacial gains in the order of 50% of the natural albedo for equator-facing systems with optimized tilt angle. In locations with low clearness index, the bifacial gain can even be higher.

At system level the bifacial gain will be reduced by 20% to 50% relative to a free-standing panel, due to increased self-shading and to limitation of the diffuse irradiance on the rear by adjacent sheds. Even though the rear side irradiance of a bifacial module is inherently inhomogeneous, due to self-shading and external mounting structures, our simulations have shown that the RSD of the total irradiance does not exceed 5%, even at very high albedo, and will usually be much smaller.

Moreover, bifacial systems do open up new opportunities, like vertical installation, not accessible to conventional PV. For vertical E/W bifacial the gain, relative to tilted monofacial, will be somewhat smaller than for equator-facing systems, but a 10%–20% gain is still possible at high latitudes and in installations at lower latitude with high albedo.

Initial simulation results on bifacial systems with tracking look very promising. Single-axis tracking should therefore be addressed more extensively, taking into account the effect of ground cover ratio on different locations, in order to explore the full potential of bifacial tracking across the globe.

The simulation methods applied in this chapter and their ongoing integration into commercially available software tools, like PVsyst, will give system designers and installers more visibility on the configuration layout of a bifacial PV plant at their specific location, the associated bifacial gain and the effect on their return on investment.

However, as the results in this chapter also make clear, it should be noted that exact bifacial yield predictions require good local meteorological data, especially the fraction of diffuse irradiance and data on the albedo, including its seasonal variation and dependence on weather conditions. In this respect, the prediction of the bifacial yield is more challenging than for conventional PV systems. Future work to extend simulation capabilities should address the spatial variation for the modules at the edges of a row, going beyond the concept of "unlimited sheds." Furthermore simulations should not be restricted to the optical capture of front and rear side irradiance but also include the thermal and electrical aspects of the increased irradiance on bifacial modules.

Last but not least experimental bifacial system data remain the cornerstone in further development and validation of extended simulation tools.

Glossary of terms

Azimuth	Deviation of the projection on a horizontal plane of the normal to the surface of the module from the local meridian
BGE	Bifacial gain: the energy gain of a bifacial PV system over a the energy output of a monofacial system at the same location
Bifaciality factor	The ratio of the efficiency of a module when measured at 25°C, at 1 Sun normal irradiance from the rear over the efficiency when measured from the front
DNI	Direct normal irradiance
DHI	Diffuse horizontal irradiance
G_{front}, G_{rear}	Total irradiance on the front or the rear of a module
GCR	Ground cover ratio: ratio of the PV surface area and ground surface area of a PV system
GHI	Global horizontal irradiance
h	Ground clearance or module elevation: height of the lowest point of a module above the ground
H	Total height of the PV installation
HSAT	horizontal single-axis tracking
K_T	sky clearness/clearness index
Latitude	Angular location north or south of the equator
Module table cord length	Total length of the module or row, in the tilted direction
P	Pitch or row-to-row distance of a PV system
r	row-to-row spacing normalized by table chord length
R	Ground reflectance or albedo coefficient: the fraction of the radiation reflected from the ground
Self-shade	Shade created on the ground behind a module by direct irradiance on the front
T_{amb}	Ambient temperature
T_m	Module operating temperature
U	Effective total heat transfer coefficient of the module
α_f,α_r	Net absorption coefficient: $\alpha_i = 1 - R_i - T_i - \eta_i$, with R_i, T_i, and η_i the reflection, transmission, and efficiency of side i
β	Tilt angle of the module

References

[1] NASA Earth Observations, https://neo.sci.gsfc.nasa.gov/view.php?datasetId=MCD43C3_M_BSA., last access on April 24, 2018.

[2] NASA Visible Earth catalog, https://visibleearth.nasa.gov/view.php?id=60636, last access on April 24, 2018.

[3] M. Chiodetti, 'Bifacial PV Plants: Performance Model Development and Optimization of Their Configuration', Master Thesis, *KTH Royal Institute of Technology*, 2015.

[4] M. Chiodetti, A. Lindsay, P. Dupeyrat, *et al.*, 'PV bifacial yield simulation with a variable albedo model', 32nd EUPVSEC, München, Germany, Europe, 2016.

[5] R. E. Dickinson, 'Land surface processes and climate: surface albedos and energy balance', in Advances in Geophysics Theory of Climate, 25th ed., B. Saltzman, Ed., pp. 305–53, Elsevier (1983); Chapter Supplement C.

[6] Y. K. Chieng and M. A. Green, 'Computer simulation of enhanced output from bifacial photovoltaic modules', Progress in Photovoltaics 1 (1993) 293.

[7] P. Dupeyrat, C. Lucas, A. Lindsay, A. Plotton, and K. Radouane, 'Investigations on albedo dependency of bifacial PV yield', 29th EUPVSEC, Amsterdam, Germany, Europe, 2014.

[8] U. A. Yusufoglu, T. H. Lee, T. M. Pletzer, *et al.*, 'Simulation of energy production by bifacial modules with revision of ground reflection', Energy Procedia 55 (2014) 389–95.

[9] U. A. Yusufoglu, T. M. Pletzer, L. J. Koduvelikulathu, C. Comparotto, R. Kopecek, and H. Kurz, 'Analysis of the annual performance of bifacial modules and optimization methods', IEEE Journal of Photovoltaics 5 (2015) 320–8.

[10] I. Shoukry, J. Libal, R. Kopecek, E. Wefringhaus, and J. Werner, 'Modelling of bifacial gain for stand-alone and in-field installed bifacial PV modules', Energy Procedia 92 (2016) 600–8.

[11] B. Y. H. Liu and R. C. Jordan, 'The interrelationship and characteristic distribution of direct, diffuse and total solar radiation', Solar Energy 4 (1960) 1–19.

[12] X. Sun, M. R. Khan, C. Deline, and M. A. Alam, 'Optimization and performance of bifacial solar modules: a global perspective', Journal of Applied Energy 212 (2018) 1601–10 doi.org/10.1016/j.apenergy.2017.12.041

[13] S. Wang, O. Wilkie, J. Lam, *et al.*, 'Bifacial photovoltaic systems energy yield modelling', Energy Procedia 77 (2015) 428–33.

[14] L. Kreinin, N. Bordin, A. Karsenty, A. Drori, D. Grobgeld, and N. Eisenberg, 'PV module power gain due to bifacial design. Preliminary experimental and simulation data', 35th IEEE Photovoltaic Specialists Conference, 2010, pp. 002171–5.

[15] C. Deline, S. MacAlpine, B. Marion, F. Toor, A. Asgharzadeh, and J. S. Stein, 'Assessment of bifacial photovoltaic module power rating methodologies-inside and out', IEEE Journal of Photovoltaics 7 (2017) 575–80.

[16] B. Marion, S. MacAlpine, C. Deline, *et al.*, 'A practical irradiance model for bifacial PV modules', IEEE 44th Photovoltaic Specialists Conference (PVSC), 2017.

[17] M. R. Khan, A. Hanna, X. Sun, and M. A. Alam, 'Vertical bifacial solar farms: physics, design, and global optimization', Applied Energy 206 (2017) 240–8.

[18] PVSyst, http://files.pvsyst.com/help/bifacial_system_2d.htm, last access on April 24, 2018.

[19] E. Molin, B. Stridh, A. Molin, and E. Wäckelgård, 'Experimental yield study of bifacial PV panels in nordic conditions', 33rd EUPVSEC, Amsterdam, 2017.

[20] J. E. Castillo-Aguilella and P. S. Hauser, 'Bifacial photovoltaic module best-fit annual energy yield model with azimuthal correction', IEEE 43rd Photovoltaic Specialists Conference (PVSC), 2016, pp. 3109–12.

[21] M. Kutzer, A. Fülle, A. Jahnke, *et al.*, 'Ertragssteigerung durch bifaciale Modultechnologie', Symposium Photovoltaische Solarenergie, 2018.

[22] M. Suri, T. Huld, E. Dunlop, M. Albuisson, and L. Wald, 'Online data and tools for estimation of solar electricity in Africa: the PVGIS approach', 21st EUPVSEC, Dresden, 2006. Accessible online at http://re.jrc.ec.europa.eu/pvg_tools/en/tools.html, last access on April 24, 2018.

[23] T. Huld, M. Šúri, and T. Cebecauer, 'Performance of single-axis tracking photovoltaic systems in Europe', Photovoltaics International 6 (2009).

[24] F. Bizarri, 'La Silla PV plant: Innovative bifacial PV plant at La Silla observatory in Chile', BIFI Workshop, Konstanz, 2017.

[25] A. Lindsay, M. Chiodetti, D. Binesti, *et al.*, 'Modelling of Single-Axis Tracking Gain for Bifacial PV Systems', 32nd EUPVSEC, München, Germany, Europe, 2016.

[26] S. Guo, T. M. Walsh, and M. Peters, 'Vertically mounted bifacial photovoltaic modules: a global analysis', Energy 61 (2013) 447–54.

[27] M. Ito and E. Gerritsen, 'Geographical Mapping of the Performance of Vertically Installed Bifacial Modules', 32nd EUPVSEC, München, Germany, Europe, 2016.

[28] G. J. M. Janssen, B. B. Van Aken, A. J. Carr, and A. A. Mewe, 'Outdoor performance of bifacial modules by measurements and modelling', Energy Procedia 77 (2015) 364–73.

[29] J. Appelbaum, 'Bifacial photovoltaic panels field', Renewable Energy 85 (2016) 338–43.

[30] H. K. Hajjar, F. A. Dubaikel, and I. M. Ballard, 'Bifacial photovoltaic technology for the oil and gas industry', 2015 Saudi Arabia Smart Grid (SASG), 2015.

[31] M. R. Vogt, H. Holst, M. Winter, R. Brendel, and P. P. Altermatt, 'Numerical modeling of c-Si PV modules by coupling the semiconductor with the thermal conduction, convection and radiation equations', Energy Procedia 77 (2015) 215–24.

[32] B. B. van Aken, 'Bifacial modules: hot or cool?', BIFI Workshop, Konstanz, 2017.

[33] G. J. M. Janssen, R. S. R. Gali, K. M. de Groot, A. J. Carr, B. B. Van Aken, and I. G. Romijn, 'Impact of inhomogeneous irradiance at the rear of bifacial panels on modelled energy yield', 33rd EUPVSEC, Amsterdam, 2017.

Chapter 9

Summary and outlook

Radovan Kopecek[1] and Joris Libal[1]

9.1 Summary

The extremely hot summer in 2018 could have a similar consequence for the coal power plants as the Fukushima accident for nuclear power stations in 2011. The lack of cooling water had the consequence that many coal power plants had to be shut down during summer and more electricity from renewables was fed into the grid. Due to hotter and hotter summers the coal power plants have to be step by step replaced by renewables: this was also requested by the EU commission after the summer 2018. This will speed up PV installations in EU and world-wide again.

We have summarised in our bifacial PV book that, in order to bring new photovoltaic (PV) technology into the market – even if it is only an evolutionary technology – much more has to be considered and worked on than just high power and low costs. A big challenge is how to make the technology bankable and how to reach and to convince the end customers. Even if the lowest bid in PV history was submitted by EDF/Masdar of 1.79 USct/kWh in 2017 (first bid below 2 USct/kWh) [1] using bifacial HSAT technology in Saudi Arabia's tender, it was not accepted by the customer – most likely due to the lack of financing institutions and accordingly due to a negative outcome of its bankability evaluation. However, this bid made everybody 'nervous' who still did not have bifaciality on the roadmap. From then on many big electrical companies involved in PV were making their own studies on this technology and were hiring R&D institutes and other technical advisors to help them out with yield simulations and related studies. The largest systems in 2017 were the 53 MWp fixed tilt system from Yingli (Figure 9.1) and the 71 MWp HSAT system from SPIC (Figure 9.2) [2]. There are big plans for 2018 to install several 100 MWp systems, e.g. in Mongolia by Yingli based on nPERT technology [3] and in Taiwan by NSP based on bifacial PERC+ technology [4].

However, bankability still remains an issue. Therefore, setting standards, as well as create easy, understandable and comfortable simulations that are validated for their accuracy by a sufficient amount of case studies (field data), are very

[1]International Solar Energy Research Center Konstanz e.V., Germany

Figure 9.1 53 MWp bifacial PV system from Yingli Solar. Largest bifacial fixed tilt system in 2017

Figure 9.2 71 MWp bifacial PV system from SPIC Solar. Largest bifacial HSAT system in 2017 [2]

important issues. The chapters of this book describe step-by-step the technological, economical and commercial status of bifacial technology and sketch the future variety and fields of applications.

There is a huge variety of solar cell and module technologies such as p-type and n-type devices, both sided contacted cells, rear contact cells, half cell technology, double glass modules and transparent back sheet technology, encapsulant-free modules, multi busbar interconnections and many others. What technology will have the highest impact in the bifacial field is not easy to predict. Sure is that since the beginning of 2016 innovations have penetrated the PV market and because of its huge production volume even niche technologies will have a huge impact. The PV era just started and – thanks to its low cost and easy applicability – this technology will be, just like Internet, spread all around the word in the coming years. And will make also poor countries independent on energy imports.

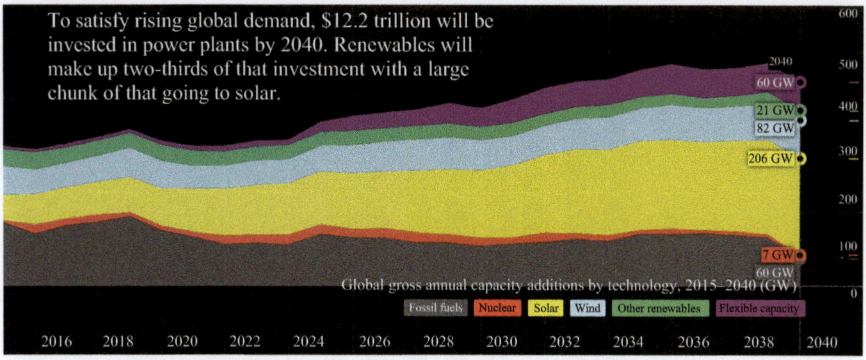

Figure 9.3 Energy outlook from Bloomberg from 2015 showing a 200 GW PV Market from 2040 on [5]

9.2 Outlook

9.2.1 Growth of PV

Few years back, nobody would have expected that in year 2017 100 GWp of PV systems will be installed worldwide. Prediction by e.g. Bloomberg showed an optimistic scenario already in 2015 [5], however, still underestimating the power of PV to a large extent (Figure 9.3).

Bloomberg's newest studies show that PV is becoming cheaper than coal energy [6] and that we will have an accumulated 1 TWp market already from 2021 on with a yearly volume of 200 MWp already in that year: so 20 years earlier as predicted by Bloomberg in the 2015 study. This shows the unleashed dynamics of PV technology. In addition, also the movement from standard cell and modules towards advanced technologies was not predicted to go that fast—as we will summarise in the next sections. The ITRPV underestimated the power of PERC.

9.2.2 Predictions of new cell and module technologies

The ITRPV roadmap [7] predicted a fast penetration of new solar cell technologies into the PV market and that from 2021 on 'PERX' (solar cells with passivated rear side, e.g. nPERT or pPERC) solar cells will dominate the market. New studies show however that this will happen already in 2018. PERC will become mainstream [8] and new technologies such as PERT, low cost IBC and HJT will follow (Figure 9.4). Also the module market will be very dynamic in terms of new technologies – e.g. double glass modules and half-cell technologies will become main stream in the next years – and all these innovation are another argument for going bifacial (Figure 9.4).

At the PV Cell Tech show in 2018 in Malaysia it was presented and forecasted by PV Tech that the low efficiency solar cell market, which is mostly consisting of Al BSF cells, will disappear fully from the 'PV map' in 2022 and from then on the market will be composed of rear side passivated high efficiency devices only (Figure 9.4).

Every cell manufacturer has its own high-efficiency road map [10] and the solar cell market will be a mixture of PERXs, HJTs, IBCs – with and without

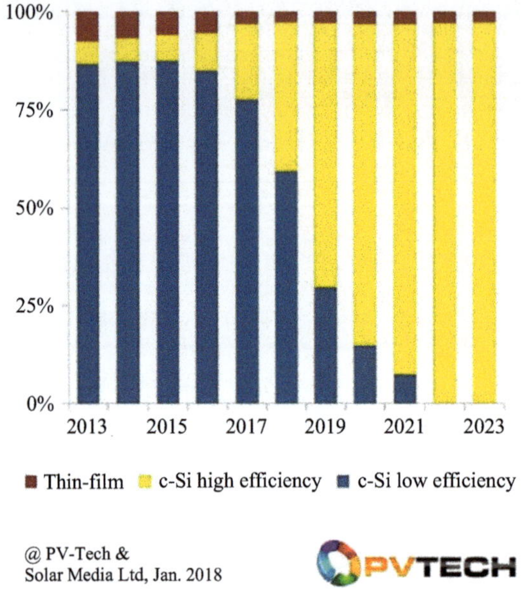

Figure 9.4 Prediction of market share for thin film and c-Si technology divided in low (Al-BSF) and high (PERX, HJT, IBC) efficiency [9]

passivated contacts. Because of the more advanced rear side, all there cell concepts are easy to be made bifacial.

9.2.3 Future of bifacial PV

'Bifacial mono c-Si photovoltaics will win.' This is not only a slogan of one big Chinese company LONGi that has this vision [11] – the entire PV community is starting to believe this since 2017/2018. At the SNEC Show in Shanghai in April 2017 and at the Intersolar Europe in Munich in June 2017, the shows were flooded by bifacial modules and the leading PV companies had promoted this vision. Some of them are even dreaming of highest efficiency cost effective n-type bifacial disruptive technologies, such as devices with carrier selective contacts leading to efficiencies of above 25% – similar to what Sunpower and Panasonic are doing since years. The only difference is that the costs must be comparable to standard solar cells.

We believe that much faster than from 2028 on, as predicted by ITRPV [7], more than 40% of the modules on the market will be based on 'real bifacial technologies'. According to Bloomberg, PV will become the most important energy source. Figure 9.3 shows the predicted evolution of the most prominent energy sources, where PV will be the most important one from 2021 on – whereas fossil fuel and nuclear energy will slowly disappear. As mentioned, in the past all predictions of volume and costs for PV were highly underestimated and this is also the case with the Bloomberg study as we reached the 100 GWp market already in 2017. Therefore, we believe that the impact of PV will be even higher than predicted in

Figure 9.3 and bifaciality will have a 100 GWp market share in less than 5 years from now.

We hope that this book brought you this promising bifacial technology closer and that you gained knowledge in the above-mentioned fields, so that you can go ahead and make use of bifaciality yourself. Then we, all authors of this bifacial book, fulfilled our mission.

References

[1] 'Abu Dhabi plant to produce region's cheapest electricity from solar.' [Online]. Available from http://www.thenational.ae/business/energy/abu-dhabi-plant-to-produce-regions-cheapest-electricity-from-solar [Accessed 24 Apr. 2018].

[2] '71 MW Bifacial Solar Project in China World's largest Bifacial Solar Power Project Installed in China's Qinghai Province.' taiyangnews, 01 Oct. 2018. [Online]. Available from http://taiyangnews.info/technology/20-mw-bifacial-solar-system-in-china/ [Accessed 24 Apr. 2018].

[3] J.S. Hill, 'Yingli Solar Begins Construction of 100 Megawatt Top Runner Project in Inner Mongolia,' CleanTechnica, 20 Nov. 2017. [Online]. Available from https://cleantechnica.com/2017/11/20/yingli-solar-announces-begins-construction-of-100-mw-top-runner-project-in-inner-mongolia/ [Accessed 24 Apr. 2018].

[4] M. Osborne, 'Neo Solar Power's mono PERC bifacial modules selected for 100 MW PV plant in Taiwan,' PV Tech. [Online]. Available from https://www.pv-tech.org/news/neo-solar-powers-mono-perc-bifacial-modules-selected-for-100mw-pv-plant-in [Accessed 24 Apr. 2018].

[5] '2015: New energy outlook.' [Online]. Available from https://www.bnef.com/dataview/new-energy-outlook/index.html [Accessed 24 Apr. 2018].

[6] 'Solar Energy 2017.' [Online]. Available from https://www.bloomberg.com/quicktake/solar-energy [Accessed 24 Apr. 2018].

[7] 'International Technology Roadmap for PV.' [Online]. Available from http://www.itrpv.net/ [Accessed 24 Apr. 2018].

[8] 'Panel predictions 2018.' [Online]. Available from https://www.pv-magazine.com/magazine-archive/panel-predictions-2018/ [Accessed 24 Apr. 2018].

[9] 'Gigawatt cell producers to explain technology expansion plans at PV CellTech 2018.' [Online]. Available from https://www.pv-tech.org/news/gigawatt-cell-producers-to-explain-technology-expansion-plans-at-pv-celltec [Accessed 24 Apr. 2018].

[10] 'PV CellTech 2018.' [Online]. Available from https://www.pv-tech.org/topics/pv-celltech-2018 [Accessed 24 Apr. 2018].

[11] 'Intersolar Europe 2017: Bifacial will be mainstream in two years says LONGi.' [Online]. Available from https://www.pv-tech.org/news/intersolar-europe-2017-bifacial-will-be-mainstream-in-two-years-says-longi [Accessed 24 Apr. 2018].

Index